W0042029

Lecture Notes in Control and Information Sciences

Edited by M. Thoma

Vol. 43: Stochastic Differential Systems
Proceedings of the 2nd Bad Honnef Conference
of the SFB 72 of the DFG at the University of Bonn
June 28 – July 2, 1982
Edited by M. Kohlmann and N. Christopeit
XII, 377 pages. 1982.

Vol. 44: Analysis and Optimization of Systems
Proceedings of the Fifth International
Conference on Analysis and Optimization of Systems
Versailles. December 14–17, 1982
Edited by A. Bensoussan and J. L. Lions
XV, 987 pages, 1982

Vol. 45: M. Arató
Linear Stochastic Systems
with Constant Coefficients
A Statistical Approach
IX, 309 pages. 1982

Vol. 46: Time-Scale Modeling of Dynamic Networks
with Applications to Power Systems
Edited by J. H. Chow
X, 218 pages. 1982

Vol. 47: P. A. Ioannou, P. V. Kokotovic
Adaptive Systems with Reduced Models
V, 162 pages. 1983

Vol. 48: Yaakov Yavin
Feedback Strategies for Partially
Observable Stochastic Systems
VI, 233 pages, 1983

Vol. 49: Theory and Application of Random Fields
Proceedings of the IFIP-WG 7/1
Working Conference
held under the joint auspices of the
Indian Statistical Institute
Bangalore, India, January 1982
Edited by G. Kallianpur
VI. 290 pages. 1983

Vol. 50: M. Papageorgiou
Applications of Automatic Control Concepts
to Traffic Flow Modeling and Control
IX, 186 pages. 1983

Vol. 51: Z. Nahorski, H.F. Ravn, R.V.V. Vidal
Optimization of Discrete Time Systems
The Upper Boundary Approach
V, 137 pages 1983

Vol. 52: A. L. Dontchev
Perturbations, Approximations and Sensitivity Analysis
of Optimal Control Systems
IV, 158 pages. 1983

Vol. 53: Liu Chen Hui
General Decoupling Theory of Multivariable
Process Control Systems
XI, 474 pages. 1983

Vol. 54: Control Theory for Distributed
Parameter Systems and Applications
Edited by F. Kappel, K. Kunisch,
W. Schappacher
VII, 245 pages. 1983.

Vol. 55: Ganti Prasada Rao
Piecewise Constant Orthogonal Functions
and Their Application to Systems and Control
VII, 254 pages. 1983.

Vol. 56: Dines Chandra Saha, Ganti Prasada Rao
Identification of Continuous
Dynamical Systems
The Poisson Moment Functional
(PMF) Approach
IX, 158 pages. 1983.

Vol. 57: T. Söderström, P. G. Stoica
Instrumental Variable Methods
for System Identification
VII, 243 pages. 1983.

Vol. 58: Mathematical Theory of
Networks and Systems
Proceedings of the MTNS-83 International
Symposium
Beer Sheva, Israel, June 20–24, 1983
Edited by P. A. Fuhrmann
X, 906 pages. 1984

Vol. 59: System Modelling and Optimization
Proceedings of the 11th IFIP Conference
Copenhagen, Denmark, July 25-29, 1983
Edited by P. Thoft-Christensen
IX, 892 pages. 1984

Vol. 60: Modelling and Performance
Evaluation Methodology
Proceedings of the International Seminar
Paris, France, January 24–26, 1983
Edited by F. Bacelli and G. Fayolle
VII, 655 pages. 1984

Vol. 61: Filtering and Control of Random
Processes
Proceedings of the E.N.S.T.-C.N.E.T. Colloquium
Paris, France, February 23–24, 1983
Edited by H. Korezlioglu, G. Mazziotto, and
J. Szpirglas
V, 325 pages. 1984

For information about Vols. 1–42 please contact your bookseller or Springer-Verlag.

Lecture Notes in Control and Information Sciences

Edited by M. Thoma and A. Wyner

100

J. P. Zolésio (Editor)

Boundary Control and Boundary Variations

Proceedings of the IFIP WG 7.2 Conference
Nice, France, June 10-13, 1987

Springer-Verlag
Berlin Heidelberg GmbH

Series Editors
M. Thoma · A. Wyner

Advisory Board
L. D. Davisson · A. G. J. MacFarlane · H. Kwakernaak
J. L. Massey · Ya Z. Tsypkin · A. J. Viterbi

Editor of Conference Proceedings of the series:
Computational Techniques in Distributed Systems IFIP-WG 7.2
Irena Lasiecka
Dept. of Applied Mathematics
Thornton Hall
University of Virginia
Charlottesville, VA 22903
USA

Editor
J. P. Zolésio
Laboratoire de Mathematiques
Departement de Mathematiques
Université de Nice
Parc Valrose
F-06034 Nice Cedex

ISBN 978-3-540-18546-8

Library of Congress Cataloging in Publication Data
IFIP W.G. 7.2 Conference (1986: Nice, France)
Boundary control and information sciences; 100)
1. Control theory – Congresses.
2. Boundary value problems – Congresses.
I. Zolésio, J. P. II. Title. III. Series.
QA402.3.I454 1986 629.8'312 87-28679
ISBN 978-3-540-18546-8 ISBN 978-3-540-48015-0 (eBook)
DOI 10.1007/978-3-540-48015-0

This work is subject to copyright. All rights are reserved, whether the whole or part of the material is concerned, specifically the rights of translation, reprinting, re-use of illustrations, recitation, broadcasting, reproduction on microfilms or in other ways, and storage in data banks. Duplication of this publication or parts thereof is only permitted under the provisions of the German Copyright Law of September 9, 1965, in its version of June 24, 1985, and a copyright fee must always be paid. Violations fall under the prosecution act of the German Copyright Law.

© Springer-Verlag Berlin Heidelberg 1988
Originally published by Springer-Verlag Berlin Heidelberg New York in 1988

2161/3020-543210

PREFACE

This volume comprises the proceedings of the Working Conference "Boundary variations and boundary control" held in Nice (France), June 10-13, 1986 space. The Conference was organized for the Working Group 7.2 (Computational methods for control systems described by partial differential equation) of Technical Committe 7 (Modelling and Optimization Techniques) of the International Federation for Information Processing (IFIP).

The organizing Committe consisted of the following members :

J.P. MARMORAT (Ecole des Mines, Sophia Antipolis)
L. PASSERON (Aérospatiale, Cannes)
M. SOULI (Département de Mathématiques, Nice)
C. TRUCHI (Ecole des Mines, Sophia-Antipolis)
J.P.ZOLESIO (Département de Mathématiques, Nice and C.N.R.S Montpellier).

The International Committe of the Working Group 7.2 consists of the following members :

A. BERMUDEZ, Univ. Santiago de Compostela, Spain
A. BUTKOWSKI, Control Institut, Moscow
R. CURTAIN, Univ. of Groningen, Netherlands
G. DA PRATO, Scuola Normale, Pisa, Italy.
R.GLOWINSKI, INRIA, Paris.
K. HOFFMAN, Univ. of Augsburg, Germany
W. KRABS, Technische Hochschule, Darmstadt, Germany
I. LASIECKA (CHAIRMAN), Univ. of Virginia, U.S.A.
J. LIONS, Collège de France and CNES, Paris, France
U. MOSCO, Univ. of Rome, Rome, Italy
O. PIRONNEAU, INRIA, Paris, France
J.P. YVON, INRIA, Paris, France
J.P. ZOLESIO, Univ. of Nice and CNRS Montpellier, France.

It was a great pleasure for me to welcome 45 invited participants from 9 different contries whose research involves the use of boundary techniques in optimization problems governed by partial differential equations. The aim of this Conference was to stimulate exchange of Ideas between the group working on shape optimization (including free boundary problems) and the group working on boundary control of hyperbolic systems (including stabilization). An important remark is that if one considers a dynamical system governed by linear elasticity the choice of Lagrangian coordinate leads to discuss boundary conditions, or boundary control (for example to stabilize) , while the choice of Eulerian coordinates lead to a moving boundary and moving domain Ω_t. This remark challenges us to consider the domain (or its boundary) as a control.

This Conference was sponsored by :

- Département de Mathématiques, Université de Nice.
- Ecole Nationale Supérieure des Mines de Paris (Centre de Mathématiques) Sophia-Antipolis.
- Aérospatiale, Cannes-La-Bocca.

The collaboration of these three Institutions working on stabilization of flexible structures was mainly initiated by Professor M.C. DELFOUR during the year 1984 that he spent in Sophia-Antipolis. My thanks go to the 33 authors of the contributions contained in this volume.

March 1987 J.P. ZOLESIO

CONTENTS

Towards a multipurpose optimal shape design computer code

G. Arumugam (INRIA)
O. Pironneau (Université Paris 6 and INRIA)

Abstract :

Optimal shapes of distributed systems can be obtained by the techniques of optimal control and numerical analysis. However these shapes are often unfeasible because they violate constraints which were not thought of at first. Thus optimal design requires constant changes of criteria and constants functionals. In this paper we propose a few solutions to minimize code developments under such conditions.

I. Introduction :

When an optimal shape design problem is well posed, and when its state equation is a PDE which is easy to solve numerically then the optimal shape can be found in general with good precision, even in 3-D (see [1], [2], [3], [4]). However to our experience, optimal shape design problems are rarely well posed in the sense that the optimal shape usually does not satisfy the engineer who has set up the problem. This is because many constraints so obvious to him were not included or because the criteria proposed should really have been a weighted averge of several criteria ; for instance off design point performance are usually important also but difficult to include in a criteria. Thus in practice there is a long dialog between the engineer, the optimal controler, and the computer before a good formulation of the problem is found. And this dialog requires long and expensive code developements.

2. Objective :

Airplane design can be optimized in several places such as (see fig. 1)

. Airfoil shape or wing shape

. Engine/wing attachment shape

. Nozzle shape.

While the general natural criteria are drag reduction and lift control, these are not used in pratice because they would require to use the full compressible Navier-Stokes equations for the State. Instead the potential inviscid approximation is used :

(1) $u = \nabla\phi$ (velocity)

(2) $e = (1 - |\nabla\phi|^2)^{1/\gamma-1}$ (density), $\nabla.(\rho u) = 0$

(3) $p = (1 - |\nabla\phi|^2)^{\gamma/\gamma-1}$ (pressure)

and the viscous drag is assumed to be proportional to the maximum of the pressure on the part of the body to be optimized or proportional to the gradient of p. Other criteria have been suggested such as the position of the shock on the wing, the jump of the velocity across the shock... all non differentiable, be it noted.

The constraints are

. smooth radius of curvature

. solidity of the structures

. fabrication feasibility and cost

. minimum lift, Joukowski conditions, off design perfomances...

The challenge is to design a 2-D code for shape optimization that would accomodate all the above criteria and constraints, be extendable to 3-D and as much as possible independant of the geometry.

3. Methodology :

The traditional approach for such problem is to parameterise the unknown boundary S by a small number of parameters $\{\alpha_1,\ldots,\alpha_N\}$ and do a trial/error or parabolic fit optimization [5] with respect to $\{\alpha_1,\ldots,\alpha_2\}$ i.e. if J is the criterium, this amount to solve $\min_{\alpha_i \varepsilon A} J(\alpha_i)$ by using finite difference approximation of the derivatives

(4) $J'_{\alpha_i} \simeq [J(\alpha_i+\delta\alpha_i) - J(\alpha_i)]/\delta\alpha_i \qquad i=1,\ldots,N$

If N is grater than, say 5, this method is too expensive because to compute $J(\alpha_i)$ one must solve PDE once so each computation of (4) requires N solutions of the DPE.

The standard method of Optimal Control (see [1] for example) for solving a model problem like

(5) $\min_{S\varepsilon Y} \{J(\phi,S) : A(S,\phi) = 0\}$

where A represente the PDE, ϕ its solution and S the unknown boundary, one proceeds as follows :

1. Discretize (5) ; for example with the FEM (5) becomes

(6) $$\min_{q^j \varepsilon Q} \{J_h(\phi_i, q^j) : A_h(q^j, \phi_i) = 0\}$$

where ϕ_i denotes the values of ϕ at the node q^i.

2. Compute

(7) $$(\nabla_j J_h)_\ell = \frac{\partial J_h}{\partial \phi_i} \frac{\partial \phi_i}{\partial q^j_\ell} + \frac{\partial J_h}{\partial q^j_\ell}$$

where $\partial\phi_i/\partial\phi^j_\ell$ is computed by solving

(8) $$\frac{\partial A_h}{\partial \phi_i} \frac{\partial \phi_i}{\partial q^j_\ell} + \frac{\partial A_h}{\partial q^j_\ell} = 0$$

3. Compute the gradients of the contraints $G_h(\phi_i, q^j)$ by the same method.

4. Update the triangulation by the contrainted optimization iterative algorithm ; usually like

(9) $$q^j \to q^j + \rho\,(\nabla_j J_h + \mu \nabla_j G_h)$$

We propose to keep the same methodology but do delay as much as possible specific choices of J and G. On the contrary A and A_j are assumed to be given and fixed.

4. Incompressible flows

At low mach number (2) may be approximated by a Laplacien. Thus a fairly general problem is

(10) $$\min_{S \varepsilon J} \{E(S,\phi,\nabla\phi) : -\Delta\phi = 0 \text{ in } \Omega \quad \phi|_{\Gamma_1} = \phi_1, \frac{\partial\phi}{\partial n}\Big|_{\Gamma_2} = g\}$$

(11) $$S = [S \varepsilon \Sigma : F(S,\phi) \le 0\}$$

A Lagrange Finite Element discretization yields

(12) $$\min_{q^i \varepsilon Q} \{E(q_i,\phi_j,(\nabla\phi)_k) : \int_\Omega \nabla\phi_h . \nabla w^i \, dx = \int_{\Gamma_2} g w^i \, d\gamma : \forall i \varepsilon I$$
$$\phi_h = \sum_{j \varepsilon I} \phi_j w^j + \phi_1\}$$

where w^j denotes the basic functions of the Finite Element space, $(\nabla\phi)_k$ the values of $\nabla\phi$ at point x^k and I the indices of the nodes which are not on Γ_1.
Similary S yields

$$(13)\quad Q = \{q^i \varepsilon P : F(q^i, P^i_j, (\nabla\phi)_k) \le 0\}$$

where P is the set of points giving $S\varepsilon\Sigma$ and such that the triangulation is admissible.
Of course (12) is of the form

$$(14)\quad \min \{J(q^i) : q^j \varepsilon Q\}$$

and with self explainatory short hand notation one has

$$(15)\quad J'_{,q^i_\ell} = E_{,q^i_\ell} + E_{,r_j}\, \phi_{j,q^i_\ell} + E_{,s_k}\, (\nabla\phi)_{k,q^i_\ell}$$

(E,q, E,r, E,s are the derivatives of E(q, r, s)

Since

$$(16)\quad \phi_h(x) = \Sigma\, \phi_j\, w^j(x), \quad \int_\Omega \nabla\phi_h\, \nabla w^j dx = \int_{\Gamma_2} g w^j\, d\gamma$$

we have

$$(17)\quad (\nabla\phi)_{k,q^i_\ell} = \Sigma\, \phi_{j,q^i_\ell}\, \nabla w^j(q^k) + \phi_j (\nabla w^j)_{k,q^i_\ell}$$

but (see [1,p12])

$$(18)\quad w^j_{,q^i_\ell} = - w^i(x)\, \frac{\partial w^j}{\partial x_\ell}$$

so, with summation over repeated indices, we have

$$(19)\quad J'_{,q^i_\ell} = E_{,q^i_\ell} - E_{,j_k} - \frac{\partial\phi_h}{\partial x_\ell}(x)\, w^i(x^k) + (E_{,r_j} - E_{,s_k}\, \nabla w^j(x^k))\phi_{j,q^i_\ell}$$

To get rid of ϕ_{j,q^i_ℓ} let us differentiate (16). Here for the sake of clarity we assume that g, Γ_2 are fixed and :

$$(20)\quad \int_\Omega \nabla(\Sigma\phi_{j,q^i_\ell} w^j + \Sigma\, \phi_j w^j{}_{,q^i_\ell})\nabla w^j + \nabla\phi_h\, \nabla w^j{}_{,q^i_\ell} + \int_{\Omega,q^i_j} \nabla\phi_h\, \nabla w^j = 0$$

or, using (18) and [1, Proposition 3 p 102]

$$(21)\quad \int_\Omega \nabla(\Sigma\phi_{n,q^i_\ell} w^n - \frac{\partial\phi_h}{\partial x_\ell}\nabla w^i)\nabla w^j = \int_\Omega \nabla\phi_h.\nabla w^i \frac{\partial w^j}{\partial x_\ell} - \nabla\phi_h.\nabla w^j \frac{\partial w_i}{\partial x_\ell}$$

So if we define the adjoint state P_h by

$$(22)\quad \int_\Omega \nabla P_h \nabla w^j = E,r_j - E,s_k . \nabla w^j(x^k) \qquad \forall_j \in I$$

Then

$$(23)\quad J',q^i_\ell = E,q^i_\ell - E,s_k \frac{\partial\phi_h}{\partial x_\ell}(x,k) + \int_\Omega [\nabla w^i . \nabla P_h \frac{\partial\phi_h}{\partial x_\ell} + \nabla\phi_h . \nabla w^i \frac{\partial P_h}{\partial x_\ell} - \nabla\phi_h . \nabla P \frac{\partial w^i}{\partial x_\ell}]\, dx$$

5. Implementation

To compute (23) we must specify E. This is done by a FORTRAN Function statement ; with linear elements (NQ vertices, NT triangles) :

```
FUNCTION E(Q, F, GF)
DIMENSION Q (2, NQ, F(NQ), GF(3,NT)
E =
RETURN
END
```

To compute E,q^i_ℓ , E,r_j , E,s_k there are 3 possibilities

1. The user provides similar functions
2. Symbolic manipulation programs like MACSYMA does it
3. Finite difference approximations are used ; for instance

$$(24)\quad E,r_j\ (q,r,s) \simeq (E(q,r_j + \delta r_j, s) - E(q,r,s))/\delta r_j$$
with $\delta r_j = 10^{-6}$.

We have chosen the third way because 1 is too constraining 2 is not portable and 3 is not so expensive. It requires N function evaluation of E ; Solutions of (18) and (22) always require O(N) operations also. For example with 90 vertices and, 140 triangles, 30 iterations of a gradient method requires 3 minutes with method 1 and 4 minutes with method 3.

When the mesh is changed one must make sure that the distribution of vertices stay reasonable, that the triangles do not flip over or become too flat... There are 4 methods

1. Attach the motion of linear nodes to the motion of boundary nodes by having a family of lines on which these moves (figure 2)

2. Use an automatic mesh generation (AMG) and find the displacement of the inner nodes δq^j in term of the outer motions δq^i :

$$(25)\quad \delta q^j = T_{ij}\ \delta q^i$$

by running AMU twice, once with q^i and once with $q^i+\delta q^i$.

3. leave all inner nodes move freely within control regions defined from their neighours (Figure 3)

4. Find a general rule that links δq^j with δq^i.
In all cases except method 3 we have a relation like (25) and the total gradient of J with respect to q^i is $J,q^i_\ell + T_{ij}\ J,q^j_\ell$.
While method 1 is certainly the best, it is the worst from the programming point of view, because it stepends very much on the geometry of the problem Method 2 is simple but expensive.
Method 3 is all right if one is near enough to the solution, but not otherwise, because δq^i is often very much bigger than δq^j, so outernodes move fast, bump into inner nodes and then every thing moves very slowly.

To implement method 4 we followed Marrocco [6]

$$(26)\quad \delta q^j = \frac{\Sigma \alpha_{ji} w_i \delta q^i}{\Sigma_i \alpha_{ji} w_i}$$

with $\alpha_{ij} = |q^i - q^j|^{-\beta}$, $\beta = 1.6$ and (see figure 4)

$$(28)\quad w_j = \frac{1}{2}\left(|q^{j+1}-q^j| + |q^j - q^{j-1}|\right)$$

It is proposed by the fact that inner points near to boundary point should have the same speed and by the fact that the density of boundary point should not affect the formula. However much a formula turns out to be quite sensitive to β.

Finally to prevent oscillations of the boundary due to the discretization we link the motions of every other boundary nodes to its two neighbour (Figure 5)

$$(29)\quad \delta q^j = (\ell_2\ \delta q^{j-1} + \ell_1 \delta q^{j+1}) \ /\ (\ell_1+\ell_2) \qquad j = 2\ ,\ 4,\ 6\ldots$$

6. Numerical results

We have tested the method on Problem 10 with

$$E = \int_D |\nabla\phi - \frac{1}{|D|}\int_D \nabla\phi \, dx|^2 \, dx \text{ or } \int_D |\nabla\phi - u_d|^2 \, dx \tag{30}$$

modified either by changing the position of D or by adding penalization terms coming from desired constraints.

Experiment 1 (Figure 6 and 7)

Objective : compare the method with the classical method as used by Angrand [7]. Here u_d is given corresponding to $\nabla\phi_d$ for a given nozzle and the objective is to recover the nozzle that gave ϕ_d. Performances are excellent.

Experiment 2 (Figure 8 and 9)

Objective : use the first form of E in (30) and check the results. Notice that the smoothness of Γ is not satisfactory. Thus Penalization of the angles of the boundary was added to the criteria.

Experiment 3 (Figure 10, 11, 12)

Objective : test the performance of the method on a wide range of different domains.

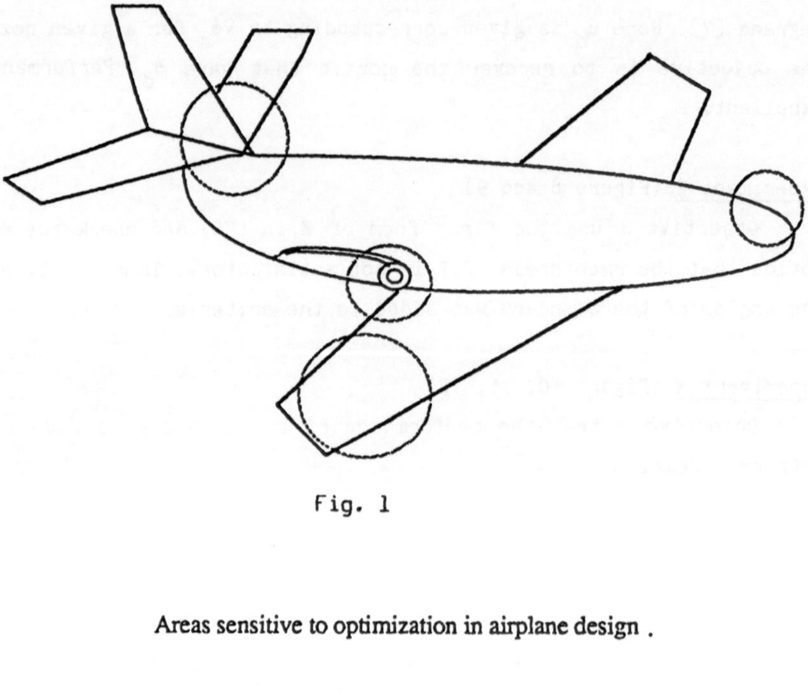

Fig. 1

Areas sensitive to optimization in airplane design .

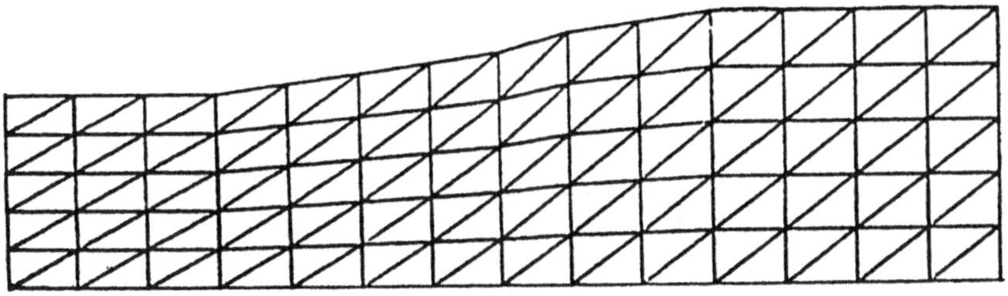

Fig. 2

Each inner node move vertically to the motion of the curved boundary nodes

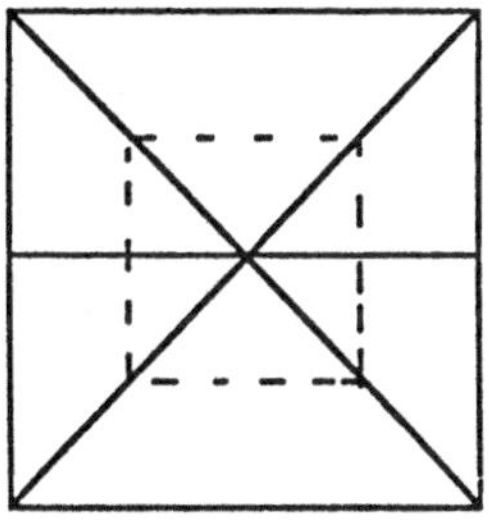

fig. 3

Each node is assigned to move inside the dotted region.

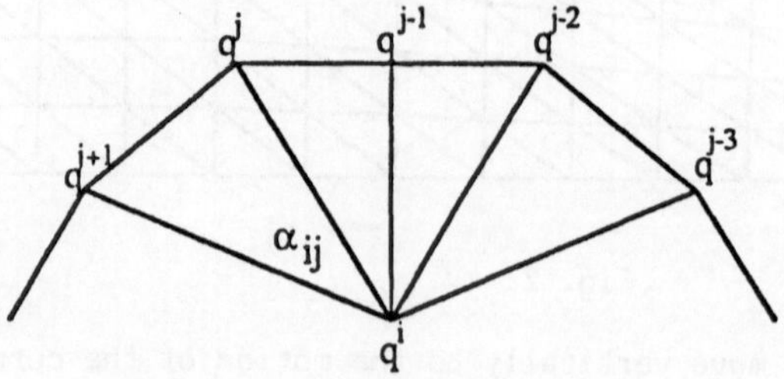

fig.4

The motion of interior node is linked with the boundary nodes.

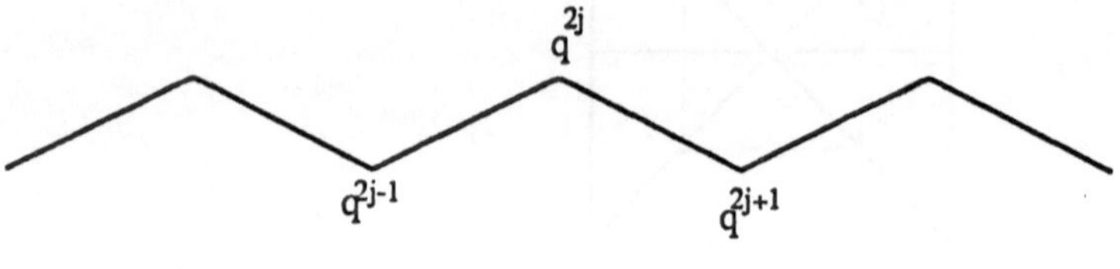

fig. 5

The motion of every other node is constrained to the motion of other nodes.

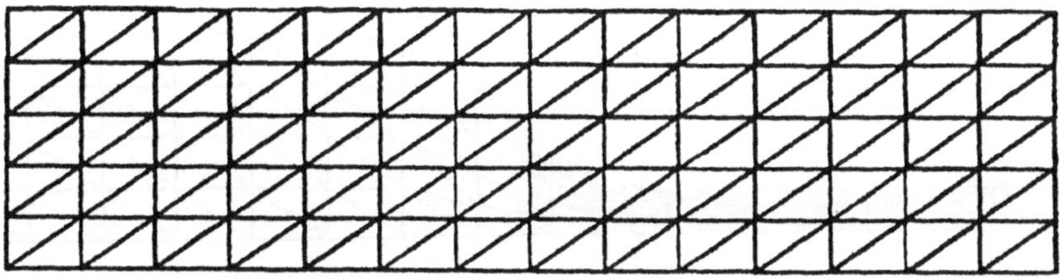

a. Initial domain

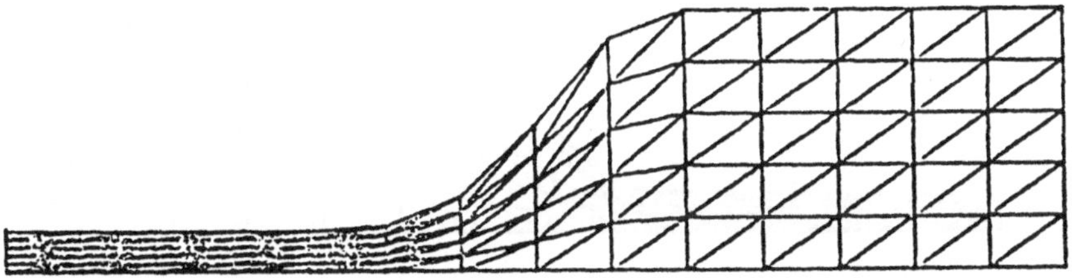

b. Final domain

Fig. 6

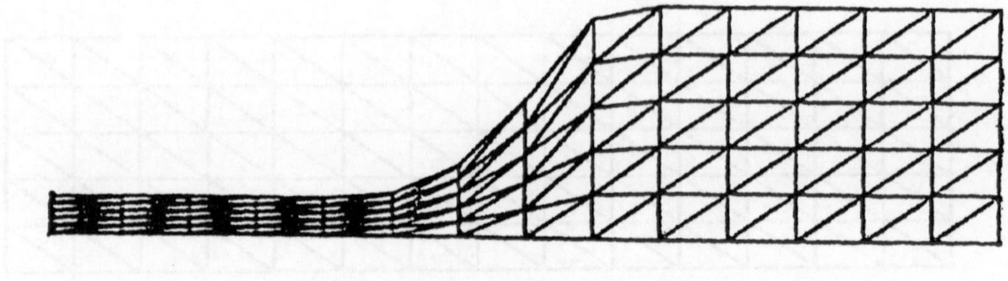

a. Initial domain

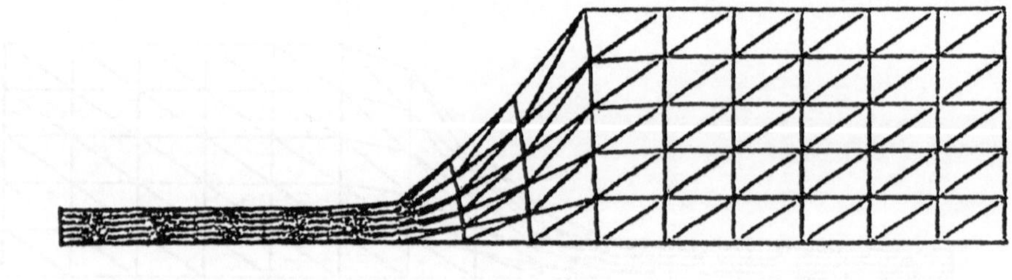

b. Final domain

Fig. 7

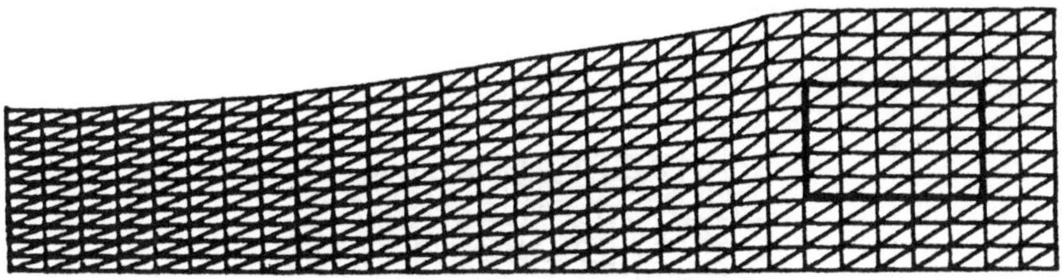

a. Initial domain

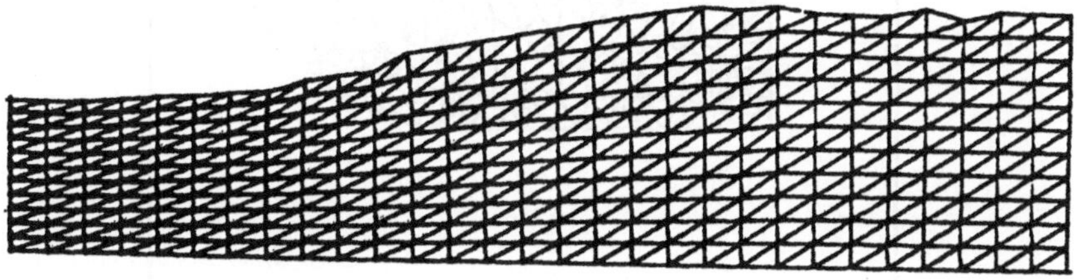

b. Final domain

Fig. 8

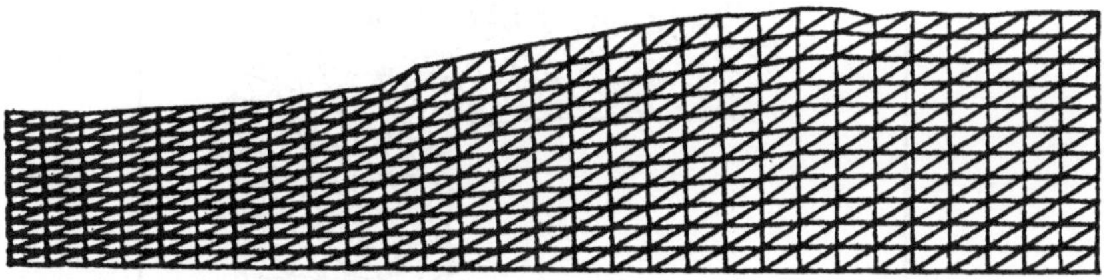

Fig. 9

Final domain after adding penalty term to the cost function

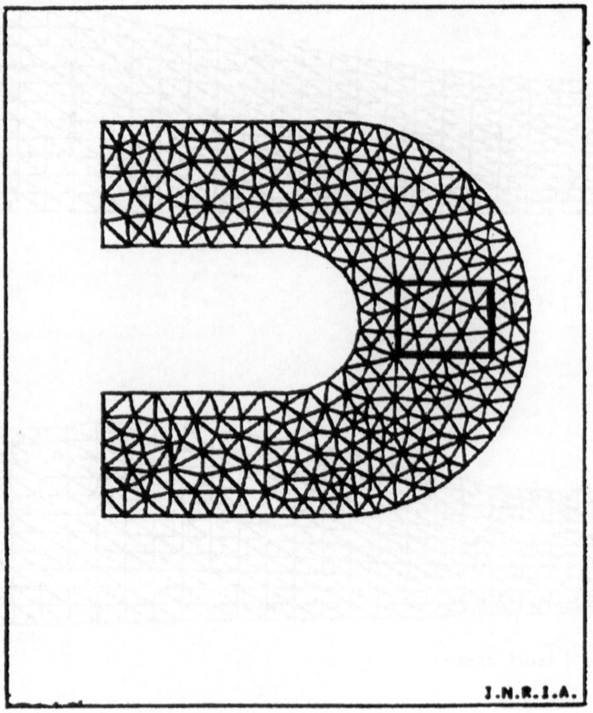

a. Initial domain

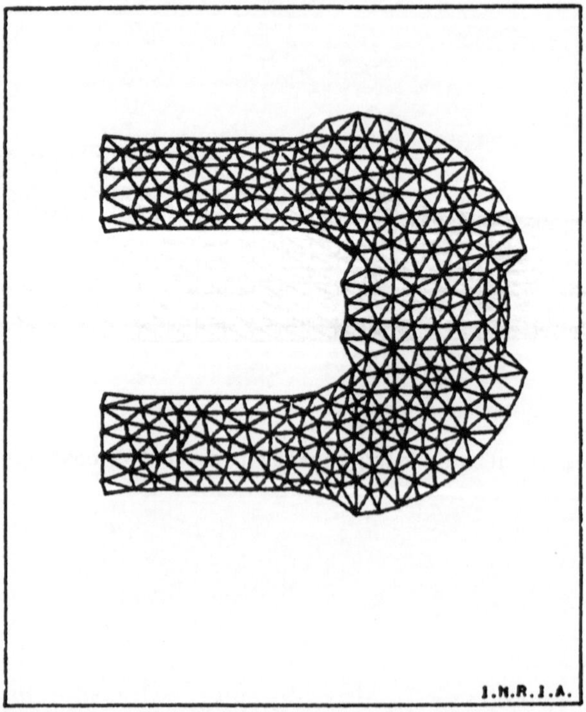

Fig. 10 b. Final domain

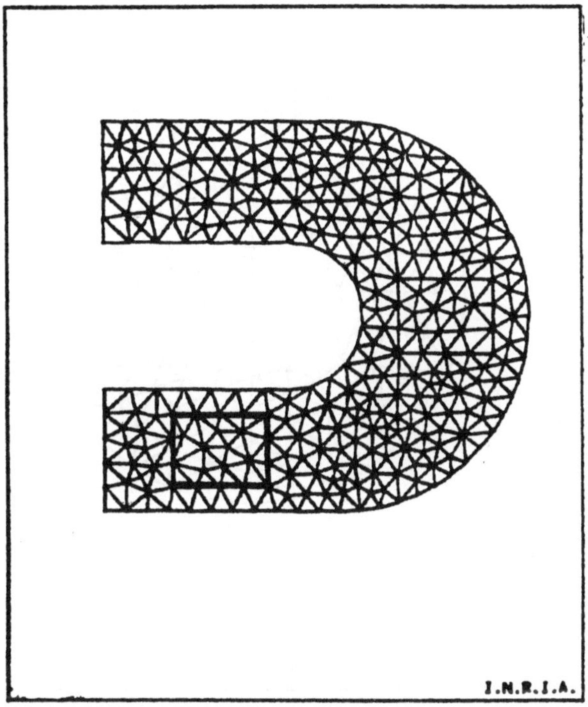

a. Initial domain

Fig. 11 b. Final domain

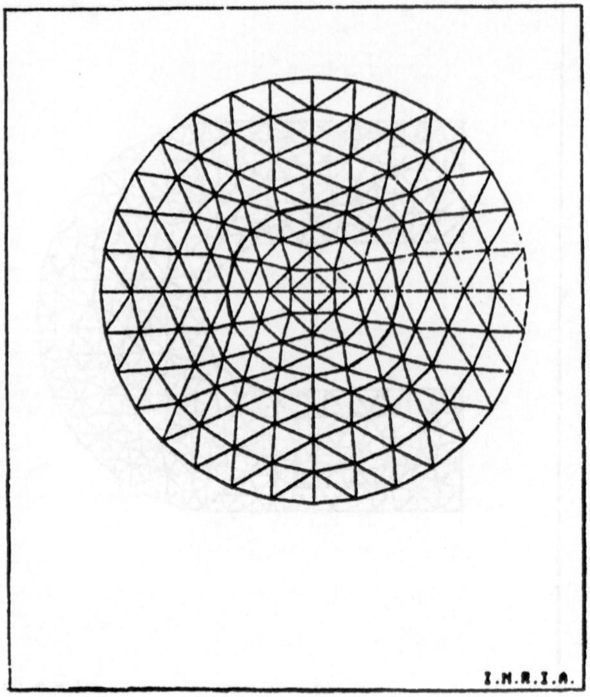

a. Initial domain

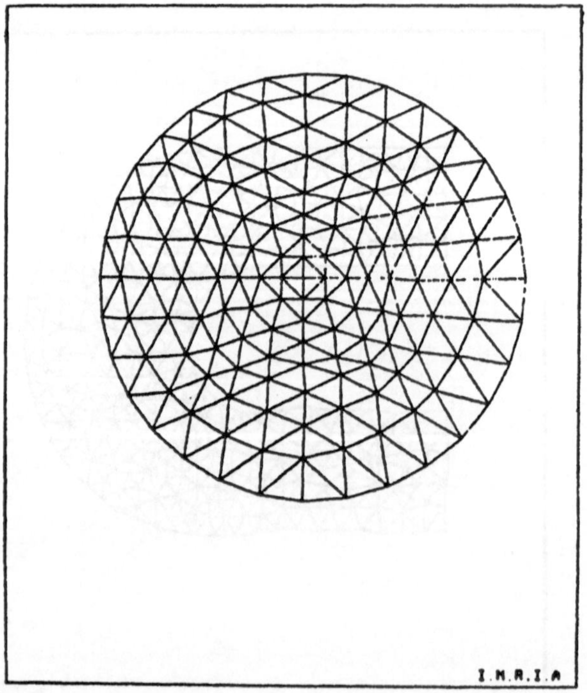

Fig. 12 b. Final domain

REFERENCES

(1) O. PIRONNEAU : Optimal shape design for elliptic Systems, Springer-Verlag, Springer Series in Computational Physics (1983).

(2) N.V. BANICHUK : Problems and Methods of Optimal Structural design, Plenum Press New York and London 1983.

(3) E.J. HAUG, K. CHOI, V. KANKOV : Design sensitivity analysis of structural systems (To appear).

(4) E.J. HAUG, J. CEA : Optimization of distributed parameter Structures, Nato advanced study institute series N° 49, 50, Volume 1, 2 (1981).

(5) E. POLAK : Computational methods in Optimization, Academic, New York (1971).

(6) A. GERDOLLE, A. MARROCCO, S. MARTIN : Diffusion d'impuretés avec oxydation, Computing Methods in Applied Sciences and Engineering VI, R. Glowinski and J.L. Lions (Eds.) (1984).

(7) F. ANGRAND : Methods numériques pour des problèmes de conception optimale en Aérodynamiques, Thèse de 3ème cycle, 1980.

STABILITY ENHANCEMENT OF FLEXIBLE STRUCTURES

BY NONLINEAR BOUNDARY-FEEDBACK CONTROL

A.V. Balakrishnan
LABORATORY FOR FLIGHT SYSTEMS RESEARCH
6750 BOELTER HALL
LOS ANGELES, CALIFORNIA 90024

ABSTRACT

We establish strong stability for a class of nonlinear boundary feedback controllers using an abstract wave-equation formulation of a beam stabilization problem arising in the control of flexible structures in space.

Research supported in part under AFOSR grant #83-0318, Applied Math Division.

1. Introduction

The problem of active feedback stabilizing flexible structures is of recent interest especially for deployment in space [1]. A significant feature of this application is the need for "robust" controllers - whose design does not require precise knowledge of system parameters -- see [2]. Although a class of such linear stabilizing control laws is given in [2], it would appear that to generate the control effort necessary, the actuators (reaction jets, for example), would need to be nonlinear: linear for small amplitudes but nonlinear (saturating-type) for large amplitudes.

Another feature of the space application is the uncertainty in the damping parameters (even including the model). The strong stabilizability using linear controllers requires controllability and this was established in [2] neglecting damping; but the proof of controllability when damping is present is unclear. In this paper we show that for a reasonable damping model the robust linear controller still yields strong stability under a natural extension of a condition involving the undamped modes used in [2]. Under the asumption that there is nonzero natural damping, however small, we prove that we have strong stability for a class of nonlinear feedback control laws which are linear for small amplitudes and can saturate for large amplitudes.

We only treat the abstract wave-equation version of the problem: the reduction of the physical model (biharmonic beam equations with delta-function controls) to the abstract form is given in [3]. We only note here one feature of the problem: the differential operators do not involve any "boundary conditions": and although the control is exerted on the boundary, the formulation does not fall into the class of boundary-control problems treated by

Lasiecka and Triggiani [4] so that in particular the results therein are not directly applicable.

Finally we note that our results may be regarded as an extension of the Benchimol-Slemrod result [5] (see Levan [6] for a recent treatment) to a class of nonlinear controllers but without invoking controllability -- albeit in a particular case. Our proof, although totally elementary, relies heavily on the eigenfunction decomposition of the generator. The problem statement and the main results are in Section 2.

2. General Results

Let H denote a separable Hilbert space and let us consider the following canonical abstract differential equation characterizing the response $x(t)$ of a flexible structure to the applied input $u(t)$:

$$M\ddot{x}(t) + D\dot{x}(t) + Ax(t) + Bu(t) = 0 \tag{2.1}$$

where M is a self-adjoint positive definite (zero is in the resolvent spectrum) linear bounded operator mapping H into H;

A is a self-adjoint nonnegative definite closed linear operator with dense domain and with compact resolvent; we shall (for simplicity) also assume that zero is in the resolvent set.

D is a self-adjoint nonnegative definite closed linear operator whose domain includes that of $\sqrt{A}$; and

B is a finite-dimensional linear operator mapping R^m into H.

The next step is to introduce the "energy norm" space, which we shall denote by $\mathcal{H}_E$. On the product space

$$\mathcal{D}(\sqrt{A}) \times \mathcal{H} \tag{2.2}$$

we can introduce the "energy" inner product:

$$[Y,Z]_E = [\sqrt{A}\, y_1, \sqrt{A}\, z_1] + [My_2, z_2] \tag{2.3}$$

where

$$Y = \begin{vmatrix} y_1 \\ y_2 \end{vmatrix} \quad ; \quad Z = \begin{vmatrix} z_1 \\ z_2 \end{vmatrix} \tag{2.4}$$

and complete the space in this inner product to yield $\mathcal{H}_E$. To avoid confusion we shall use a subscript E to denote the inner product in $\mathcal{H}_E$. Let $\mathcal{A}$ denote the operator

$$\mathcal{A} = \begin{vmatrix} 0 & I \\ -M^{-1}A & -M^{-1}D \end{vmatrix} \tag{2.5}$$

with domain:

$$\mathcal{D}(\mathcal{A}) = \left[Y \;\middle|\; Y = \begin{vmatrix} y_1 \\ y_2 \end{vmatrix} ; \; \begin{matrix} y_1 \in \mathcal{D}(A) \\ y_2 \in \mathcal{D}(\sqrt{A}) \end{matrix} \right] . \tag{2.6}$$

We shall now show that $\mathcal{A}$ is closed.

$$\begin{vmatrix} y_{1,n} \\ y_{2,n} \end{vmatrix} = Y_n \in \mathcal{D}(\mathcal{A}) \; ; \qquad Y_n \to Y = \begin{vmatrix} y_1 \\ y_2 \end{vmatrix}$$

$$\begin{vmatrix} z_{1,n} \\ z_{2,n} \end{vmatrix} = Z_n = \mathcal{A}Y_n \; ; \qquad Z_n \to Z = \begin{vmatrix} z_1 \\ z_2 \end{vmatrix}$$

where

$$z_{1,n} = y_{2,n} \tag{2.7}$$

$$Mz_{2,n} = -Ay_{1,n} - Dy_{2,n} \tag{2.8}$$

and

$$\|z_n - z\|_E^2 = \|\sqrt{A}\,(z_{1,n} - z_1)\|^2 + [(M(z_{1,n} - z_n), (z_{2,n} - z_n)]$$

Since $(\sqrt{A})$ has a bounded inverse, it follows that

$$z_{1,n} = (\sqrt{A})^{-1}(\sqrt{A}\, z_{1,n})$$

converges. But by (2.7)

$$y_{2,n} = z_{1,n}$$

so that

$$\{y_{2,n}\} \quad \text{and} \quad \{\sqrt{A}\, y_{2,n}\}$$

are Cauchy. Now since domain of D includes that of $\sqrt{A}$

$$D(\sqrt{A})^{-1}$$

is linear bounded, and hence

$$Dy_{2,n} = D(\sqrt{A})^{-1}(\sqrt{A}\, y_{2,n})$$

converges. Since the left side of (2.8) is Cauchy, this implies that

$$\{Ay_{1,n}\}$$

is Cauchy, and of course $\{y_{1,n}\}$ is Cauchy. Hence we see that $\mathcal{A}$ is closed.

Since

$$\mathcal{A}^* = \begin{bmatrix} 0 & -I \\ M^{-1}A & -M^{-1}D \end{bmatrix}$$

with same domain as $\mathcal{A}$, and for Y in $\mathcal{D}(\mathcal{A})$:

$$[\mathcal{A}Y, Y]_E + [\mathcal{A}^*Y, Y]_E = -2[Dy_2, y_2] \tag{2.9}$$

where

$$Y = \begin{vmatrix} y_1 \\ y_2 \end{vmatrix}$$

we see that $\mathcal{A}$ and $\mathcal{A}^*$ are dissipative, and hence [7] $\mathcal{A}$ generates a strongly continuous contraction semigroup which we shall denote $S(t)$, $t \geq 0$. Let $\mathcal{A}_0$ denote the "undamped" generator:

$$\mathcal{A}_0 = \begin{bmatrix} 0 & I \\ -M^{-1}A & 0 \end{bmatrix}$$

with domain

$$= \left[Y \,\middle|\, Y = \begin{vmatrix} y_1 \\ y_2 \end{vmatrix}, \ y_1 \in \mathcal{D}(A) \right] .$$

Let

$$\mathcal{D} = \begin{bmatrix} 0 & 0 \\ 0 & M^{-1}D \end{bmatrix} .$$

We note that $\lambda > 0$ belongs to the resolvent of $\mathcal{A}_0$ and that $R(\lambda, \mathcal{A}_0)$ is compact. From the easily verified resolvent equation:

$$R(\lambda, \mathcal{A}) = R(\lambda, \mathcal{A}_0) [I + \mathcal{D}R(\lambda, \mathcal{A})] \tag{2.10}$$

it follows that $R(\lambda, \mathcal{A})$ is compact, since $\mathcal{D}R(\lambda, \mathcal{A})$ is bounded. Rewriting (2.1) as

$$\dot{Y}(t) = \mathcal{A}Y(t) + \mathcal{B}u(t) \tag{2.11}$$

where

$$Y(t) = \begin{vmatrix} x(t) \\ \dot{x}(t) \end{vmatrix}$$

$$\mathcal{B}u = \begin{vmatrix} 0 \\ -M^{-1}Bu \end{vmatrix}$$

we see that (2.11) has the "mild" solution:

$$Y(t) = S(t)\,Y(0) + \int_0^t S(t-\sigma)\,\mathcal{B}u(\sigma)\,d\sigma \qquad (2.12)$$

Let $S_0(\cdot)$ denote the semigroup generated by $\mathcal{A}_0$. Let $\{\phi_k\}$ denote the eigenvectors of

$$M^{-1}A\,\phi_k = \omega_k^2\,\phi_k$$

orthonormalized so that

$$[M\phi_k, \phi_j] = \delta_j^k$$

and such that ω_k^2 are monotone increasing. Note that these are the undamped or "natural" modes.

We shall say that a (time-invariant) "feedback" control

$$u(t) = K(Y(t)) \qquad (2.13)$$

where $K(\cdot)$ maps $\mathcal{H}_E$ into R^m, "stabilizes" the system (2.11) if

$$Y(t) = S(t)\,Y(0) + \int_0^t S(t-\sigma)\,\mathcal{B}K(Y(\sigma))\,d\sigma$$

has a unique strongly continuous solution such that it is globally stable. That is to say

$$\|Y(t)\|_E \to 0 \qquad \text{as} \quad t \to \infty$$

for every initial $Y(0)$.

We begin with linear controllers.

THEOREM 2.1. Suppose

$$[\phi_k, (D + BB^*)\phi_k] \neq 0 \tag{2.14}$$

for every k. Then the linear feedback control

$$u(t) = -P\mathcal{B}^*Y(t) \tag{2.15}$$

where P is positive definite and has a bounded inverse, yields global asymptotic stability.

Conversely, (2.14) is necessary if these controls are to yield global stability.

Proof. Without loss of generality we may take $P = I$. To begin with we assume that the strong version of (2.11) is satisfied:

$$M\ddot{x}(t) + D\dot{x}(t) + Ax(t) + Bu(t) = 0 \ . \tag{2.16}$$

Let

$$E(t) = \frac{1}{2} \|Y(t)\|_E^2$$

$$= \frac{1}{2} ([\sqrt{A}\,x(t), \sqrt{A}\,x(t)] + [M\dot{x}(t), \dot{x}(t)]) \ . \tag{2.17}$$

Then

$$\frac{d}{dt} E(t) = -[D\dot{x}(t), \dot{x}(t)] - [\dot{x}(t), Bu(t)] \tag{2.18}$$

$$= -[(D + \lambda(t)BB^*)\dot{x}(t), \dot{x}(t)] \tag{2.19}$$

where

$$\lambda(t) = \frac{[u(t), B^*\dot{x}(t)]}{\|B^*\dot{x}(t)\|^2} \tag{2.20}$$

$$= 0 \qquad \text{if} \quad \|B^*\dot{x}(t)\| = 0 \ .$$

From (2.18) we obtain that

$$E(t) \text{ is monotone nonincreasing in } t$$

and that

$$E(0) - E(\infty) = \int_0^\infty [D\dot{x}(t), \dot{x}(t)]\, dt + \int_0^\infty \lambda(t)[B^*\dot{x}(t), B^*\dot{x}(t)]\, dt \ . \tag{2.21}$$

Hence if we consider first the choice

$$u(t) = B^*\dot{x}(t) \tag{2.22}$$

or equivalently

$$\lambda(t) = 1$$

we have correspondingly

$$\int_0^\infty [D\dot{x}(t), \dot{x}(t)]\, dt + \int_0^\infty [B^*\dot{x}(t), B^*\dot{x}(t)]\, dt = E(0) - E(\infty) < \infty \ . \tag{2.23}$$

To prove global stability, it is enough to show that

$$E(\infty) = 0$$

in (2.21), or in (2.23). Next let

$$a_k(t) = [x(t), M\phi_k]$$

so that we have the "modal" expansion:

$$x(t) = \sum_1^\infty a_k(t)\, \phi_k \tag{2.24}$$

and

$$E(t) = \frac{1}{2}\sum_1^\infty \left(\omega_k^2 a_k(t)^2 + \dot{a}_k(t)^2\right) \ . \tag{2.25}$$

Using this expansion in (2.16) we have

$$\ddot{a}_k(t) + \omega_k^2 a_k(t) = -[D\dot{x}(t), \phi_k] - [B^*\dot{x}(t), B^*\phi_k] \tag{2.26}$$

or,

$$\ddot{a}_k(t) + \{[D\phi_k,\phi_k] + [B^*\phi_k, B^*\phi_k]\}\,\dot{a}_k(t) + \omega_k^2 a_k(t) = -f_k(t) \tag{2.27}$$

where

$$f_k(t) = [D\dot{x}(t), \phi_k] - \dot{a}_k(t)[D\phi_k, \phi_k] + [B^*\dot{x}(t), B^*\phi_k] - \dot{a}_k(t)[B^*\phi_k, B^*\phi_k] \;.$$

Suppose now that for some k:

$$D\phi_k = 0\;; \qquad B^*\phi_k = 0\;.$$

Then

$$x(t) = a_k(t)\phi_k$$

is a solution of (2.17), provided only that

$$\ddot{a}_k(t) + \omega_k^2 a_k(t) = 0$$

and hence $|a_k(t)|$ does <u>not</u> go to zero as $t \to \infty$; this takes care of the necessity condition. Assume then that

$$2\sigma_k = [D\phi_k, \phi_k] + [B^*\phi_k, B^*\phi_k]$$

is nonzero. Then (2.27) can be "solved" to yield (assuming small damping for simplicity):

$$a_k(t) = a_k\, e^{-\sigma_k t} \cos \lambda_k t + b_k\, e^{-\sigma_k t}\, \frac{\sin \lambda_k t}{\lambda_k} - \int_0^t W(t-s)\, f_k(s)\, ds \tag{2.28}$$

where

$$W(t) = \frac{e^{-\sigma_k t} \sin \lambda_k t}{\lambda_k} ;$$

$(-\sigma_k \pm i\lambda_k)$ are the roots of

$$s^2 + 2\sigma_k s + \omega_k^2 = 0 .$$

Now because:

$$|[D\dot{x}(t), \phi_k]|^2 \leq [D\dot{x}(t), \dot{x}(t)] [D\phi_k, \phi_k]$$

it follows from (2.23) that

$$\int_0^\infty |[D\dot{x}(t), \phi_k]|^2 dt < \infty .$$

Hence a little analysis shows that

$$\int_0^t W(t-s) [D\dot{x}(s), \phi_k] ds \to 0 \qquad \text{as} \quad t \to \infty .$$

By a similar reasoning, also

$$\int_0^t W(t-s) [B^*\dot{x}(s), B^*\phi_k] ds \to 0 \qquad \text{as} \quad t \to \infty .$$

Hence it follows that

$$\varepsilon(t) = a_k(t) - 2\sigma_k \int_0^t W(t-s) \dot{a}_k(s) ds \tag{2.29}$$

where

$$\varepsilon(t) \to 0 \qquad \text{as} \quad t \to \infty .$$

By integration by parts we obtain that

$$\varepsilon(t) = a_k(t) + a_k(0) W(t) + 2\sigma_k \int_0^t \dot{W}(t-\sigma) a_k(\sigma) d\sigma .$$

Hence

$$a_k(t) + 2\sigma_k \int_0^t \dot{W}(t-\sigma)\, a_k(\sigma)\, d\sigma = \beta(t) \tag{2.30}$$

where

$$\beta(t) \to 0 \qquad \text{as} \quad t \to \infty .$$

Solving (2.30) (using elementary Laplace transform techniques) we obtain that

$$a_k(t) \to 0 \qquad \text{as} \quad t \to \infty .$$

Differentiating (2.28) and proceeding in a similar way we can also show that

$$\dot{a}_k(t) \to 0 \qquad \text{as} \quad t \to \infty .$$

Let

$$\Phi_k^+ = \begin{vmatrix} \phi_k \\ i\omega_k\phi_k \end{vmatrix}$$

$$\Phi_k^- = \begin{vmatrix} \phi_k \\ -i\omega_k\phi_k \end{vmatrix} .$$

Then

$$\lim_{t\to\infty} [Y(t), \Phi_k^+] = \lim_{t\to\infty} [Y(t), \Phi_k^-] = 0$$

for each k, which is enough to imply that $Y(t)$ converges weakly to zero. Next we note that

$$Y(t) = S_B(t)\, Y(0)$$

where $S_B(t)$ is the (strongly continuous contraction) semigroup generated by

$$A - BB^*$$

which has a compact resolvent. Hence weak stability implies strong stability [7]:

$$\|Y(t)\|_E = \|S_B(t)\, Y(0)\|_E \to 0 \,. \tag{2.31}$$

In particular we have (2.23) that

$$\|Y(0)\|_E^2 = \int_0^\infty [D\dot{x}(t), \dot{x}(t)]\, dt + \int_0^\infty \|B^*\dot{x}(t)\|^2\, dt$$

or,

$$\|Y(0)\|_E^2 = \int_0^\infty [DY(t), Y(t)]_E\, dt + \int_0^\infty \|B^*Y(t)\|^2\, dt \,. \tag{2.32}$$

Finally given any arbitrary initial condition $Y(0)$, we can find an approximating sequence $\{Y_n(0)\}$ in the domain of A such that (2.31) holds for the corresponding solution $Y_n(t)$; and the result follows from the estimate:

$$\|Y(t)\|_E \le \|S_B(t)\, Y_n(0)\|_E + \|Y_n(0) - Y(0)\|_E \,.$$

REMARK 1. We note that if $D = 0$, our condition (2.14) is equivalent to requiring $(A \sim B)$ controllability (see [2]). Hence we would have strong stability by a more general argument due to Benchimol [5]. If $B = 0$, then condition (2.14) implies that $D\phi_k \neq 0$ for any k, and we are then proving strong stability for the semigroup $S(t)$.

<u>Nonlinear Controllers</u>

Let us now go on to consider a class of nonlinear controllers. Thus we shall consider where $K(\cdot)$ in (2.13) is given by

$$K(Y) = f(B^*y_2)$$

where

$$Y = \begin{vmatrix} y_1 \\ y_2 \end{vmatrix},$$

and $f(\cdot)$ maps R^m into R^m, and satisfies the following conditions:

i) $[f(u), u] > 0$ for $u \neq 0$

ii) $\|f(u)\| \leq \lambda\|u\|$

iii) $f(\cdot)$ is Lipschitz.

A typical example of $f(\cdot)$ is

$$f(u) = v; \qquad u = \{u_i\}; \qquad v = \{v_i\}$$

$$v_i = \gamma_i \tan^{-1} \mu_i u_i; \qquad \gamma_i, \mu_i > 0$$

THEOREM 2.2. Suppose

zero is in the resolvent set of D (2.14a)

for every k. Then the feedback control

$$u(t) = -B^* f(B^*\dot{x}(t)) \qquad (2.33)$$

where

$$Y(t) = \begin{vmatrix} x(t) \\ \dot{x}(t) \end{vmatrix}$$

yields asymptotic global stability.

Proof. Under condition (2.14a), the semigroup $S(\cdot)$ generated by A

in $\mathcal{H}_E$ is strongly stable. By virtue of the Lipschitz conditions on $f(\cdot)$, existence and uniqueness of solution for each $t > 0$ is immediate, and the solution is given by:

$$Y(t) = S(t)\,Y(0) + \int_0^t S(t-\sigma)\,\mathcal{B}\,f(B^*\dot{x}(\sigma))\,d\sigma, \qquad 0 < t . \tag{2.34}$$

Next we shall show that

$$\int_0^\infty \|f(B^*\dot{x}(\sigma))\|^2\,d\sigma < \infty . \tag{2.35}$$

For this purpose let us assume first that the initial condition $Y(0)$ is in the domain of the generator $\mathcal{A}$, so that we have

$$M\ddot{x}(t) + D\dot{x}(t) + Bf(B^*\dot{x}(t)) + Ax(t) = 0 . \tag{2.36}$$

Defining the energy again as

$$E(t) = \frac{1}{2}\|Y(t)\|_E^2$$

we have that

$$\dot{E}(t) = -[D\dot{x}(t), \dot{x}(t)] - [\dot{b}(t), f(\dot{b}(t))]$$

where

$$b(t) = B^*x(t) .$$

Since

$$[\dot{b}(t), f(\dot{b}(t))] \geq 0 ,$$

we have that:

$$\int_0^\infty [D\dot{x}(t), \dot{x}(t)]\,dt < \infty$$

and since we can write

$$(\sqrt{D})^{-1} (\sqrt{D}\,\dot{x}(t)) = \dot{x}(t)$$

it follows that

$$\int_0^\infty \|\dot{x}(t)\|^2 \, dt < \infty .$$

But this implies that

$$\int_0^\infty \|B^*\dot{x}(t)\|^2 \, dt < \infty$$

and hence also

$$\int_0^\infty \|f(B^*\dot{x}(t))\|^2 \, dt < \infty$$

by virtue of our assumptions on $f(\cdot)$.

We now proceed as in the proof of the preceding theorem. Writing

$$2\sigma_k = [D\phi_k, \phi_k]$$

we have:

$$\ddot{a}_k(t) + \omega_k^2 a_k(t) + 2\sigma_k \dot{a}_k(t) = -f_k(t) \tag{2.37}$$

where $a_k(t)$ is as before, and

$$f_k(t) = [D\dot{x}(t), \phi_k] - \dot{a}_k(t)[D\phi_k, \phi_k] + [f(B^*\dot{x}(t)), B^*\phi_k]$$

since

$$\int_0^\infty [D\dot{x}(t), \phi_k]^2 \, dt + \int_0^\infty [f(B^*\dot{x}(t)), B^*\phi_k]^2 \, dt < \infty ,$$

as before, we obtain

$$\varepsilon(t) = a_k(t) + a_k(0)\, w(t) + 2\sigma_k \int_0^t \dot{w}(t-\sigma)\, a_k(\sigma)\, d\sigma$$

where $\varepsilon(t) \to 0$ as $t \to \infty$ and the rest of the arguments follow. Hence

$$a_k(t) \to 0 \qquad \text{as } t \to \infty$$

$$\dot{a}_k(t) \to 0 \qquad \text{as } t \to \infty$$

for each k. Hence it follows that

$$Ax(t) \to 0$$

$$A\dot{x}(t) \to 0 .$$

Since A^{-1} is compact, it follows that $x(t)$ converges strongly to zero (in $\mathcal{H}$). Similarly

$$A\dot{x}(t) \to 0$$

implies that $\dot{x}(t)$ converges strongy to zero (in $\mathcal{H}$). Hence $Y(t)$ converges weakly (in $\mathcal{H}_E$) to zero. Now

$$[D\dot{x}(t), \phi_k] = [\dot{x}(t), D\phi_k]$$

implies that

$$D\dot{x}(t) \to 0 .$$

Hence

$$\mathcal{A}Y(t) \to 0 .$$

since $\mathcal{A}$ has a compact resolvent it follows that $Y(t)$ converges strongly (In $\mathcal{H}_E$) to zero.

Next we relax our assumption regarding the initial condition $Y(0)$. We begin by proving Lipschitz continuity of the solution with respect to the initial condition. Thus let $Y(t)$ denote the solution with initial condition $Y(0)$ and let

$$Y(t) = M(t)(Y(0)) .$$

Then let $Y_1(0), Y_2(0) \in \mathcal{H}_E$ and

$$M(t)(Y_2(0)) - M(t)(Y_1(0)) = S_D(t)(Y_2(0) - Y_1(0))$$

$$+ \int_0^t S_D(t-\sigma)\left(Bf(B^*Y_2(\sigma)) - Bf(B^*Y_1(\sigma))\right) d\sigma .$$

Let

$$m(t) = \|Y_2(t) - Y_1(t)\|_E .$$

Then in view of our assumptions on $f(\cdot)$, we have

$$m(t) \leq m(0) + \gamma\int_0^t m(\sigma)\, d\sigma ; \qquad \gamma \geq 0$$

and hence by the usual analysis

$$m(t) \leq e^{\gamma t}\, m(0)$$

yielding Lipschitz continuity. The continuity yields in turn

$$\|Y(t)\|_E^2 = 2E(t) = \|Y(0)\|_E^2 - \int_0^t [D\dot{x}(s), \dot{x}(s)]\, ds$$

$$- \int_0^t [\dot{b}(s), f(\dot{b}(s))]\, ds .$$

and hence we obtain:

$$\int_0^\infty \|D\dot{x}(t)\|^2\, dt + \int_0^\infty [\dot{b}(t), f(\dot{b}(t))]\, dt < \infty .$$

Next we need to establish (2.37). But this follows readily from the fact (2.11) holds in the weak sense. Hence we obtain that $Y(t)$ converges strongly to zero in $\mathcal{H}_E$ for any initial $Y(0)$ in $\mathcal{H}_E$.

REMARK. It would be of interest to establish strong stability under the weaker condition (2.14). We note that the condition (2.14) is again obviously necessary; indeed

$$D\phi_k = 0 \quad \text{and} \quad B^*\phi_k = 0$$

implies that

$$x(t) = a(t)\,\phi_k$$

is a solution of (2.35) with

$$\ddot{a}(t) + \omega_k^2\, a(t) = 0 .$$

REFERENCES

[1] SCOLE Workshop Proceedings, 1984. Compiled by L.W. Taylor. NASA Langley FRC, Hampton, Virginia.

[2] A.V. Balakrishnan: On a Large Space Structure Control Problem, Proceedings of the IFIP Working Conference on Control of Systems Governed by Partial Differential Equations, Gainesville, Florida, 1986. (To be published.)

[3] A.V. Balakrishnan: A Mathematical Formulation of the SCOLE Control Problem, Part 1, NASA CR-172581.

[4] I. Lasiecka and R. Triggiani: A Cosine Operator Approach to Modelling $L_2(0,T; L_2(\Gamma))$ Boundary Input Hyperbolic Equations, Applied Math and Optimization, Vol. 7, No. 1 (1981), pp. 35-93.

[5] C.D. Benchimol: Feedback Stabilization in Hilbert Spaces, Applied Math and Optimization, Vol. 4, No. 3 (1978), pp. 225-248.

[6] N. Levan: Stability Enhancement by State Feedback, in Proceedings of the IFIP Working Conference on Control of Systems Governed by Partial Differential Equations, Gainesville, Florida, Febraury 1986.

[7] A.V. Balakrishnan: Applied Functional Analysis, 2nd edition. Springer-Verlag, 1981.

STATIONARY AND MOVING FREE BOUNDARY PROBLEMS

RELATED TO THE CAVITATION PROBLEM

G. Bayada * M. Chambat ** M. El Alaoui Talibi *

The cavitation in lubrication has been examined for over a century but physical understanding is still incomplete. Here we concern ourselves with gaseous cavitation in both steady state and unsteady operating conditions. Visual observations show that the lubricant film does not cover the whole surface Ω of the lubricated mechanism, such as journal bearing or seal, and a free boundary appears between a full film region and a cavitating region, filled with an air-fluid blend.

A rigorous approach would be to consider the Stokes or Navier Stokes equation in the full three dimensional space and to introduce the real free boundary problem. However, as the gap between the two surfaces of the lubricated mechanism is less that 12^{-5}m, a two-dimensional approximation of the equations describing the flow is used in practice: the two-dimensional Reynolds equation for the pressure. This approximation is very similar to the one of the thin plate equations. A mathematical study of the transition between Stokes and Reynolds equations can be found in [BC1] for the steady-state problem but the occurence of the cavitation has not been taken into account in this paper.

The relation ship between the three-dimensional real cavitation and the two-dimensional cavitation which is the one studied in this paper is still an open problem.

* Centre de Mathématiques, Insa Lyon - 403, 69621 Villeurbanne Cedex

** LAN, Université Lyon I - Bât. 101, 69622 Villeurbanne Cedex

A review of the mathematical problems involved in cavitation modelling and the related physical views appears in [B2] [BC2] for steady-state problems. The problem has been very often formulated as a variational inequation problem [CI] but it has been shown that this kind of models are not satisfactory regarding output or input flow values. A new model has been proposed [ELR] [FLO] by introducing a new variable θ together with the classical pressure p:

$$\theta = 1 \qquad \text{where } p > 0 \text{ in the full film region}$$

$$0 \leq \theta \leq 1 \qquad \text{where } p = 0 \text{ in the cavitating region}$$

So θ acts as a saturation function.

Let us consider first the steady state case. We define the mass flow in the (x_1, x_2) plane.

$$\vec{F} = h^3 \vec{\nabla}_x p - \theta h\vec{u} \tag{1}$$

where $\vec{u}$ is a given velocity vector. The mass flow conservation law $\operatorname{div}(\vec{F}) = 0$ alows us to obtain the following strong formulation:

$$\operatorname{div}(h^3 \vec{\nabla} p - h\vec{u}) = 0 \qquad \theta = 1 \qquad \text{if } p > 0 \tag{2}$$

$$\operatorname{div}(\theta h\vec{u}) = 0 \qquad 0 \leq \theta \leq 1 \qquad \text{if } p = 0 \tag{3}$$

$$p = 0, \quad h^3 \frac{\partial p}{\partial n} = (1 - \theta)\, h\, \vec{u}.\vec{n} \quad \text{on the free boundary} \tag{4}$$

It appears that the values of the data p or/and θ on the boundary of the area Ω play a major role in the existence and uniqueness of the solution. This is the case if we consider the starvation problem in which the natural boundary condition is

$$p = 0 \qquad \text{on } \partial\Omega \tag{5}$$

Clearly $p \equiv 0$ is a solution of the problem. In order to obtain a well posed problem, it is necessary to add to (5) another condition on θ.

We present in the following section a new method to study the corresponding free boundary problem which does not require any assumption on the free boundary but the regularity and allows us to obtain an uniqueness theorem. The third section is devoted to the study of the full moving boundary problem. The gap h between the two surfaces is a function both of space and time. The mass flow is:

$$\vec{F} = (h^3 \vec{\nabla}_x p - \theta h\vec{u}, \theta h) \tag{6}$$

The mass flow conservation law induces the following equations:

$$\mathrm{div}_x (h^3 \vec{\nabla}_x p - h\vec{u}) = \frac{\partial h}{\partial t}, \qquad \theta = 1 \qquad \text{if } p > 0 \tag{7}$$

$$\frac{\partial \theta h}{\partial t} + \mathrm{div}_x(\theta\, h\vec{u}) = 0 \qquad 0 \leq \theta \leq 1 \qquad \text{if } p = 0 \tag{8}$$

$$p = 0, \quad h^3 \frac{\partial p}{\partial n} = (1-\theta)\, h\, (\vec{u}.\vec{n} - q_n) \text{ on the free boundary} \tag{9}$$

We give two existence theorems for this problem and we briefly present some numerical results.

1 - The steady-state starvation problem.

We consider as a model problem a journal bearing $\Omega =]0,2\pi[\times]0,1[$ with a supply line $\Gamma_0 = \{0\} \times]0,1[$ located at the maximum gap. $\vec{u} = (1,0)$ is the constant velocity of the running shaft and the gap function is $h(x) = 1 + \alpha \cos(x_1)$ where $0 < \alpha < 1$.

We find a splitting of Ω in Ω_+ in which (2) is satisfied and Ω_0 in which (3) reduces now to:

$$\frac{\partial}{\partial x_1} (\theta\, h) = 0 \tag{3'}$$

Condition (4) reduces to:

$$p = 0, \qquad h^3 \frac{\partial p}{\partial n} = (1 - \theta)\, h \cos(\vec{n}, \vec{x}_1) \text{ on } (\Sigma) \tag{4'}$$

We introduce also

$$\Gamma_{00} = \Gamma_0 \cap (\bar{\Omega}_0) \text{ and } \Gamma_{0+} = \Gamma_0 \cap [\bar{\Omega}_+]$$

The physical situation is illustrated by the figure 1.

To obtain a variational formulation for the problem we define $\theta_0(x_2)$ in $L^\infty(\Gamma_0)$ such that:

$$\theta_0(x_2) = \theta\,(0,x_2) \text{ on } \Gamma_{00}$$

$$\theta_0(x_2) = 1 - h^2(0) \frac{\partial p}{\partial x_1} \quad \text{on } \Gamma_{0+}$$

LEMMA 1: We have $0 \leq \theta_0(x_2) \leq 1$ a.e.

Proof: As $p > 0$ on $\Omega+$ and $p = 0$ on Γ_0, we deduce $\theta_0(x_2) \leq 1$.

Let us consider now the new function:

$$\tilde{p} = p - \int_0^{x_1} 1/h^2(\xi)\, d\xi$$

We easily obtain by the maximum principle that $\dfrac{\partial \tilde{p}}{\partial x_1} \leq 0$ on Γ_{0+} which ends the proof.

We are able to give a precise formulation of the free boundary problem in which θ_0 acts like a parameter linked to the amount of fluid in the supply line Γ_0.

PROBLEM (P_0): Find $p \in H_0^1(\Omega)$, $\theta \in L^\infty(\Omega)$, $\theta_0 \in L^\infty(\Gamma_0)$ such that:

$$a(p,\varphi) = \int_\Omega \theta h \frac{\partial \varphi}{\partial x_1}\, dX + \int_{\Gamma_0} \theta_0 h \varphi\, dx_2 \qquad \forall\, \varphi \in V \tag{10}$$

$$H(p) \leq \theta \leq 1, \quad 0 \leq \theta_0 \leq 1, \quad p \geq 0 \tag{11}$$

where H is the Heaviside function, $a(p,\varphi)$ the bilinear form

$$a(p,\varphi) = \int_\Omega h^3\, \nabla p\, \nabla\varphi\, dX ,$$

and $\quad V = \{\varphi \in H^1(\Omega),\ \varphi = 0 \quad \text{on } \partial\Omega - \Gamma_0\}$.

As a first step, we consider a relaxed problem (P) which is obtained up to (P_0) by searching p in V and no more in $H_0^1(\Omega)$.

THEOREM 1: For each θ_0, $\theta_0 \geq 0$, there exists (p,θ) such that (p,θ,θ_0) is a solution of the relaxed problem (P) and $p \in V \cap C^0(\Omega)$.

Proof: We introduce for a little parameter ϵ the following approximated problem:

PROBLEM (P_ϵ): Find $p_\epsilon \in V$ and $\theta_0 \in L^\infty(\Gamma_0)$ such that:

$$a(p_\epsilon,\varphi) = \int_\Omega H_\epsilon(p_\epsilon)\, h \frac{\partial \varphi}{\partial x_1}\, dX + \int_{\Gamma_0} \theta_0 h \varphi\, dx_2 \qquad \forall\, \varphi \in V \tag{12}$$

$$0 \leq \theta_0 \leq 1, \quad p_\epsilon \geq 0 \tag{13}$$

where $H_\epsilon(t) = \{0 \text{ if } t \leq 0,\ t/\epsilon \text{ if } 0 \leq t \leq \epsilon ,\ 1 \text{ if } t \geq \epsilon\}$

For each θ_0 in $L^\infty(\Gamma_0)$, we use the Schauder fixed point theorem to prove the existence of p_ϵ in V satisfying (12). Condition (13.2) is fulfilled by choosing $\varphi = \overline{p}_\epsilon$ as a test function and using the assuption $\theta_0 \geq 0$. Moreover, we have:

$$\|p_\epsilon\|_V \leq C^{st} \text{ and } |H_\epsilon(p_\epsilon)|_{L^\infty} \leq 1.$$

So letting ϵ tend to zero, we obtain the existence of p and θ satisfying (12) (11). Now from (12) and choosing φ in $\mathcal{D}(\Omega)$, we write

$\operatorname{div}(h^3 \nabla p) = \dfrac{\partial \theta h}{\partial x}$ in $\mathcal{D}'(\Omega)$ which gives the regularity of p.

THEOREM 2: If $0 \leq \theta_0 \leq h(\pi)/h(0)$, there exists an obvious solution for problems (P) and (P_0).

Proof: It is easy to check that:
$(p=0,\ \theta = \theta_0(x_2)\ h(x_1)\ /\ h(0),\ \theta_0(x_2))$ is solution of (12).
The condition on θ_0 ensures that (11.2) is fulfilled.

THEOREM 3: If $\theta_0 > h(\pi)/h(0)$, the non cavitating area $\Omega+$ is non empty and is a connected one.

Proof: a detailed proof can be found in [B1] We recall here the lines of the proof.
We define first

$$\Omega_1 = \{(x_1, x_2)\ /\ 0 < x_1 < \pi,\ h(x_1) < \theta_0(x_2)\ h(0)\}$$

and we prove that p is > 0 on Ω_1.
This is a consequence of the fact that if $p(\tilde{x}_1, \tilde{x}_2) > 0$ with $\tilde{x}_1 < \pi$, then $p(x_1, \tilde{x}_2) > 0$ for $\tilde{x}_1 < x_1 < \pi$.

This last property is proved by contradiction using the continuity of p and the maximum principle with $\dfrac{\partial h}{\partial x_1} < 0$ if $x_1 < \pi$

At this step, we get that the free boundary in the area $\{x_1 < \pi\}$ is necessary a reformation boundary for the pressure.
In the second step, we prove in a similar way that if $p(\tilde{x}_1, \tilde{x}_2)$ is > 0 and $\tilde{x}_1 > \pi$, then $p(x_1, \tilde{x}_2) > 0$ for $\pi < x_1 < \tilde{x}_1$.

Remarks: - In this proof, we assume that $\Omega+$ and Ωo are smooth enough so that Green theorem may be used.

- this theorem proves that the conjecture used in [BC3] on the position of the built-up and built-down free boundaries is valid.

THEOREM 4: If $\theta_0 > h(\pi)/h(0)$, there exists a unique pair (p,θ) such that (p,θ,θ_0) is solution of (P).

Proof: See also [B1]. We use a method initialy proposed in [CH] for the dam problem by introducing the equivalent problem:

Find $p \in V$, $\theta \in L^\infty(\Omega)$, $\theta_0 \in L^\infty(\Gamma_0)$

$$a(p,\varphi) \leq \int_\Omega \theta h \frac{\partial\varphi}{\partial x_1} dX + \int_{\Gamma_0} \theta_0 h\varphi \, dx_2$$

$$\forall\varphi\ ,\ \varphi \in H^1(\Omega) \quad \varphi = 0 \ \text{ on } \Gamma_3,\ \varphi \geq 0 \text{ on } \Gamma_1.$$

where $\Gamma_3 = 2\pi \times (0,1)$ and $\Gamma_1 = \partial\Omega - \Gamma_0 - \Gamma_3$.

An identification procedure is conducted up to the $\partial\Omega_1 \cap \Gamma_1$ boundary where we are sure that two possible solutions have the same value and are strictly positive in a neighbourhood of this boundary.

We define now an auxiliary problem (PM) whose solution lies in $H^1(\Omega)$ and which will act as an upper bound for the family of solutions related to the problem P_0.

We denote $C = \{M(x_1, x_2) \quad x_1 < \pi\}$

$D = \{M(x_1, x_2) \quad x_1 > \pi\}$

PROBLEM (PM): Find $P_m \in H^1(\Omega)$, $\theta_m \in L^\infty(D)$ such that:

$$a(p_m,\varphi) = \int_C h \frac{\partial\varphi}{\partial x_1} dX + \int_D \theta_m h \frac{\partial\varphi}{\partial x_1} dX \qquad \forall\ \varphi \in H_0^1(\Omega) \quad (14)$$

$$H(p_m) \leq \theta_m \leq 1, \qquad p_m \geq 0$$

THEOREM 5: There exists a solution (p_m, θ_m) to (PM) and we can define $\theta_{0m} \in L^\infty(\Gamma_0)$ such that $(p_m, \theta_m, \theta_{0m})$ is a solution to the initial problem (P_0).

Proof: The existence of a solution for (PM) is given, as in the theorem 1 by substituting $H_\epsilon(p_\epsilon)$ for θ_m and using the Shauder fixed point theorem. Now we show that p_m is strictly positive on C. Suppose by contradiction $p_m = 0$ on an open set ω in C, (14) implies:

$$\frac{\partial h}{\partial x_1} = 0 \qquad \text{on } \omega$$

which is impossible.

So we can extend θ_m by 1 on the whole Ω.

We define as in lemma 1 the input flow θ_{0m} by

$$(1-\theta_{0m})\, h(0) = h^3(0) \frac{dp_m}{dx_1} \qquad \text{on } \Gamma_0$$

To prove that $(p_m, \theta_m, \theta_{0m})$ is a solution of (P_0), it suffices now to compute $a(p_m,\varphi)$ with φ in V instead of H^1 and to use the strong formulation associated with (14).

THEOREM 6: For each θ_0 such that $h(\pi)/h(0) \leq \theta_0 \leq \theta_{0m}$, a.e. x_2, the solution of the problem (P) is also a solution of (P_0).

Proof: The only thing to prove is that the solution p of (P) lies not only on V but also in $H_0^1(\Omega)$.

We use the monotonicity of the solutions of (P) towards θ_0 by way of their approximations p_ϵ and $p_{m\epsilon}$:

For each $\varphi \in V$, we obtain:

$$a(p_\epsilon - p_{\epsilon m}, \varphi) = \int_\Omega [H_\epsilon(p_\epsilon) - H_\epsilon(p_{m\epsilon})] h \frac{\partial \varphi}{\partial x_1}\, dX + \int_{\Gamma_0} (\theta_0 - \theta_{0m}) h \varphi\, dx_2$$

But $\theta_0 \leq \theta_{0m}$, so:

$$a(p_\epsilon - p_{\epsilon m}, \varphi) \leq \int_\Omega [H_\epsilon(p_\epsilon) - H_\epsilon(p_{m\epsilon})] h \frac{\partial \varphi}{\partial x_1}\, dX$$

We choose now as test function $\varphi = (p_\epsilon - p_{m\epsilon} - \delta)^+ / (p_\epsilon - p_{m\epsilon})$ where δ is a little parameter.

Letting δ tend to zero, we obtain [CH]:

$$p_\epsilon \leq p_{m\epsilon} \quad \text{which implies} \quad p \leq p_m$$

As p_m lies in $H^1(\Omega)$, we have also $p_{\Gamma_0} = 0$, so $p \in H_0^1(\Omega)$

Remarks: - Numerical results about the influence of θ_0 may be found in [B2],

- The function p_m is nothing else than the classical Reynolds solution widely used in the lubrication area. It satisfies the variational inequality [BC2]:

Find p in $K = \{\varphi \in H^1(\Omega),\ \varphi \geq 0 \text{ p.p.}\}$ such that

$$a(p, \varphi - p) \geq - \int_\Omega \frac{\partial h}{\partial x_1} (\varphi - p)\, dx \qquad \forall \varphi \in K$$

2 - The moving boundary problem.

We consider as a model problem a radial face seal [TF] (Fig. 2) where the gap h is a function both of the space and of the time. Equations (7), (8) and (9) can be reduced into a single one.

$$\frac{\partial}{\partial t} \theta h - \nabla \cdot (h^3 \nabla P) + \nabla \cdot (\theta h \vec{u}) = 0 \quad \text{in} \quad \mathscr{D}'(Q_T) \qquad (15)$$

where $Q_T = \Omega \times]0,T[$.

This two phases problem is quite different from the usual classical Stefan problems:

- θ is not a strictly monotone function of P,
- each unknown appears only in one phase,
- the equations are of different nature in each phase.

We give now a variational formulation of the problem (7), (8), (9).

PROBLEM ($\mathscr{P}$): Find $P \in L^2(0, T; H^1(\Omega))$, $\theta \in L^\infty(Q) \times H^1(0,T; H^{-1}(\Omega))$ such that:

$$P \geq 0 \quad \text{and} \quad \theta \in H(P) \qquad (16)$$

$$P = 0 \text{ on } \Sigma_i,\ P = P_a \text{ on } \Sigma_e \qquad (17)$$

$$\int_Q -\theta h \frac{\partial \varphi}{\partial t} + h^3 \nabla P . \nabla \varphi - \theta h\, \vec{V}.\overrightarrow{\nabla \varphi} = 0 \qquad \forall \varphi \in H_0^1(Q\ ) \qquad (18)$$

$$\theta(0,x) = \theta_0(x) \text{ in } H^{-1}(\Omega) \tag{19}$$

where:

- $\partial\Omega = \Gamma_e \cup \Gamma_i$, $\Sigma_i = \Gamma_i \times]0,T[$, $\Sigma_e = \Gamma_e \times]0,T[$.

- H is the Heaviside graph,

- P_a **is the given supply pressure,** $P_a \geq 0$.

In the following $\bar{P}_a$ **means a function in** $H^1(\Omega)$ **such that** $\bar{P}_{a|\Gamma_e} = P_a$ **and** $\bar{P}_{a|\Gamma_i} = 0$

2.a — A first existence theorem by elliptic regularisation.

In order to solve the problem $(\mathscr{P})$**, we introduce a family of problem** $\mathscr{P}_\epsilon$ **where the Heaviside graph is approximated by** C^∞ **functions** H_ϵ **while a parabolic regularization is introduced in the equation (18), such that:**

$$H_\epsilon \in C^\infty(\mathbb{R})\ ; \quad 0 \leq H_\epsilon \leq 1 \quad \text{and } H'_\epsilon \geq 0 \tag{20}$$

$$H_\epsilon(0) = 0 \quad \text{and} \lim_{\epsilon \to 0} \inf \{\xi > 0 : H_\epsilon(\xi) = 1\} = 0 \tag{21}$$

$$\lim_{\epsilon \to 0} L_\epsilon . \sqrt{\epsilon} = 0 \quad \text{where } L_\epsilon = \sup \{H'_\epsilon(\xi) : \xi \geq 0\} \tag{22}$$

We consider also a family $\bar{P}_{a\epsilon}$ **satisfying:**

$$\bar{P}_{a\epsilon|\Gamma e} = P_a\ , \quad \bar{P}_{a\epsilon|\Gamma i} = 0 \quad \text{and } ||\bar{P}_{a\epsilon}||_{H^1(Q)} \leq C \tag{23}$$

$$H_\epsilon(\bar{P}_{a\epsilon}) \xrightarrow[\epsilon \to 0]{} \theta_0 \quad \text{in } \mathscr{D}'(Q) \tag{24}$$

Remark: **For the existence of** H_ϵ **and** $\bar{P}_{a\epsilon}$ **see [ELA].**

For $\epsilon \in]0,1[$ **and** H_ϵ **and** $P_{a\epsilon}$ **satisfying the previous assumptions we define the problem** $(\mathscr{P}_\epsilon)$**:**

PROBLEM $(\mathscr{P}_\epsilon)$**:**

Find $P_\epsilon \in H^1(Q)$, $P_\epsilon \geq 0$ **in** Q, $P_\epsilon = \bar{P}_{a\epsilon}$ **on** $\partial_0 Q$ **such that:**

$$\int_Q \epsilon \frac{\partial P_\epsilon}{\partial t} \frac{\partial \varphi}{\partial t} + h^3\, \nabla P_\epsilon\, \nabla\varphi - h\, H_\epsilon(P_\epsilon)\,(\vec{V}.\nabla\varphi + \frac{\partial \varphi}{\partial t})$$

$$+ \int_\Omega h(T)\ H_\epsilon(P_\epsilon(T))\ \varphi(T) = 0 \tag{25}$$

$$\forall \varphi \in W = \{\varphi \in H^1(Q) \ / \ \varphi|_{\partial_0 Q} = 0\}$$

$$\partial_0 Q = \Sigma_1 \cup \Sigma_e \cup \Omega \times \{0\}$$

THEOREM 7: For every $\epsilon \in]0,1[$, there exists a unique solution P_ϵ of problem $(\mathcal{P}_\epsilon)$, such that:

$$\int_Q |\nabla P_\epsilon|^2 + \epsilon \left| \frac{\partial P_\epsilon}{\partial t} \right|^2 \leq C \tag{26}$$

$$\int_\Omega \varphi|\epsilon \frac{\partial P_\epsilon}{\partial t}(0,x)|^2 \leq C(\varphi)\{(1+L_\epsilon)\sqrt{\epsilon} + \int_\Omega \epsilon\varphi|\nabla \bar{P}_{a\epsilon}|^2\} \quad \forall P \in \mathcal{D}^+(\Omega) \tag{27}$$

C is a constant independent of ϵ and P_ϵ.
L_ϵ is the constant in (22).

Proof: For existence we use the Shauder fixed point theorem. Estimate (26) is obtained by choosing a test function $\varphi = P_\epsilon - \bar{P}_{a\epsilon}$ and by using assumptions (20), (21) on H_ϵ.

Now we obtain estimate (27) by choosing as a test function $\varphi \dfrac{\partial P_\epsilon}{\partial t}$ in the elliptic equation obtained from (25) and by integrating on Q_T.

THEOREM 8: The problem $(\mathcal{P})$ has at least one solution.

Proof: From theorem 7 and by letting ϵ tend to zero, we obtain a subsequence of (P_ϵ), $(H_\epsilon(P_\epsilon))$ and (P,θ) such that:

$$P_\epsilon \longrightarrow P \text{ in } L^2(0,T;\ H^1(\Omega)) \text{ weak} \tag{28}$$

$$H_\epsilon(P_\epsilon) \longrightarrow \theta \text{ in } L^\infty\text{-weak* and } L^2 \text{ weak} \tag{29}$$

$$\sqrt{\epsilon}\, P_\epsilon \longrightarrow 0 \text{ in } L^2(Q) \text{ weak} \tag{30}$$

In order to show that $\theta \in H(P)$, we use assumption (21) on H_ϵ to obtain:

$$\lim_{\epsilon \to 0} \int_Q h\, P_\epsilon\, (1 - H_\epsilon(P_\epsilon)) = 0$$

and similarly to [GIL] we define:

$$W^{\epsilon} = \epsilon \frac{\partial p_{\epsilon}}{\partial t} - h\,(H_{\epsilon}(P_{\epsilon}))$$

which enables us to use the theorem 5.1 in [LI] and we obtain:

$$W^{\epsilon} \longrightarrow \theta h \text{ in } L^2(0,T;\, H^{-1}(\Omega));$$

Now we have,

$$\lim_{\epsilon \to 0} \int_Q h\, P_{\epsilon}\, (1 - H_{\epsilon}(P_{\epsilon})) = \int_Q h\, P\, (1-\theta) = 0$$

The function W^{ϵ} is also used to show that $\theta(0) = \theta_0$ in $H^{-1}(\Omega)$: by integrating by parts (18), (25) and using (28), (29) and (30) to obtain:

$$w^{\epsilon}(0) \longrightarrow -\theta h(0) \text{ in } H^{-1}\text{-weak.}$$

Therefore

$$W^{\epsilon}(0) = \epsilon \frac{\partial p_{\epsilon}}{\partial t}(0) - h(0)\, H_{\epsilon}(\bar{P}_{a\epsilon})$$

So (24) and (27) induces:

$$W^{\epsilon}(0) \longrightarrow -\, h(0)\, \theta_0 \qquad \text{in } \mathcal{D}'(\Omega)$$

and we have $\theta(0) = \theta_0$ in $H^{-1}(\Omega)$.

Time periodic problem;

By the same technic one can show that the following time periodic problem has at least one solution when the graph is T-periodic in time:

Find $P \in L^2(0,T;\, H^1(\Omega))$, $\theta \in L^{\infty}(Q) \cap H^1(0,T;\, H^{-1}(\Omega))$ such that:

$P \geq 0$ et $\theta \in H(P)$ a.e. in Q,

$P = 0$ on Σ_i; $P = P_a$ on Σ_e,

$$\int_Q -\theta h\, \partial_t \varphi + h^3 \nabla P\, \nabla \varphi - h\vec{V}.\nabla \varphi = 0, \quad \forall\, \varphi \in V_{Per},$$

$$\theta(0,x) = \theta(T,x) \qquad \text{in } H^{-1}(\Omega)$$

where $V_{Per} = \{\varphi \in H^1(0,T;\, H^1(\Omega)),\ \varphi$ is T-periodic in time$\}$.

2.b — A second existence theorem by a half-time discretization.

We introduce a family of approximation problems by a time-discretization with the step $k = T/n$.

PROBLEM $\mathcal{P}_N$:

For $n = 1, \ldots, N$, find (P^n, θ^n) such that:

$$P^n - \bar{P}_a \in H^1(\Omega), \quad \theta^n \in L^\infty(\Omega), \qquad (31)$$

$$\theta^n \in H(P^n) \quad \text{a.e. in } \Omega, \qquad (32)$$

$$\frac{h^n\theta^n - h^{n-1}\theta^{n-1}}{k} - \nabla \cdot ((h^n)^3 \nabla P^n) = -\nabla \cdot (\vec{V}\, \theta^{n-1} h^{n-1}), \qquad (33)$$

$$\theta^0(x) = \theta_0(x) \qquad (34)$$

THEOREM 9: The problem $\mathcal{P}_N$ has a unique solution:

The problem $(\mathcal{P}_N)$ will be solved step by step. So, at each time step we prove the existence of (31) (32) (33) by the following proposition:

LEMMA 2: Let (P,θ) be a solution of problem (31)-(33) then $w = P-\bar{P}_a$ is a solution of the variational inequality of second kind:

$$w \in H_0^1(\Omega),$$

$$k \int_\Omega h^3 \nabla w\, \nabla(\varphi - w) + J(\varphi) - J(w) \geq \langle f, \varphi - w \rangle_{H^{-1}\, H_0^1} \qquad (35)$$

If w is the solution of (35) then there exists only one θ such that $(w+\bar{P}_a;\theta)$ is the solution of (31)-(33), J is the functional defined on $L^2(\Omega)$ by:

$$J(\varphi) = \int_\Omega h(x)\, \Phi\, (\varphi + \bar{P}_a)\, dx$$

$\Phi(x) = x^+$ and $H = \partial\Phi$ where $\partial\Phi$ is the sub-differential of Φ,

Proof:

— The proof of the first part of the lemma is based on the sub-differential inequality:

$$\theta \in H(P) \Longleftrightarrow \Phi(\lambda) - \Phi\, P(x) \geq \theta(x)\, (\lambda - P(x)) \quad \forall \lambda \in \mathbb{R} \text{ and a.e. in } \Omega.$$

— The second part is obtained via the existence of a Lagrange multiplier for (35) which has one and only one solution by classical existence theorem for variational inequality of second kind [GLO].

A priori estimates:

LEMMA 3: – Let us assume that there exist $P^0 \in L^2(\Omega)$ such that $\theta^0 \in H(P^0)$ then we have:

$$k \sum_{n=1}^{N} \|P^n\|^2_{H^1(\Omega)} \leq C$$

$$k \sum_{n=1}^{N} \left\| \frac{h^n \theta^n - h^{n-1}\theta^{n-1}}{k} \right\|_{H^{-1}(\Omega)} \leq C$$

If $P^0 \in H^1(\Omega)$ and $\nabla . \vec{V} = 0$ we have:

$$\|P^n\|_{H^1(\Omega)} \leq C$$

$$\left\| \frac{h^n \theta^n - h^{n-1}\theta^{n-1}}{k} \right\|_{H^{-1}(\Omega)} \leq C$$

where C is a constant independant of k.

Proof: We use the inequality:

$$(\theta^n - \theta^{n-1})\, P^n \geq 0 \qquad \text{a.e. in } \Omega \qquad \forall\, n \geq 1.$$

and we choose the test functions:

$$\varphi = P^n - \overline{P}_a \quad \text{and } \varphi = P^n - P^{n-1}$$

Now, we introduce the functions defined on [0,T] from the sequences P^n and θ^n by:

$$\tilde{\theta}_k(t,x) = \frac{\theta_k^{n+1}(x) - \theta_k^n(x)}{k}\,(t-nk) + \theta_k^n(x)$$

$$P_k(t,n) = P_k^{n+1}(x) \quad \text{if } t \in [nk,(n+1)k].$$

Lemma 2 induces the estimates:

$$\|\tilde{\theta}_k\|_{L^\infty(Q) \cap H^1(0,T;\, H^{-1}(\Omega))} \leq C$$

$$\|P_k\|_{L^2(0,T;H^1(\Omega))} \leq C$$

So, there exists a sub-sequence of $\tilde{\theta}_k$ and P_k such that

$$\tilde{\theta}_k \longrightarrow \theta \quad \text{in } L^{\infty}(Q) \text{ weak}^* \text{ and } H^1(0,T,H^{-1}(\Omega)), \text{ weak,}$$

$$P_k \longrightarrow P \quad \text{in } L^2(0,T;\ H^1(\Omega)) \text{ weak.}$$

We have also $\theta \in H(P)$ by showing that:

$$\int_Q \Phi(V) - \Phi(P) \geq \int_Q \theta(V-P) \qquad \forall\ V \in L^2(Q)$$

The property $P \geq 0$ is obtained by choosing a test function $\Psi = P_\epsilon$ solution of the differential equation defined in $H^1_0(\Omega)$ by:

$$- \epsilon \frac{dP_\epsilon}{dt} + P_\epsilon = P^- \quad \text{in } [0,T]$$

$$P_\epsilon(T) = 0$$

and letting ϵ tend to zero

2.c – *Numerical results*: We present here typical results obtained by a slightly different approximation, where we have to solve at each time step a variational inequality of the first kind instead of the second kind [see ELA]. The function h(x,t) is given by:

$$h(x,y,t) = h_0 + y\ [\chi_2 \sin(x-\omega t) - \chi_1 \sin x]$$

(x,y) are the polar coordinates and χ_1, χ_2 geometrical data.

Figure (3) gives the variation of $w(t) = \int_\Omega P(x,t)\,dx$ wich is the load, and figure (4) gives the localization of the cavitated area at three particular time steps.

REFERENCES

[B1] G.BAYADA - "Localisation de la zone de coïncidence pour les problèmes à frontière libre décrits par des équations à coefficients discontinus", Publication L.A. 740 (Ex SANTI) n°29 - 1984

[B2] G. BAYADA - "Variational formulation and associated algorithm for the starved finite journal bearing", Journal of Lubrication Technology, ASME, Vol. 105, Juillet 1983, pp. 453-457.

[BC1] G. BAYADA, M. CHAMBAT - "The transition between the Stokes equation and the Reynolds equation: a mathematical proof", J. Appl. Math. Opt. 14, 1986, pp. 73-93.

[BC2] G. BAYADA, M. CHAMBAT - "Sur quelques modélisations de la zone de cavitation en lubrification hydrodynamique", A paraître au Journal. Méca. Th. et App., 1986.

[BC3] G. BAYADA, M. CHAMBAT - "Non linear variational formulation for a cavitation problem in lubrication", Journal of Mathematical Analysis and Application", Vol. 90, N°2, 12/1982, pp. 286-298.

[Ci] C. CIMATTI - "On a problem of the theory of lubrication governed by a variational inequality", Appli. Math. Opt. (2) 3, 1977, pp. 227-242.

[CH] J. CARILLO MENENDEZ, M. CHIPOT - "Sur l'unicité du problème de l'écoulement à travers une digue", C.R. Acad. Sc. Paris, T292, 1981, pp. 191-194.

[ELA] EL ALAOUI TALIBI - "Sur une problème à frontière libre associé à la modélisation de la cavitation en mécanique des films minces", Thèse 1986, Université Lyon I.

[ELIOC] C.M. ELLIOT, J.R. OCKENDON - "Weak and variational method for moving boundary problem", RNM, Pitman, 1981.

[ELR] G. ELROD, M.L. ADAMS - "A computer program for cavitation and starvation problem", Proc. Cavitation and related phenomena in lubrication, D. Dowson; M. Godet, C.M. Taylor, Mech. Eng. Pub. Ltd 1975, Londres.

[FLO] L. FLOBERG, J. JACKOBSON - "The finite journal bearing considering vaporization", Trans. Chalmers University, Sweden, V. 190, 1957.

[GIL] G. GIRALDI - "A new approach to evolution free-boundary problems" - C.P.D.E. 4 (10) 1979, pp. 1092-1122.

[GLO] R. GLOWINSKI - Cours D.E.A. "Inéquations variationnelles", 1982-1983, Paris VI.

[TF] A. TOURNERIE, J. FRENE - "Computer modelling of the functioning modes of non-contacting face-seals", Tribology International, 17, 4, 1984, pp. 179-184.

[LI] J.L. LIONS - "Quelques méthodes de résolutions de problèmes aux limites non linéaires", Dunod, 1969.

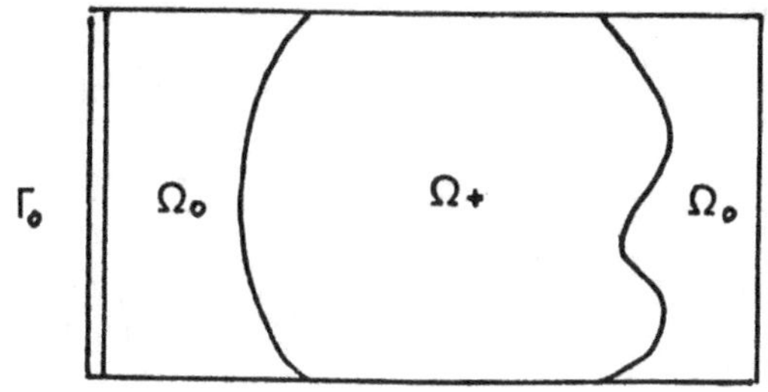

FIG 1 JOURNAL BEARING

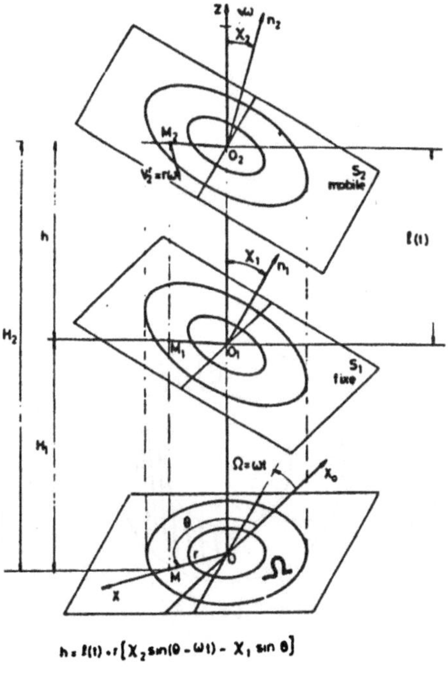

FIG 2 SEAL GEOMETRY

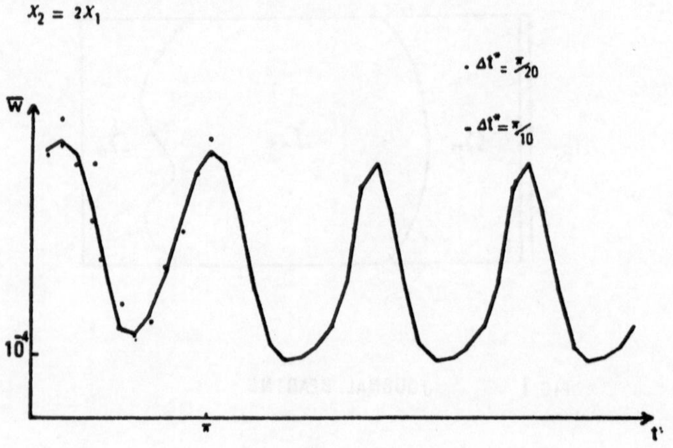

FIG3 LOAD VARIATION

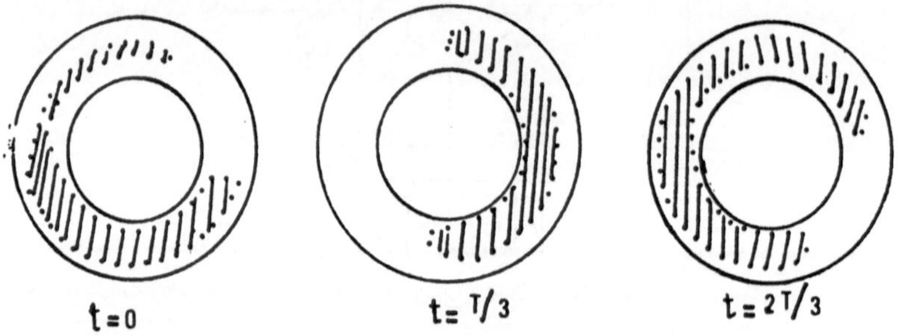

FIG 4 LOCATION OF THE CAVITY FOR VARIOUS TIMES

On Optimal Design of Activily Controlled Distributed Parameter Structures

by

Martin P. Bendsøe
Mathematical Institute
The Technical University of Denmark
Building 303
DK-2800 Lyngby, Denmark.

Abstract. We consider problems of simultaneous optimal design of controls and structure for an active control of a flexible structure. Problem formulations and relevant objectives for modal control of distributed structures or large scale structures are discussed. As an example of optimizing the performance of a simple modal control system, we optimize the design of a cantilevered beam together with the feedback gains and actuator positions of its controllers. Example problems of minimizing the spill-over of control energy into uncontrolled modes due to active damping of a number of lower modes are considered as well.

1. INTRODUCTION.

In the case that a given flexible structure is to be equipped with means for active control, and where those means are comprised of specified control devices, an optimal control strategy might be established in general on a rational basis by the use of techniques from the field of optimal control ([1], [2]). However, it should be expected that considerable improvement in the performance of the system can be obtained by a simultaneous design of the structure, the control system and the control strategy. Optimal positioning of actuators and sensors has received much attention in the literature, and the objective of the optimization is most often minimization of control effort ([3], [4]) or the maximization of energy dissipation due to active damping ([5]). Criteria related more directly to structural performance also have been taken into account ([6], [7]). Recently, the design of the structure itself has been included as an unknown in studies of the optimal control of flexible structures ([8], [9], [10]), and in this paper we will address the problem formulation and solution procedure for this more general combined optimal design of structure, control strategy and control system. The emphasis will be on distributed parameter stuctures and large scale structures, where control is implemented via a reduced order model (Refs. [8], [9] use full order models, [10] a reduced order model).

In this paper we provide a short introduction to the basic types of problems that arises when considering optimal structural design for control. The paper addresses problems for design of distributed structures with active controllers, where the local influence of actuators and local measurements of sensors plays a major role, and relevant criteria and design problems are formulated.

In this paper the interplay between structural design and control design is emphasized. For further information on control methods for structural control the reader is directed to Ref. [11] and the list of references therein.

2. ACTIVE CONTROL DAMPING OF A DISTRIBUTED PARAMETER STRUCTURE.

In the control of large scale structures and distributed structures, interesting design criteria arise due to the difficulties in implementing distributed controls, and the commonly used technique of modal control gives rise to design problems that are very similar to the well-known problem of eigenfrequence optimization for distributed structures (cf. Ref. [23]). To illustrate this we consider an active control of a cantilevered beam by a finite number m of actuators, distributed along the length of the beam. The motion of the beam is given by the solution $w(x,t)$ of the partial differential equation

$$\rho a(x) \frac{\partial^2 w(x,t)}{\partial t^2} + \xi\rho a(x) \frac{\partial w(x,t)}{\partial t} + \\ + \frac{\partial^2}{\partial x^2}\left(Ea(x)^2 \frac{\partial^2}{\partial x^2} w(x,t)\right) = \sum_{i=1}^{m} b_i(x) f_i(t). \tag{1}$$

(Initial conditions, Boundary conditions)

Here ρ is the mass density of the beam material, ξ is the damping coefficient, and $a(x)$ is the area of the cross-section which are taken to be geometrically similar with second area monent given by $k\cdot a(x)^2$. The constant k multiplied by Young's modulus for the beam material is denoted by E in (1). The actuator forces are $f_i(t)$, with actuator influence functions $b_i(x) \in L^2(0,L)$. We will seek to design the beam so that an active damping of the beam is optimal in some sense, and we use the varying cross-sectional area $a(x)$ as a distributed design variable.

The literature on control and optimal control of distributed system is abundant, so as examples of treatments on the optimal control aspects we refer to Refs. [12] - [14], and for a survey on the issues of controllability etc. of distributed systems we refer to Ref. [15]. We note here that in (1) we do not have distributed controls as the actuator influence functions are, a priori, fixed and cannot be made to vary with time, so many earlier treatments of design for optimal distributed control do not apply here. This restriction on the form of the controls means that in general the system (1) will not be controllable, in the sense that specific, desired dynamics cannot always be obtained. This means that optimal control problems for (1) should employ functionals that do not assume controllability. In the case of a quadratic functional, optimal positioning of the actuators as well as optimal positioning of sensors (in case of feedback control) has been considered in e.g. Ref. [13]. In practise the distributed nature of the problem of controlling for example a beam is often circumvented by introducing a reduced order model, i.e. a finite dimensional approximation of (1), and then using this model for the control design. Such reduced order models are conveniently constructed by introducing modal coordinates, and the controls are then designed from a model that takes a number of important modes into consideration (Refs. [1] and [16] - [21]). These reduced order models will often be controllable and thus desired dynamics of the modes considered can be obtained. The use of reduced order models also play an important role for the control of large scale structures ([11]). The use of reduced order models based on modal data require good knowledge of the modal data for the system, and even though control design in most cases will be robust with respect to the modal data ([22]), the modal data may be very dependent on a precise knowledge of the properties of the system.

The analysis which follows can be extended to any linear, distributed structure for which the governing equations can be written as

$$\frac{\partial^2 w}{\partial t^2} + \zeta \frac{\partial w}{\partial t} + Aw = B\, f(t) \tag{2}$$

with $w(t)$ in some suitable Hilbert space, A a strongly elliptic operator and with B denoting the control force operator. Equation (2) holds for external damping, while for a viscous (internal) damping our model should be

$$\frac{\partial^2 w}{\partial t^2} + \zeta A \frac{\partial w}{\partial t} + Aw = B\, f(t)\ . \tag{3}$$

and such a model leads to quite different optimal designs.

For the analysis we introduce modal coordinates for the system (1), and write the solution $w(x,t)$ of (1) as

$$w(x,t) = \sum_{i=1}^{\infty} u_i(t)\varphi_i(x) \tag{4}$$

where the orthonormal mode shapes $\varphi_i \in H^2(0,L)$ are given by

$$\frac{d^2}{dx^2}\left(E\,a(x)^2\,\frac{d^2}{dx^2}\varphi_i(x)\right) = \lambda_i\,\rho\,a(x)\varphi_i(x) \tag{5}$$

$$\int_0^L \rho\,a(x)\varphi_i(x)\varphi_j(x)\,dx = \delta_{ij} \tag{6}$$

corresponding to the natural eigenfrequencies $\omega_i = \sqrt{\lambda_i}$, ordered according to $0 < \lambda_1 \leq \lambda_2 \leq \dots$ .The modal coordinates $u_i(t)$ then satisfy the infinite system of ordinary differential equations:

$$\ddot{u}_i(t) + \xi\,\dot{u}_i(t) + \lambda_i\,u_i(t) = \sum_{j=1}^{m} B_{ij}\,f_j(t) \tag{7}$$

where $B_{ij} = \int_0^L \varphi_i(x)b_j(x)\,dx$, with actuator influence functions $b_j \in L^2(0,L)$. The equations for the modal coordinates are coupled due to the control forces and the local influence of the actuators. In the following we will only consider simple control design, so we choose to use the control forces to damp out the vibrations of the n lower modes, using a velocity feedback. We thus set

$$f_j(t) = G_{ji}\,\dot{u}_i \tag{8}$$

where $\underline{\underline{G}}$ is a $m \times n$ gain matrix. In this situation the dynamics of the lower modes will be governed by the equation

$$\underline{z} = (u_1,\dots,u_n\ ,\ \dot{u},\dots,\dot{u}_n)\ ,\ \frac{d\underline{z}}{dt} = \underline{\underline{A}}\,\underline{z} \tag{9}$$

with

$$\underline{\underline{A}} = \begin{pmatrix} \underline{\underline{0}} & \underline{\underline{E}} \\ -\mathrm{diag}(\lambda_1,\dots,\lambda_n) & -\mathrm{diag}(\ _1,\dots,\ _n) + \underline{\underline{B}}\,\underline{\underline{G}} \end{pmatrix} . \tag{10}$$

We have assumed that the velocities of the lower modes can be observed directly, and this assumption makes the dynamics of the lower modes independent of the uncontrolled (higher order) modes. The local influence of the actuators will, in general, make $B_{ij} \neq 0$ for almost all i and j , and thus the feedback law (8) will excite the higher modes and a spill-over of control energy into the higher modes will occur. The direct observation of modal velocities can be based on the use of special filters ([16]) or modal observers ([19]), but if this is not possible the local observations of local sensors will introduce a coupling of the higher modes into the lower modes which will further degrade the performance of the system ([16] - [21]). For the situation presented here we have the following properties:

i) With internal damping as in (2), a modal velocity feedback control can be constructed so that the stability margin of the system (9) is improved as compared to the uncontrolled system and at the same time the stability margin of the uncontrolled modes is unchanged ([16] - [18], [20]).

ii) With viscous damping as in (3), modal control can improve the stability margin of the full system (in this case the damping increases with mode number) (cf. [21]).

iii) For a fixed number of actuators, the control spill-over will decrease when the size of the model (10) is increased. (Refs. [16] - [18], [20], [21]).

For the sake of algebraic simplicity it is often convenient to consider cases where the number of actuators is chosen to be equal to the number of modes; then by choosing the feedback gain as

$$\underline{\underline{G}} = \underline{\underline{B}}^{-1}\,\mathrm{diag}(\alpha_1,\ldots,\alpha_n) \tag{11}$$

the dynamics of the controlled modes decouple, see (9) and (10), and we have independent modal space control (cf. Ref. [19]):

$$\left.\begin{aligned} &\underline{\tilde{z}}_i = (u_i,\dot{u}_i)\ , \quad \frac{d\tilde{z}_i}{dt} = \underline{\underline{A}}_i\,\underline{\tilde{z}}_i \\ &\underline{\underline{A}}_i = \begin{pmatrix} 0 & 1 \\ -\lambda_i & -\xi_i + \alpha_i \end{pmatrix} \end{aligned}\right\} \quad i = 1,\ldots,n\ . \tag{12}$$

Note that for independent modal space control, the conclusion iii) above no longer applies.

From the remarks above we see that for modal control there are basically two types of design problems one should consider:

A. Design of structure and controllers with the purpose of optimizing the dynamics of the controlled modes.

B. Design of structure and controllers with the purpose of minimizing control spill-over.

With the purpose of optimizing damping in mind we will use the energy

$$\begin{aligned} \|w(x,t)\|_E^2 &= \int_0^L \left[Ea^2\left(\frac{\partial^2 w}{\partial x^2}\right)^2 + \rho a\left(\frac{\partial w}{\partial t}\right)^2 \right] dx \\ &= \sum_{i=1}^{\infty} \left[\lambda_i u_i^2 + \dot{u}_i^2 \right] \end{aligned} \tag{13}$$

of the solution $w(x,t)$ to Eq. (1) as a performance measure. For exemplification, in a modal model with 2 modes the criteria for optimizing the dynamics of the 2 modelled modes will be

$$E_1 = \frac{1}{2} \int_0^{\infty} \underline{z}^T(t)\ \underline{\underline{Q}}_1 \underline{z}(t) dt \tag{14}$$

where

$$\underline{\underline{Q}}_1 = \begin{pmatrix} \lambda_1 & 0 & 0 & 0 \\ 0 & \lambda_2 & 0 & 0 \\ 0 & 0 & 1 & 0 \\ 0 & 0 & 0 & 1 \end{pmatrix} \tag{15}$$

while a minimization of spillover into mode no. 2 due to an active control damping of mode nr. 1 can be achieved by optimizing with respect to the criteria

$$E_2 = \frac{1}{2} \int_0^{\infty} \underline{z}^T(t)\ \underline{\underline{Q}}_2 \underline{z}(t) dt \tag{16}$$

with

$$\underline{\underline{Q}}_2 = \begin{pmatrix} 0 & 0 & 0 & 0 \\ 0 & \lambda_2 & 0 & 0 \\ 0 & 0 & 0 & 0 \\ 0 & 0 & 0 & 1 \end{pmatrix} . \tag{17}$$

Note that E_1 and E_2 are dependent on the initial condition $\underline{z}_0$, so that a design based on minimization of these functionals will depend on the choice of initial condition $\underline{z}_0$. This unfortunate situation can be rectified by considering a *worst-case* design using the functional

$$\max_{\underline{z}_0} \quad E(\underline{z}_0)$$

$$\frac{1}{2}\underline{z}_0^T \underline{\underline{Q}}\, \underline{z}_0 = 1 \tag{18}$$

This functional can be simplified by introducing a positive definite, symmetric matrix $\underline{\underline{P}}$, which is a solution of the Lyapunov equation

$$\underline{\underline{A}}^T\underline{\underline{P}} + \underline{\underline{P}}\,\underline{\underline{A}} = -\underline{\underline{Q}} \ , \tag{19}$$

so that for the solution $\underline{z} = e^{\underline{\underline{A}}t}\underline{z}_0$ to (9):

$$\frac{d}{dt}(\underline{z}^T\underline{\underline{P}}\,\underline{z}) = -\underline{z}^T\underline{\underline{Q}}\,\underline{z} \ . \tag{20}$$

Thus $E(\underline{z}_0)$ can be expressed as

$$E(\underline{z}_0) = \frac{1}{2}\int_0^\infty \underline{z}^T\underline{\underline{Q}}\,\underline{z}\,dt = \frac{1}{2}[-\underline{z}^T\underline{\underline{P}}\,\underline{z}]_0^\infty = \frac{1}{2}\underline{z}_0^T\underline{\underline{P}}\,\underline{z}_0 \ . \tag{21}$$

We can now formulate an appropriate criterium for an optimal design problem as

$$\underset{a(x);\ G_{ij}}{\text{minimize}} \left\{ \max_{\underline{z}_0} \frac{\underline{z}_0^T\underline{\underline{P}}\,\underline{z}_0}{\underline{z}_0^T\underline{\underline{Q}}\,\underline{z}_0} \right\} \tag{22}$$

where the cross-sectional area $a(x)$ of the beam as well as the feedback gains G_{ij} are used for optimizing the damping of any initial disturbance in the modelled modes. Taking the simple form of $\underline{\underline{Q}}$ into account (22) can be written simply as

$$\min\left(\text{max eigenvalue of } \sqrt{\underline{\underline{Q}}}^{-1}\,\underline{\underline{P}}\sqrt{\underline{\underline{Q}}}^{-1}\right) \ . \tag{23}$$

In Figs. 1 to 3 below we provide results for the optimal design of a cantilevered beam, using the criteria (22) for a model with 1 or 2 modes. The beam is controlled with one actuator. The reduced order models consist of the first mode or the first two modes and the design variables are the feedback gain α for the first mode, the position A of the actuator and as a distributed design variable we use the crosssectional area $a(x)$ of the beam. The cases are:

Case 1. A model with 1st. mode. We optimize the damping effect on this mode. This problem reduces to maximizing the smallest eigenvalue, λ , of the beam. (cf. Ref. [23]). Fig. 1.

Case 2. A model with 1st and 2nd modes for which we optimize the damping of these modes, for impulses in the 1st mode only. Fig. 2.

Case 3. A model with 1st and 2nd modes. We minimize the energy in the 2nd mode, due to control damping of impulses in the 1st mode only, i.e. spill-over is minimized. Fig. 3.

As an example of the full formulation of one of these problems we write the one for Case 2:

$$\underset{a(x),\alpha,A}{\text{minimize}} \left(\text{maximal eigenvalue of } \begin{pmatrix} p_{11}/\lambda_1 & p_{13}/\sqrt{\lambda_1} \\ p_{13}/\sqrt{\lambda_1} & p_{33} \end{pmatrix} \right)$$

subject to: $\underline{\underline{A}}^T \underline{\underline{P}} + \underline{\underline{P}}\,\underline{\underline{A}} = -\underline{\underline{Q}}$

$$\underline{\underline{A}} = \begin{bmatrix} 0 & 0 & 1 & 0 \\ 0 & 0 & 0 & 1 \\ -\lambda_1 & 0 & -\xi-\alpha B_1 & 0 \\ 0 & -\lambda_2 & -\alpha B_2 & -\xi \end{bmatrix}$$

$$\underline{\underline{Q}} = \begin{bmatrix} \lambda_1 & 0 & 0 & 0 \\ 0 & \lambda_2 & 0 & 0 \\ 0 & 0 & 1 & 0 \\ 0 & 0 & 0 & 1 \end{bmatrix} \tag{24}$$

$$\frac{d^2}{dx^2}\left(Ea(x)^2 \frac{d^2\varphi_i}{dx^2}\right) = \lambda_i \rho\, a(x)\varphi_i \quad , \quad i = 1,2$$

$$\int_0^L \rho a(x)\varphi_i(x)\varphi_j(x)\,dx = \delta_{ij} \quad , \; i = 1,2 \; ; \; B_i = \frac{1}{2\Delta}\int_{A-\Delta}^{A+\Delta} \varphi_i(x(dx) \; , \; i = 1,2$$

$$-\infty < \alpha_{min} \le \alpha \le \alpha_{max} < \infty \quad ; \quad 0 < A_{min} \le A \le A_{max} < L$$

$$0 < a_{min} \le a(x) \le a_{max} < \infty \; , \quad x \in [0,L] \quad ; \quad \int_0^L a(x)\,dx \le Vol$$

where we have imposed a volume constraint and minimum and maximum constraints on the design variables. The actuator influence function is an L^2-approximation of the δ-function, with width $\Delta = 0.05\,L$.

The problems above were solved by first finding an analytical expression for the functional and then minimizing this by use of a sequential quadratic optimization algorithm (the Harwell VF02AD). Sensitivities were computed using results for the sensitivity of modal data ([24]); the possibility of multi modal beam designs was ignored, which implies that the eigenvalues are differentiable with respect to $a(x)$.

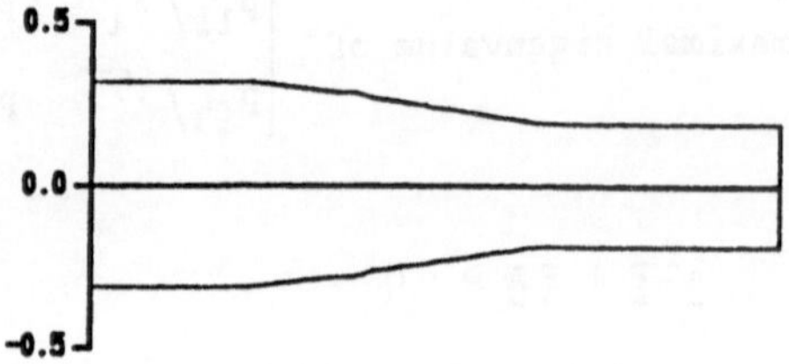

Fig. 1. Case 1: Maximization of λ_1. The eigenvalue λ_1 for this design is 3 times larger than the corresponding λ_1 for a uniform beam of the same volume.

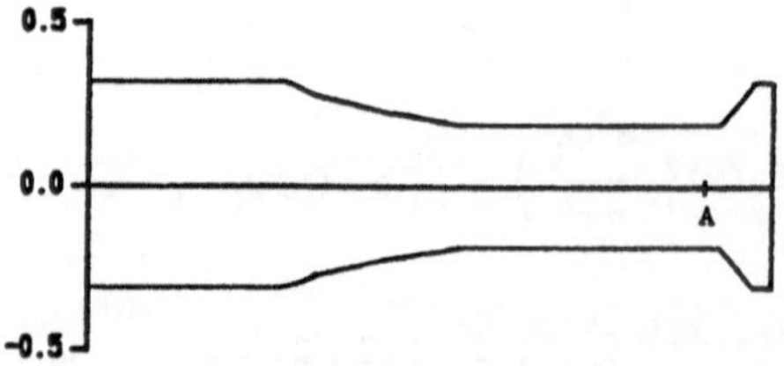

Fig. 2. Case 2: Minimization of energy in mode 1 and 2. One actuator of optimal position A, with feedback gain $\alpha = 1.48$. Value of functional is 48% of the value for a uniform design of the same volume.

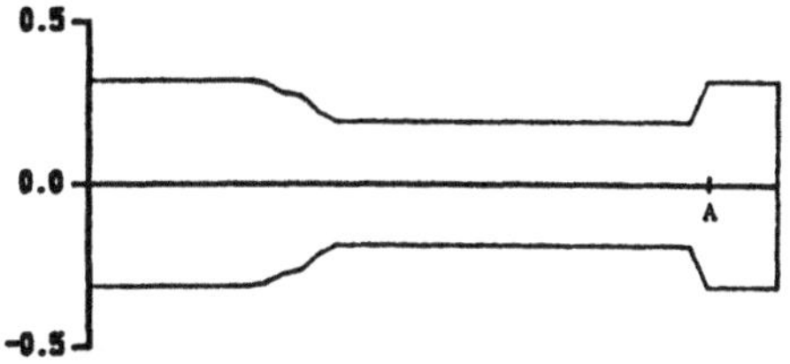

Fig. 3. Case 3: Minimization of energy in mode 2. Actuator at optimal position A, with feedback gain $\alpha = \alpha_{min} = 0.5$. Value of functional is 4.4% of the value for a uniform beam of the same volume.

For the results presented above we used a discretization for the eigenvalue problem consisting of 200 elements along the beam, while the design variable $a(x)$ was represented by a discretization into 40 elements.

We note that the maximization of the damping results in an increase of the eigenfrequency of mode 1, which is the mode with the velocity feedback, while the eigenfrequency of the second mode is decreased. This latter has the effect of building up material at the end of the beam. For the minimization of spill-over the second eigenfrequency is again decreased, and for this case the first eigenfrequency is also slightly decreased. The optimal actuator position corresponds to the location where the coupling induced by control is minimal, while the optimal feedback gain α in Case 3 is minimal in order to minimize spill-over; for Case 2 it is determined by a trade-off between maximal damping of mode 1 and a minimazation of spill-over.

The above results are for damping modelled as in (2). For viscous damping, as in Eq. (3), the optimal designs are rather different, as the increased damping resulting from increased eigenfrequencies is of overruling importance, see Figs. 4,5.

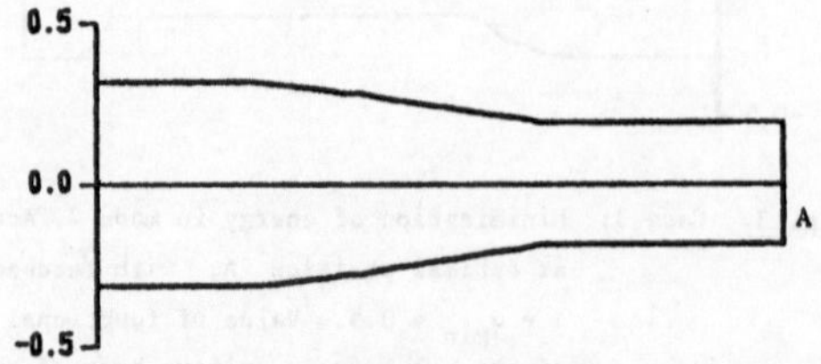

Fig. 4. Case 2 for viscous damping. $\alpha_{opt} = 1.09$, and value of functional is 48% of value for uniform beam of same volume. Design is almost the same as in Fig. 1.

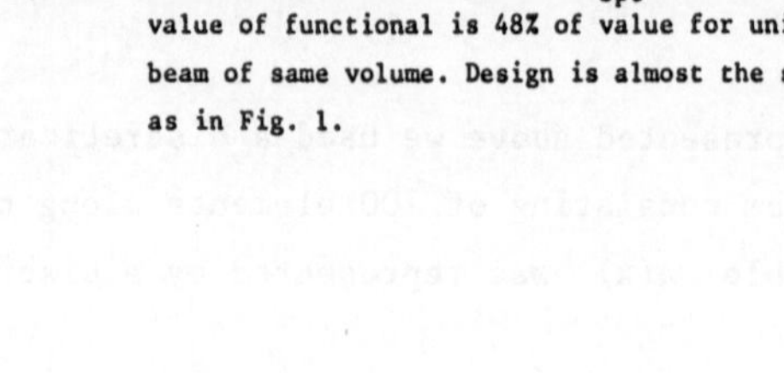

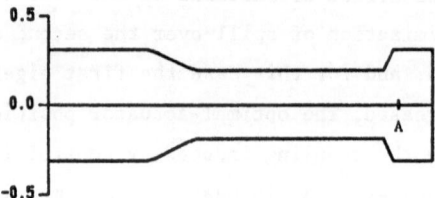

Fig. 5. Case 3 for viscous damping. $\alpha_{opt} = \alpha_{min}$, and value of functional is 3.6% of value of uniform beam of same volume.

3. MINIMIZATION OF CONTROL SPILLOVER.

The minimization of spillover considered above is only concerned with coupling of the first and second mode, and the criteria of spillover is the total energy in the second mode, due to an initial excitation of the first mode. It is, however, possible to give an estimate of the energy in all the uncontrolled modes due to initial impulses in the controlled modes and estimates of the peak energy or of the total energy can be obtained. In both cases an important measure of the coupling between the controlled and the uncontrolled modes is the factor (see [17], [25]):

$$CP = \sum_{i=1}^{M} \sum_{j=N+1}^{\infty} \left(\int_{o}^{L} \rho\, a(x) b_i(x) \varphi_j(x) dx \right)^2 \tag{25}$$

which is the sum of the square of the norms of the projections of the force influence functions $b_i(x)$ on the uncontrolled modes; the orthonormality of the modes makes it possible to compute this factor from modal data of the controlled modes only:

$$CP = \sum_{i=1}^{M} \left[\int_{o}^{L} \rho\, a(x) b_i(x)\ dx - \sum_{j=1}^{N} \left(\int_{o}^{L} \rho\, a(x) b_i(x) \varphi_j(x) dx \right)^2 \right] \tag{26}$$

and a minimization of this factor results in designs as in Fig. 6.

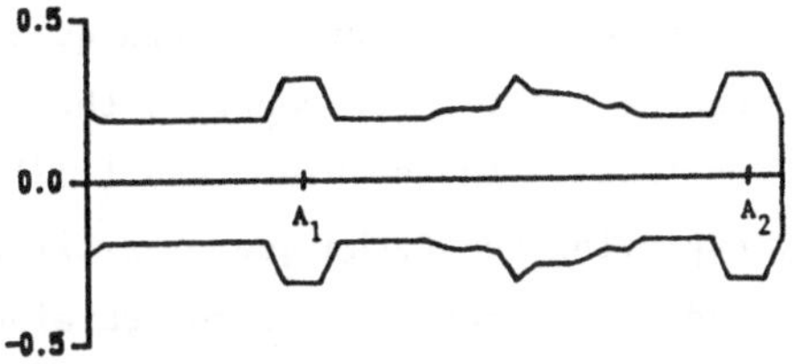

Fig. 6. Minimization of the value CP for a model with 3 modes and two actuators. Optimal positions of actuators indicated by A_i, $i = 1,2$.

The spillover is, however, not only dependent on the coupling coefficient CP above, but it also depends on the dynamics of the controlled system and the size of the control forces needed for the damping of the controlled modes. Taking this into account one can write (cf. Ref. [25]):

$$\|\underline{z}_R(t)\|_E^2 \leq M \cdot e^{-\xi t}(1-e^{-(\sigma-\xi/2)t})^2 \|\underline{z}_0\|_E^2 \qquad (27)$$

$$M = \frac{2\lambda_1+\xi}{2\lambda_1-\xi} \cdot \frac{1}{(\sigma-\xi/2)^2} \frac{1}{\mu} \cdot \rho(\underline{\underline{G}}^T G) \cdot K \cdot CP \qquad (28)$$

for the energy $\|\underline{z}_R(t)\|_E$ in the residual modes due to an excitation $\underline{z}_0$ in the controlled modes. Here σ is the stability margin of the system (10), μ is given as

$$\mu = \min\{1, \lambda_1, \ldots, \lambda_N\} \qquad (29)$$

and $\rho(\cdot)$ denotes the spectral radius of a matrix. Finally, the constant K is

$$K = \rho(\underline{\underline{D}}^T \underline{\underline{D}})\, \rho((\underline{\underline{D}}^T D)^{-1}) \qquad (30)$$

where $\underline{\underline{D}}$ is a matrix that transforms the coefficient matrix of (9) into Jordan normal form. The constant M in (28) depends only on the modal data of the controlled modes and a minimization of M will lead to a decrease in the eigenfrequencies of the controlled modes. Thus, in order not to degrade the performance of the active control damping, lower bounds on the eigenfrequencies should be imposed when spillover is minimized. Notice that M is a non-differentiable function of the eigenfrequencies of the controlled modes, so that, in general, non-differentiable optimization techniques should be used. However, for independent modal space control, the controlled modes decouple and one can restate the problem (see [25]) as a differentiable problem by employing the technique of artificial bounds (cf. Ref. [26]). Fig. 7 and 8 show optimal designs obtained for this special case, and for damping modelled as in Eq. 1.

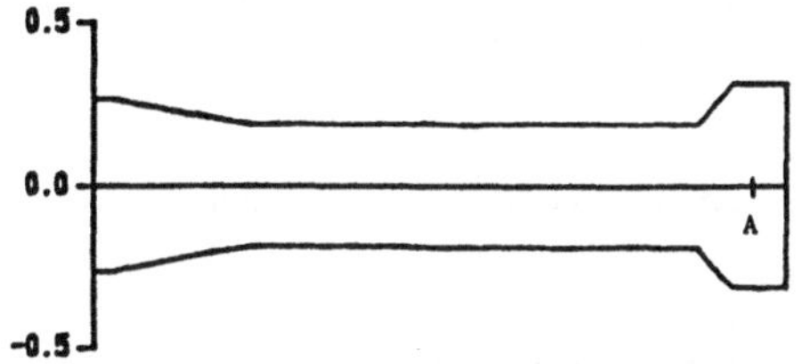

Fig. 7. Minimization of spillover coefficient M for the case of one controlled mode, with one actuator at position A. Value of functional is reduced to 42% of value for a uniform beam of the same volume.

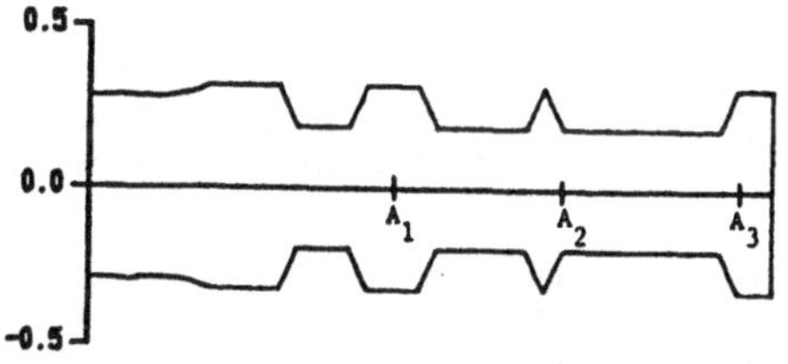

Fig. 8. Minimization of spillover coefficient M for the case of three controlled modes and three actuators, at positions A_i, $i = 1,2,3$. Value of functional is 43% of the value for a uniform beam of the same volume.

REFERENCES

[1] L. Meirovitch, L.M. Silverberg: Globally Optimal Control of Self-Adjoint Distributed Systems. Optimal Contr. Appl. Meth., Vol. 4 (1983), 365-386.

[2] M. Abdel-Rohman, H.H.E. Leipholz: Optimal Feedback Control of Elastic, Distributed Parameter Structures. Computers and Structures, 19 (1984), 801-805.

[3] O. Ibidapo-Obe: Optimal Actuator Placements for the Active Control of Flexible Structures. J. Math. Anal. Appl., 105 (1985), 12-25.

[4] M.I.J. Chang, T.T. Song: Optimal Controller Placement in Modal Control of Complex Systems. J. Math. Anal. Appl., 75 (1980), 340-358.

[5] G. Schultz, G. Heimbold: Dislocated Actuator/Sensor Positioning and Feedback Design for Flexible Structures. J. Guidance, Vol. 6 (1983), 361-367.

[6] K. Yoshida, T. Shimogo, T. Matsumoto: Optimal Design of Multiple Vibration Controllers for Elastic Beams. Bull. ISME, 27 (1984), 1974-1982.

[7] T.T. Song, M.I.J. Chang: On Optimal Control Configuration in Theory of Modal Control. in "Structural Control", H.H.E. Leipholz (Ed.), North-Holland, 1980, 723-738.

[8] A.L. Hale, R.J. Lisowski, W.E. Dahl: Optimal Simultaneous Structural and Control Design of Maneuvering Flexible Spacecraft. J. Guidance, Vol. 8 (1985), 86-93.

[9] N.S. Khot, F.E. Eastep, V.B. Venkayya: Simultaneous Optimal Structural/Control Modifications to Enhance the Vibration Control of Large Flexible Structure. Paper presented at the AIAA Guidance and Control Conference, Snowmass, Colorado, April, 1985.

[10] A. Messac, J. Turner: Dual Structural-Control Optimization of Large Space Structures, NASA Symposium on Recent Experiences in Multidisciplinary Analysis and Optimization, April 1984.

[11] G.S. Nurre, R.S. Ryan, H.N. Scofield, J.L. Sims: Dynamics and Control of Large Space Structures, J. Guidance, Control and Dynamics, Vol. 7 (1984), 514-526.

[12] J.L. Lions: Optimal Control of Systems Governed by Partial Differential Equations. Springer-Verlag, Berlin 1971.

[13] J.L. Lions: Some Aspects of the Optimal Control of Distributed Parameter Systems. SIAM, Philadelphia, 1972.

[14] N.U. Ahmed, K.L. Teo: Optimal Control of Distributed Parameter Systems. North-Holland, New York, 1981.

[15] D.L. Russell: Controllability and Stabilizability Theory for Linear Partial Differential Equations: Recent Progress and Open Questions. SIAM Review, Vol. 20 (1978), 639-739.

[16] M.J. Balas: Active Control of Flexible Systems. JOTA, Vol. 25 (1978), 415-436.

[17] M.J. Balas: Modal Control of Certain Flexible Dynamic Systems. SIAM J. Control, Vol. 16 (1978), 450-462.

[18] J.S. Gibson: An analysis of optimal modal regulation: convergence and stability. SIAM J. Control, Vol. 19 (1981), 686-707.

[19] L. Meirovitch, H. Baruh, H. Öz: A Comparison of Control Techniques for Large Flexible Systems. J. Guidance, Vol. 6 (1983), 302-310.

[20] M.J. Balas: Feedback Control of Dissipative Hyperbolic Distributed Parameter Systems with Finite Dimensional Controllers. J. Math. Anal. Appl. 98 (1984), 1-24.

[21] Y. Sakawa: Feedback Control of Second Order Evolution Equations with Damping. SIAM J. Control and Optimization 22 (1984), 343-361.

[22] A.L. Hale, G.A. Rahn: Robust Control of Self-Adjoint Distributed-Parameter Structures. J. Guidance, Vol. 7 (1984), 265-273.

[23] N. Olhoff: Maximizing Higher Order Eigenfrequencies of Beams with Constraints on the Design Geometry. J. Struct. Mech., Vol. 5 (1977), 107-134.

[24] E.J. Haug, B. Rousselet: Design Sensitivity Analysis in Structural Mechanics. II. Eigenvalue Variations. J. Struct. Mech., 8 (1980), 161-186.

[25] M.P. Bendsøe, N. Olhoff, J.E. Taylor: On the design of structure and controls for optimal performance of actively controlled flexible structures. Mechanics of Structures and Machines. (To appear.)

[26] J.E. Taylor, M.P. Bendsøe: An Interpretation for Min-Max Structural Design Problems Including a Method for Relaxing Constraints. Int. J. Solids Structures. Vol. 20 (1984), 301-314.

A DOMAIN CONTROL APPROACH TO STATE-CONSTRAINED CONTROL PROBLEMS[#]

J.F. Bonnans and V. Gaudrat

INRIA

Domaine de Voluceau

BP 105 - Rocquencourt

78153 Le Chesnay Cedex - France

C. Saguez

SIMULOG

3, Avenue du Centre

78182 Saint-Quentin en Yvelines Cedex - France

Abstract.

We consider a system governed by a family of parabolic equations. The problem is to minimize the diameter of the set of non-admissible states. We give some conditions that ensure that the problem has a solution and show how to approach the original problem by a smooth problem. We give some encouraging numerical results concerning an application of these ideas to the continuous casting problem.

Summary

This study has been supported by AFME. MM. Jolivet and Larrecq, from IRSID, are gratefully acknowledged for this useful advices.

I - INTRODUCTION

The study presented in this paper is motivated by a real-world problem : the steel continuous-casting control problem [6,7,10]. In that problem, the state of the system (i.e., the temperature of the steel) should satisfy as much as possible some constraints that ensure the good quality of the steel. However, for non-stationary processes, it may happen that no control is admissible with respect to all state constraints. Then, in the classical approach, a criterion penalizing theses constraints is introduced. But from an economic view-point, this approach is not good.

In fact we want to minimize the diameter of the domain of non-admissible states. In some other cases, it may be better to minimize the measure of the non-admissible set ; we refer to [3] for that point of view.

It is interesting to make a comparison between these non-standard criterions and the usual trick that consists in penalizing quadratically the constraints on the state. We know that, if an admissible control exists one can solve approximately the state-constrained problem by solving the penalized problem with a large penalty coefficient [8,11] (see [4] for the use of an augmented Lagrangian). However, in our case, a quadratic penalization may have the effect to enlarge the domain in which the constraints are not satisfied. Speading the defects in such a way, of course, would not be good from our point of view. A further discussion will be given in section 5.

The applications of the results of this paper to the continuous-casting process will be detailed in [1,5]. In order to simplify this paper and to focus on the main ideas, we choosed to study a problem with a simple model equation. This equation is in fact a family of parabolic equations, indexed by a parameter τ, the coupling effect being made by the control. We first suppose that the equation is linear. Then, in section 2, we formulate the problem and check the existence of an optimal control. We give a mean to approximate the original problem by a smooth problem in section 3. Section 4 is devoted to the case of a non-linear state equation (the two phases Stephan problem). We show then that the preceding results still hold if some form of compacity of the set of admissible controls holds. Some numerical results are discussed in section 5.

II - A MODEL CONTROL PROBLEM

We will define a model problem that, except for the linearity of the state equation, retains the main difficulties of the real problem. The difficulties related to nonlinearity will be discussed in section 4. Let Ω be a bounded open subset of $\mathbb{R}^n$ with smooth boundary Γ. Let $T > 0$ be given. We define

$$Q = \Omega \times (0,T), \quad \Sigma = \Gamma \times (0,T).$$

The symbol Δ will represent the Laplacian with respect to the variable x ($x \in \Omega$). The outward derivative to Ω will be denoted by n. Let us consider the system

$$\frac{\partial y}{\partial t}(\tau,x,t) - \Delta y(\tau,x,t) = 0, \quad \text{in } (0,1) \times Q,$$

$$\frac{\partial y}{\partial n}(\tau,\gamma,t) = u(\tau,\gamma,t), \quad \text{on } (0,1) \times \Sigma, \tag{2.1}$$

$$y(\tau,x,0) = y^0(x), \quad \text{for } \tau \in (0,1),\ x \in \Omega.$$

We often write $u(\tau)$, $y(\tau)$ for $u(\tau,.,.)$, $y(\tau,.,.)$. We suppose that

$$y^0(x) \text{ belongs to } L^2(\Omega). \tag{2.2}$$

Let u (the control variable) be in

$$U = L^2(0,1,L^2(\Sigma)).$$

We know [8] that for each τ, the mapping $u(\tau) \to y(\tau)$ is affine and continuous from $L^2(\Sigma)$ onto

$$W(0,T) = \{y \in L^2(0,T,H^1(\Omega))\ ;\ \frac{dy}{dt} \in L^2(0,T,H^1(\Omega)')\}.$$

Hence the mapping $u \to y_u$ (solution of (2.1)) is affine and continuous from U onto

$$Y = L^2(0,1,W(0,T)).$$

Let K (resp. Y_{ad}) be a closed convex subset of U (resp. W(0,T)). We wish to minimize some criterion J(u) as well as to satisfy the constraints

$$u \in K$$

$$y(\tau) \in Y_{ad}, \quad \text{a.e. } \tau \in (0,1) \tag{2.3}$$

Then we consider the state-constrained optimal control problem

$$\text{Min } J(u) \quad \text{s.t. } u \in K \text{ and (2.1), (2.3)} \tag{P0}$$

The study of (P0) could be made by applying some general tools (see [2] and the bibliography therein). Also, there exists efficient numerical procedures for this type of problem [4,11]. However, our concern is with a case when (P0) has no admissible state. For the reasons given in the introduction we formulate a new criterion in the following way. Define

$$\tau_1(u) = \sup\{s \in [0,T]\ ;\ y_u(\tau) \in Y_{ad}, \ \text{a.e. } 0 \le \tau \le s\},$$

$$\tau_2(u) = \inf\{s \in [\tau_1(u),T]\ ;\ y_u(\tau) \in Y_{ad}, \ \text{a.e. } s \le \tau \le 1\}.$$

For a given u in U, $\tau_1(u)$ and $\tau_2(u)$ are unambiguously defined, and $0 \le \tau_1(u) \le \tau_2(u) \le 1$. We define the problem

$$\min \tau_2(u) - \tau_1(u) + \sigma J(u), \quad \text{s.t. (2.1) and } u \in K \tag{P}$$

where σ is a strictly positive real number.

In the sequel of this paper we suppose that

J is weakly l.s.c. on U (2.4)

dom J has a non-empty intersection with K (2.5)

Hypothesis (2.4) holds for instance if J is a l.s.c. convex function.

THEOREM 1 **Under hypothesis** (2.4)(2.5), **if either** K **is bounded in** U **or** $J(u) \to +\infty$ **when** $\|u\| \to \infty$, **then problem** (P) **has (at least) one solution.** □

Proof By (2.5) we have that $\inf(P) < +\infty$. let $\{u^n\}$ be a minimizing sequence. The assumptions of the theorem imply that $\{u^n\}$ is bounded, hence has at least a weak-limit point $\bar{u}$. As K is closed and convex, $\bar{u}$ is in K, and $J(\bar{u}) \leq \liminf J(u^n)$ by (2.4). The proof will be complete if we prove that $\tau_2(u) - \tau_1(u)$ is w.l.s.c., which is done in the following lemma. □

LEMMA 1 **The mapping** $u \to \tau_2(u) - \tau_1(u)$ **is weakly l.s.c. from** U **onto** $\mathbb{R}$. □

Proof Let $\{u^n\}$ be weakly convergent in U towards u. Let (τ_1,τ_2) be a limit point of $(\tau_1(u^n), \tau_2(u^n))$. For all $\varepsilon > 0$ there exists a subsequence such that the associated states $\{y^n\}$ are in

$$C_\varepsilon = \{z \in Y \ ; \ z(\tau) \in Y_{ad}, \ \text{a.e. } 0 \leq \tau \leq \tau_1 - \varepsilon \text{ or } \tau_2 + \varepsilon \leq \tau \leq 1\}$$

Also, $y^n \to y_u$ in w-Y ; but as C_ε is closed in Y and convex, this implies that y_u is in C_ε for all $\varepsilon > 0$, which proves that $\tau_1(u) \geq \tau_1$ and $\tau_2(u) \leq \tau_2$. This proves the lemma. □

Remarks

(i) Problem (P) is non-convex. Hence the solution may be non-unique even if J is strictly convex.
(ii) Problem (P) is highly discontinuous.
(iii) If $J(u) = 0$ problem (P) can be viewed as an optimal time control problem.
(iv) Suppose that (PO) has an admissible solution and let $\{u_\sigma\}$ be a sequence of solution of (P) with $\sigma \to 0$. Then any weak limit of $\{u_\sigma\}$ is a solution of (PO).

III - A REGULARIZED PROBLEM

Even if J is smooth, the criterion of problem (P) is highly discontinuous. Hence problem (P) cannot be solved numerically by a standard descent methods. One approach is to regularize (P) in order to approximate it by a smooth problem. This might also be a mean to derive some optimality conditions for (P) ; however, we make no attempt in that direction (see [3] on this subject).

In order to approximate (P), we first reformulate it as

$$\min_{(u,\tau_1,\tau_2)} \quad \tau_2 - \tau_1 + \sigma\, J(u)$$

$$\text{s.t. } u \in K,\ 0 \le \tau_1 \le \tau_2 \le 1, \qquad (P')$$

$$y_u(\tau) \in Y_{ad}, \quad \text{a.e. } \tau \in A(\tau_1,\tau_2), \text{ and } (2.1),$$

with

$$A(\tau_1,\tau_2) = \{\tau \;;\; 0 \le \tau \le \tau_1 \text{ or } \tau_2 \le \tau \le 1\}.$$

Problem (P') contains some constraints involving y and τ_1, τ_2. Let $P_{Y_{ad}}$ be the projection onto Y_{ad} in W(0,T). For any $\varepsilon > 0$, we define the problem

$$\min \tau_2 - \tau_1 + \sigma J(u) + \frac{1}{\varepsilon}\int_{A(\tau_1,\tau_2)} \|y_u(\tau) - P_{Y_{ad}}\, y_u(\tau)\|^2 d\tau \qquad (P_\varepsilon)$$

$$\text{s.t. } u \in K,\ 0 \le \tau_1 \le \tau_2 \le 1 \text{ and } (2.1).$$

THEOREM 2 Under the assumptions of Theorem 1, problem (P_ε) has (at least) one solution. □

The proof of Theorem 2 is similar to the one of Theorem 1. Just notice that the last term of the criterion is continuous and convex, hence w.l.s.c., with respect to y.

THEOREM 3 Suppose that the assumptions of Theorem 1 hold. Let $\{u_\varepsilon, \tau_{1_\varepsilon}, \tau_{2_\varepsilon}\}$ be a sequence of solutions of (P_ε) (such a sequence exists by Theorem 2). Then $\{u_\varepsilon\}$ is bounded and any weak limit-point of $\{u_\varepsilon\}$ is a solution of (P). □

Proof For any solution $\bar{u}$ of (P) (there exists some) we have

$$\tau_{2\varepsilon} - \tau_{1\varepsilon} + \sigma J(u_\varepsilon) + \frac{1}{\varepsilon}\int_{A(\tau_{1_\varepsilon},\tau_{2_\varepsilon})} \|y_{u_\varepsilon} - P_{Y_{ad}}\, y_{u_\varepsilon}\|^2\, d\tau \qquad (3.1)$$

$$\le \tau_2(\bar{u}) - \tau_1(\bar{u}) + \sigma J(\bar{u}).$$

By the boundedness of K or the coercivity of J, $\{u_\varepsilon\}$ is bounded. Let u be a weak limit point of u_ε, and τ_1, τ_2 be limit points of τ_{1_ε}, τ_{2_ε}. From the w.l.s.c. of J and the above relation we deduce that

$$\tau_2 - \tau_1 + J(u) \le \tau_2(\bar{u}) - \tau_1(\bar{u}) + J(\bar{u}).$$

As K is weakly closed, u is in K. In addition, from (3.1) we deduce that for some C > 0 :

$$\int_{A(\tau_{1_\varepsilon},\tau_{2_\varepsilon})} \|y_{u_\varepsilon} - P_{Y_{ad}}(y_{u_\varepsilon})\|^2 \le C\varepsilon.$$

For all $\alpha > 0$, there exists a subsequence still indexed by ε such that

$$B_\alpha = \{\tau \; ; \; 0 \leq \tau \leq \tau_1 - \alpha \text{ or } \tau_2 + \alpha \leq \tau \leq T\}$$

is included in $A(\tau_{1_\varepsilon}, \tau_{2_\varepsilon})$, hence

$$\int_{B_\alpha} \| y_{u_\varepsilon} - P_{Y_{ad}}(y_{u_\varepsilon}) \|^2 \, d\tau \leq C\varepsilon.$$

This implies by passing to the limit that $y_u(\tau)$ is in Y_{ad}, a.e. in B_α, for all $\alpha > 0$; hence $\tau_1 \leq \tau_1(u)$ and $\tau_2(u) \leq \tau_2$ and u is a solution of (P). □

Remark The quadratic penalization to the distance to one set is continuously differentiable (in Hilbert spaces). Hence, if J is smooth, problem (P_ε) can be solved by a gradient and projection method. From a practical point of view, we may notice that a projection in W(0,T) is not easy to compute. Hence for practical reasons it may be better to choose some other spaces. For instance, if Y_{ad} is the set of non-negative function it is better to choose the space $L^2(Q)$ because the projection involves then only to take the positive part of y ; however, these points do not modify deeply an analysis.

IV - DEALING WITH NONLINEARITY

We shall now see how a nonlinearity in the state equation modifies our analysis. Essentially, we will see that some kind of compacity of the controls with respect to τ is needed and, interestingly, this will be related to the numerical results. We assume that the equation is of the following form

$$\begin{aligned} &\frac{\partial w}{\partial t}(\tau,x,t) - \Delta y(\tau,x,t) = u(\tau,x,t) \quad \text{in } (0,1) \times Q, \\ &y(\tau,\gamma,t) = 0 \quad \text{on } (0,1) \times \Sigma, \\ &w(\tau,x,0) = w^0(x,\tau) \quad \text{on } (0,1) \times \Omega \;, \\ &w(\tau,x,t) \in \beta(y(\tau,x,t)) \quad \text{in } (0,1) \times Q \;, \end{aligned} \tag{4.1}$$

with β maximal monotone from $\mathbb{R}$ onto $\mathbb{R}$, everywhere defined, and such that for some $\alpha > 0$:

$$(\beta(y) - \beta(z))(y-z) \geq \alpha(y-z)^2 \text{ for any } y,z \text{ in } \mathbb{R} \tag{4.2}$$

Remark The assumptions on β hold in particular for the two-phases Stefan problem in which $\beta(y)$ is defined by

$$\beta(y) = \begin{cases} y \text{ if } y < 0, \\ [0,1] \text{ if } y = 0, \\ 1+y \text{ if } y > 0. \end{cases}$$

Remark In our application (the steel casting problem) the control is made through the boundary rather than in all the domain. We prefer to study equation (3.1), however, in order to use some results of Z. Meike and D. Tiba [9].

We suppose that

$$w^0(x,\tau) \text{ is in } L^2(\Omega)$$
$$\beta^{-1}(w^0(x,\tau)) \text{ is in } H_0^1(\Omega). \tag{4.3}$$

We denote by w-V the weak topology of a Banach spaces V.

We introduce the spaces

$$Y = \{y \in L^\infty(0,T,H_0^1(\Omega)) \; ; \; \frac{dy}{dt} \in L^2(0,T,L^2(Q))\},$$

$$W = \{w \in L^\infty(0,T,L^2(Q)) \; ; \; \frac{dw}{dt} \in L^2(0,T,H^{-1}(\Omega))\}.$$

THEOREM 4 (Z. Meike, D. Tiba [9]). **We suppose that (4.2) and (4.3) hold. Let τ be given in (0,1) ; then the mapping $u(\tau) \to (y(\tau),w(\tau))$ is continuous from $w\text{-}L^2(Q)$ onto $(w\text{-}Y) \times (w\text{-}W)$ and the following inequality holds :**

$$\|y(\tau)\|_Y + \|w(\tau)\|_W \le C(1 + \|u(\tau)\|_{L^2(Q)}),$$

where C does not depend on u and τ. □

A convenient tool for the study of system (4.1) will be the following topology.

DEFINITION **Let V be a Banach space and $Z = L^2(0,1,V)$. We will say that $\{z^n\}$ converges weakly-punctually in Z towards z (in brief $z^n \to z$ wp-Z) if $\|z^n\|$ is bounded and**

$$z^n(\tau) \to z(\tau) \text{ in w-V, a.e. } \tau \text{ in } (0,1). \quad □$$

LEMMA **The weak-punctual topology is stronger that the weak-topology.** □

Proof Suppose that $\{z^n\}$ w.p. converges towards z in Z. As $L^\infty(0,1,V')$ is densely imbedded in Z', in order to prove that $z^n \to z$ in w-Z, it is sufficient to prove that (we denote by <.,.> the duality product between V and V')

$$\int_0^1 \langle z^n,w\rangle \to \int_0^1 \langle z,w\rangle \text{ for any } w \text{ in } L^\infty(0,T,V').$$

Define $f^n(\tau) = \langle z^n(\tau), w(\tau)\rangle$. Then, by the definition of the w.p. topology :

$$f^n(\tau) \to f(\tau) = \langle z(\tau),w(\tau)\rangle, \text{ a.e. } \tau \text{ in } (0,1).$$

On the other hand,

$$\int_0^1 (f^n(\tau))^2 \, d\tau \le \int_0^1 \|z^n(\tau)\|_V^2 \, \|w(\tau)\|_{V'}^2,$$
$$\le \|w\|^2_{L^\infty(0,T,V')} \, \|z^n\|_Z^2,$$

hence f^n is bounded in $L^2(0,1)$. Then, by a corollary of Egorov's Theorem, we deduce that $f^n \to f$ in w-$L^2(0,1)$, and in particular

$$\int_0^1 f^n(\tau)\, d\tau \to \int_0^1 f(\tau) d\tau$$

which is the desired relation. □

We define $U = L^2([0,1] \times Q)$.

THEOREM 5 **The mapping** $u \to (y,w)$ **(solution of** (4.1)) **is continuous from** wp-U **onto** wp-$L^2(0,1,Y) \times$ wp-$L^2(0,1,W)$. □

Proof Suppose that $\{u^n\}$ w.p. converges towards u in U, and denote (y^n,w^n) (resp. y,w) the state associated to u^n (resp. u). As $u^n(\tau) \to u(\tau)$ in w-$L^2(Q)$ a.e. τ in (0,1), we deduce from theorem 4 that

$$y^n(\tau) \to y(\tau) \text{ in w-}Y, \text{ a.e. } \tau \text{ in } (0,1),$$

$$w^n(\tau) \to w(\tau) \text{ in w-}W, \text{ a.e. } \tau \text{ in } (0,1).$$

From the inequality of theorem 4 we deduce that (y^n,w^n) is bounded in $L^2(0,T,Y) \times L^2(0,T,W)$. This implies that (y^n,w^n) w.p. converges towards (y,w). □

We now state the optimal control problem :

$$\min \tau_2(u) - \tau_1(u) + J(u), \text{ s.t. } u \in K \qquad \text{(P1)}$$

where $\tau_2(u)$ and $\tau_1(u)$ are now defined through the solution y of (4.1). The following result is a simple consequence of theorem 5.

THEOREM 6 **We suppose that assumptions** (2.4) **and** (2.5) **hold and that** K **is a** w.p. **compact subset of** U. **Then problem** (P1) **has (at least) one solution.** □

V - NUMERICAL RESULTS

We give in this section an application of our ideas to the continuous casting problem [1,5,7,10]. The state equation after using the Kirchhoff transform is

$$\partial G(y) / \partial t - \Delta y = 0 \text{ in } [0,\tau f] \times Q,$$

$$\partial y / \partial n + uy = 0 \text{ in } [0,\tau f] \times \Sigma,$$

$$y|_{t=0} = y_0 \qquad \text{in } [0,\tau f] \times \Omega$$

where G(y) is the enthalpy. The "state constraints" are

$$G(y)(\tau,x,t) \le L \text{ if } t \ge t_a,$$

$y \leq y_1$ on $[0,\tau f] \times \Sigma$,

$y(\tau,x,\phi(\tau)) \geq y_2$ on $[0,\tau f] \times \partial\Omega$,

with $\phi(\tau)$ a given function ; also the discrete derivative of y is subject to some bounds [5,7]. All these constraints are penalized in L^2 spaces. The control u (water flow) is subject to be piecewise constant, and to bounds on each componant and on the total flow at each time.

Call y_u the solution of the state equation. The problem is

$$\begin{cases} \text{satisfy as much as possible } y_u(\tau) \in Y_{ad}, \\ \text{s.t. } u \in K_0. \end{cases}$$

Denoting by $P_{Y_{ad}}$ the projection into Y_{ad} in a convenient space [5].

We formulate the classical problem

$$\min_u \frac{1}{\varepsilon} \int_0^{\tau f} \|y_u - P_{Y_{ad}} y_u\|^2 \, d\tau \qquad \text{s.t. } u \in K \tag{P1}$$

and the "new" problem

$$\min_{(u,\tau_1,\tau_2)} \tau_2-\tau_1 + \frac{1}{\varepsilon} \int_{A(\tau_1,\tau_2)} \|y_u - P_{Y_{ad}} y_u\|^2 \, d\tau \text{ s.t. } u \in K \tag{P2}$$

We have computed an approximate solution of (P1) by a gradient-projection method. For problem (P2) we designed a special algorithm for two reasons : (i) non homogeneity of (u,τ_1,τ_2) and bad conditioning of (P2) for ε small ; (ii) no meaningness of the results of the computation if (τ_1,τ_2) are far from $(\tau_1(u_1),\tau_2(u))$. For this reasons we choose an heuristical two-level algorithm : minimize with respect to u with (τ_1,τ_2) fixed, then update (τ_1,τ_2), as follows (see [5] for more details) :

ALGORITHM

step 0 Set u and ε.

$\bar{\tau}_1 \leftarrow 0$, $\bar{\tau}_2 \leftarrow \tau f$

$\tau_i \leftarrow \tau_i(u)$, $i = 1,2$.

step 1 $\overset{+}{\tau}_1 \leftarrow \max \{\tau \; ; \|y(\sigma) - P_{Y_{ad}}(y(\sigma))\| \leq \varepsilon$ **for all** $\sigma \leq \tau\}$,

$\overset{+}{\tau}_2 \leftarrow \min \{\tau \; ; \|y(\sigma) - P_{Y_{ad}}(y(\sigma))\| \leq \varepsilon$ **for all** $\sigma \geq \tau\}$.

If $\overset{+}{\tau}_2 - \overset{+}{\tau}_1 \geq \tau_2 - \tau_1$ **then**

If $\bar{\tau}_1 \geq \overset{+}{\tau}_1$ **or** $\bar{\tau}_2 \leq \overset{+}{\tau}_2$ **then stop**

else $\bar{\tau}_i \leftarrow \overset{+}{\tau}_i$, $i = 1,2$

step 2 Minimize with respect to u **; go to step 1.**

The system is subject to a decrease of the speed casting between time 80 and 120. The initial point of the optimization is an optimal control for the corresponding stationary process.

We performed 20 iterations on problem P1. Figure 1 shows that globally, the defects are strongly reduced. However, the diameter of the defective zone is not reduced as figure 2 shows. The optimal control oscillates too much in some regions but is nicely smoothed, without significant modification of the state, by adding a term that penalizes quadratically the derivative (figures 3 and 4). For problem 2, 5 major iterations on (τ_1, τ_2) have been performed. They correspond with 25 steps with respect to u. The defective zone decreases. In addition the difference with the optimal control for the stationary case is concentrated in small regions (a nice property from an engineering view-point). The control may be regularized as for problem 1 (figures 5 to 8).

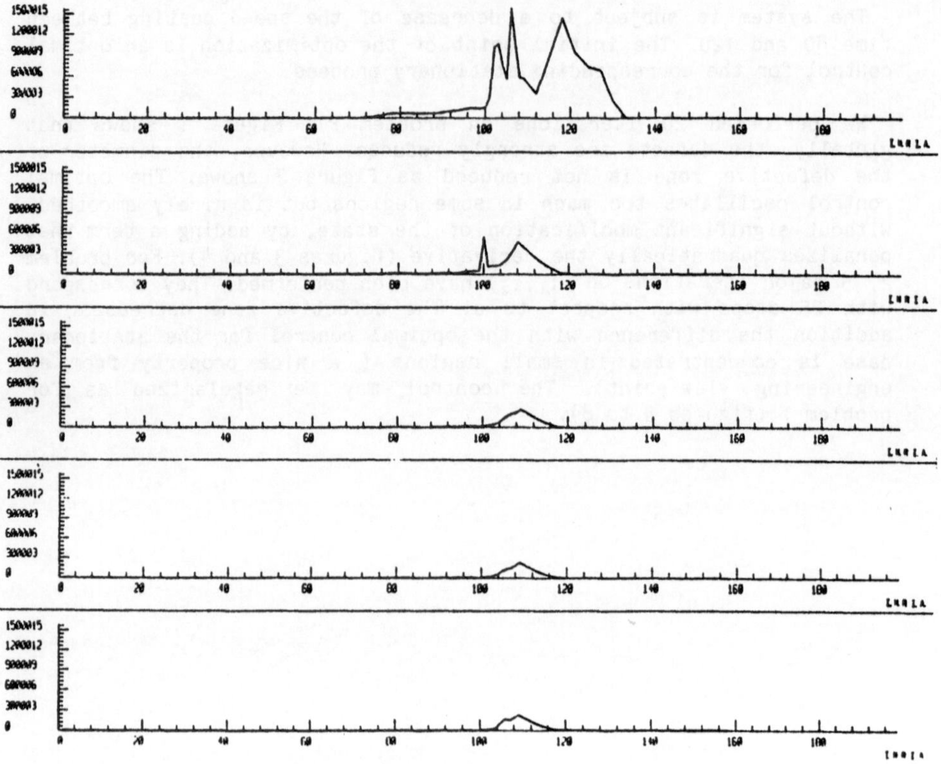

Figure 1
penalization of state constraints as a functions of τ for iterations 1,4,8,13,17 (problem 1)

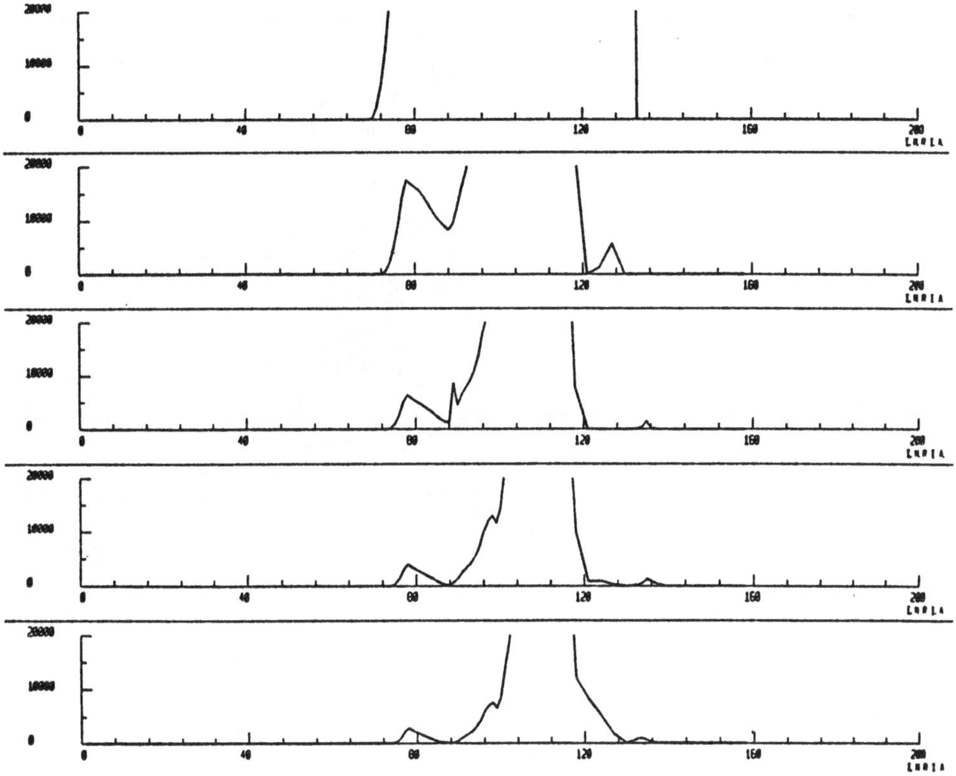

Figure 2
same as figure 1 with a different vertical scale

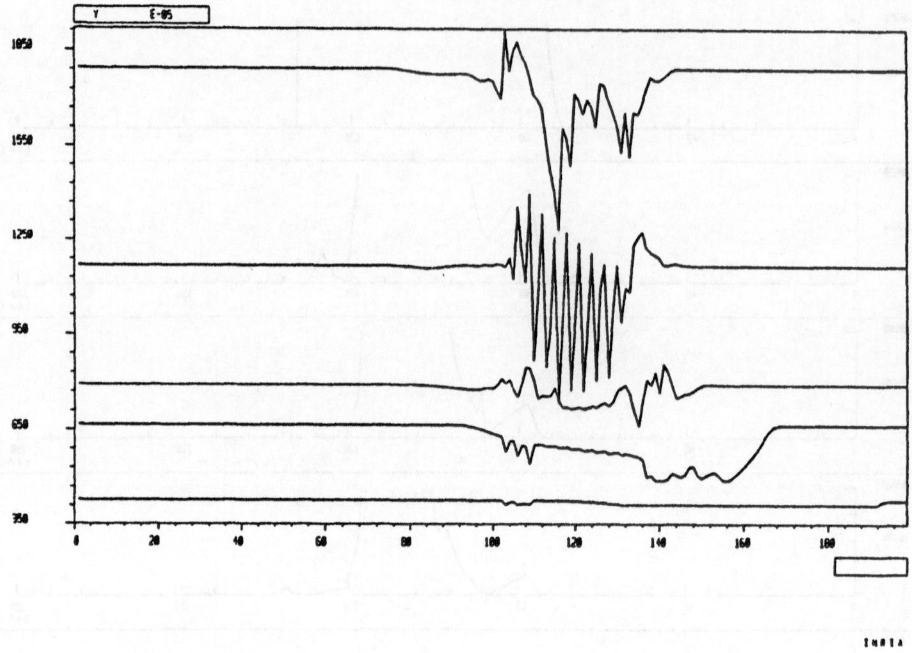

Figure 3
computed optimal control for problem 1

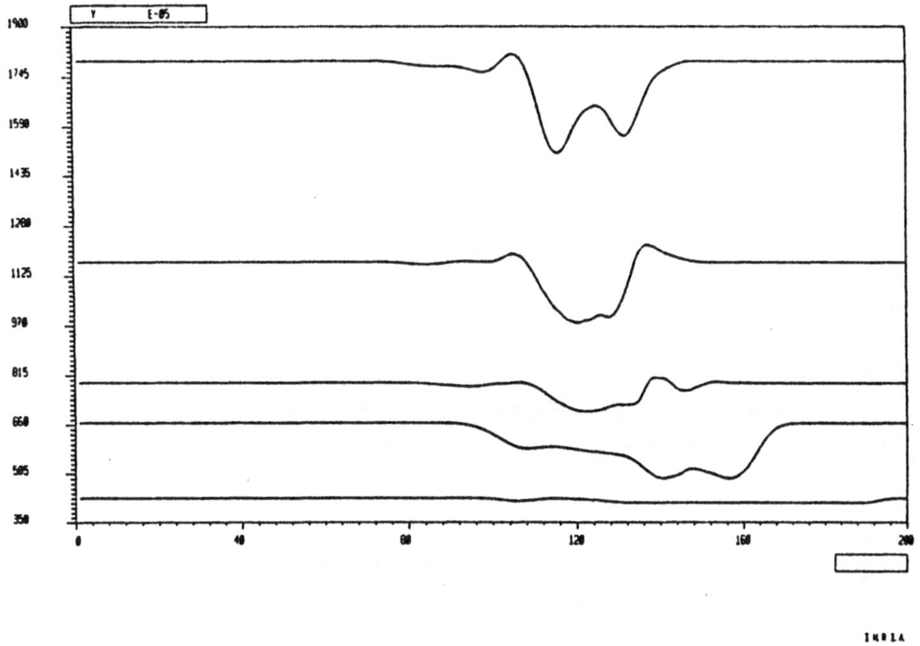

Figure 4
computed optimal control when a penalization of the derivative is added to the criterion for problem 1

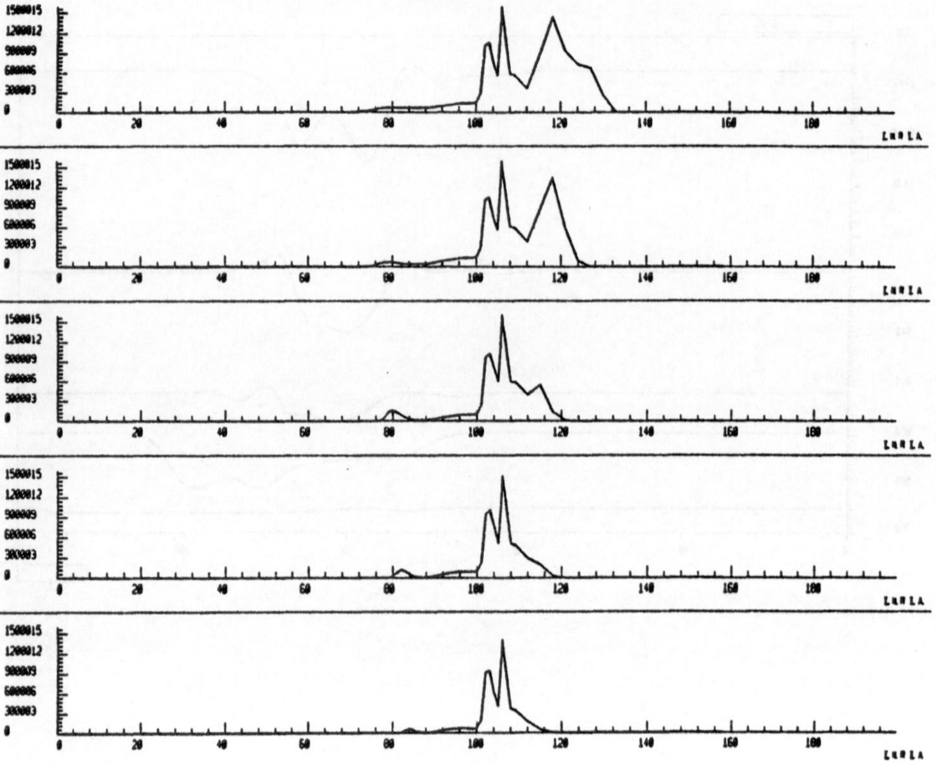

Figure 5
penalization of state constraints as a function of τ
(iterations in τ) (problem 2)

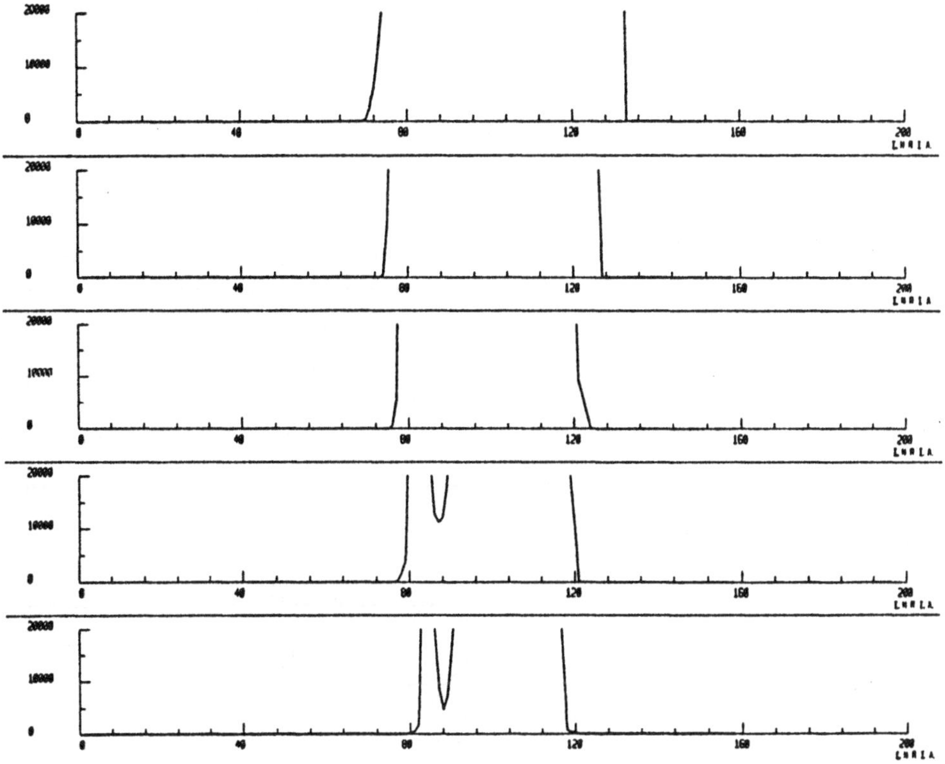

Figure 6
same as figure 5, with a different vertical scale

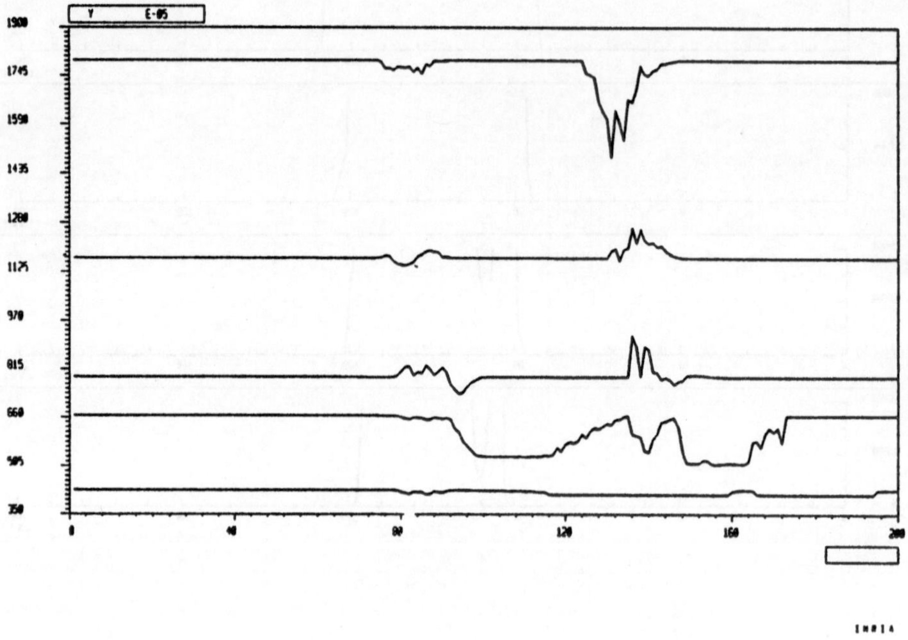

Figure 7
computed optimal control for problem 2

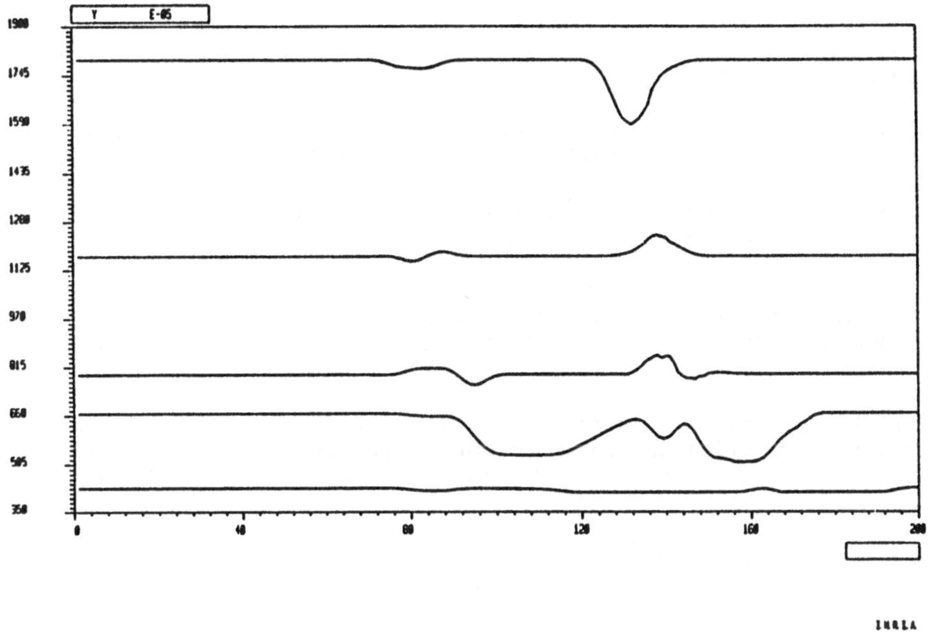

Figure 8
computed optimal control when a penalization for the derivative is added to the criterion of problem 2

REFERENCES

[1] J.F. Bonnans, V. Gaudrat, C. Saguez, J.P. Yvon (1987). **Reduction of the defective zone in the non-stationary continuous casting process.** 10th World IFAC Congress, Munich, to appear.

[2] J.F. Bonnans, E. Casas (1984). **On the choice of the function spaces for some state-constrained control problems.** Numer. Funct. Anal & Optim. 7, 333-348.

[3] J.F. Bonnans, C. Moreno, C. Saguez (1984). **Contrôle de domaines temporels.** INRIA Report n° 308.

[4] G. Di Pillo, L. Grippo (1979). **The multiplier method for optimal control problems of parabolic systems.** Appl. Math. Optim. 5, 253-269.

[5] V. Gaudrat (1987). **Quelques méthodes pour l'optimisation de la coulée continue.** Doctoral Thesis, to appear.

[6] J. Henry, M. Larrecq, J. Petegnieff, C. Saguez (1980). **Optimisation du refroidissement secondaire en coulée continue d'acier.** Worshop on "Control of metallurgical systems", Rocquencourt, France.

[7] M. Larrecq, C. Saguez, V.C. Tran, J.P. Yvon (1982). **Contrôle d'une installation de coulée continue.** Preprints of the Third IFAC Symposium on the control of distributed parameter systems, Toulouse, pp. VIII 10-16.

[8] J.L. Lions (1968). **Contrôle de systèmes gouvernés par des équations aux dérivées partielles.** Dunod, Gauthier Villars, Paris.

[9] Z. Meike, D. Tiba (1982). **Optimal control for a Stefan problem.** In Lectures Notes on Control and Information Science, n° 44, pp. 776-787, Springer Verlag, Berlin.

[10] C. Saguez (1980). **Contrôle optimal de systèmes à frontière libre.** Thèse d'état de l'Université de Technologie de Compiègne.

[11] J.P. Yvon (1970). **Application de la pénalisation à la résolution d'un problème de contrôle optimal.** Cahier de l'IRIA n° 2.

AN OPTIMIZATION PROBLEM FOR THIN INSULATING LAYERS AROUND A CONDUCTING MEDIUM

Giuseppe Buttazzo

Scuola Normale Superiore
Piazza dei Cavalieri, 7
56100 PISA (ITALY)

1. INTRODUCTION

The problem of a thin insulating layer around a conducting medium has been widely considered in the literature (see References); in this lecture I shall consider the related optimization problem, i.e. the problem of obtaining the *"best"* insulation (electrostatic, thermic, etc....) with fixed *"total insulation power"*.

The model we shall consider is the following. Let Ω be a regular bounded open subset of $\mathbb{R}^n$ (Ω is the conducting region) and let $d:\partial\Omega\to\mathbb{R}^+$ be a continuous positive function. For every $\varepsilon>0$ consider the layer Σ_ε whose thickness at thepoint $\sigma\in\partial\Omega$ is $\varepsilon d(\sigma)$

$$\Sigma_\varepsilon = \{\sigma+t\nu(\sigma) \ : \ \sigma\in\partial\Omega, \ 0<t<\varepsilon d(\sigma)\}$$

where ν denotes the outer normal versor to $\partial\Omega$, and set

$$\Omega_\varepsilon = \Omega\cup\Sigma_\varepsilon \ .$$

For the sake of simplicity, the conductivity coefficient is taken equal to 1 in Ω, whereas it is equal to ε in the layer Σ_ε. Given a function $f\in L^2(\Omega)$ (f is the density of heat sources in the thermostatic model), the temperature u_ε (set equal to zero outside Ω_ε) is the solution of the problem

$$\begin{cases} -\Delta u = f(x) & \text{in } \Omega \\ -\Delta u = 0 & \text{in } \Sigma_\varepsilon \\ u = 0 & \text{on } \partial\Omega_\varepsilon \end{cases} \tag{1.1}$$

together with the transmission conditions on $\partial\Omega$

$$\begin{cases} u^+ = u^- & \text{on } \partial\Omega \\ \varepsilon\, \partial u^+/\partial\nu = \partial u^-/\partial\nu & \text{on } \partial\Omega \end{cases} \tag{1.2}$$

where u^+, u^- are the outer and inner trace of u on $\partial\Omega$ respectively. It is easy to see that problem (1.1) with condition (1.2) can be written as a minimization problem in the following way:

$$\min \{ \int_\Omega |Du|^2\,dx + \varepsilon\int_\Omega |Du|^2\,dx - 2\int_\Omega f(x)u\,dx \ : \ u \in H^1_0(\Omega_\varepsilon) \}. \tag{1.3}$$

The study of the asymptotic behaviour (as $\varepsilon \to 0$) of u_ε has been made in several papers (see for instance [1],[3],[4],[7],[9],[10]); more precisely, the following result holds.

THEOREM 1.1. *The family* (u_ε) *tends (as* $\varepsilon \to 0$*) in* $L^2(\mathbb{R}^n)$ *to the unique solution* u_d *of the elliptic problem*

$$\begin{cases} -\Delta u = f(x) & \text{in } \Omega \\ d\, \partial u/\partial\nu + u = 0 & \text{on } \partial\Omega. \end{cases} \tag{1.4}$$

Equivalently, u_d *can be characterized as the unique solution of the minimum problem*

$$\min \{ \int_\Omega |Du|^2\,dx - 2\int_\Omega f(x)u\,dx + \int_{\partial\Omega} u^2/d\,d\sigma \ : \ u \in H^1(\Omega) \}. \tag{1.5}$$

REMARK 1.2. The limit problem (1.4) or (1.5) can be seen as a model for a conducting body whose boundary is varnished by an insulating varnish; in this way the function d can be regarded as the density of the varnish on the boundary, and the quantity $\int_{\partial\Omega} d\,d\sigma$ represents the *"total insulation power"*. Note that a different choice of the conductivity coefficient c_ε in the layer Σ_ε leads to the Dirichlet problem

$$\min \{ \int_\Omega |Du|^2\,dx - 2\int_\Omega f(x)u\,dx \ : \ u \in H^1_0(\Omega) \}$$

if $c_\varepsilon >> \varepsilon$, and to the Neumann problem

$$\min \{ \int_\Omega |Du|^2\,dx - 2\int_\Omega f(x)u\,dx \ : \ u \in H^1(\Omega) \}$$

if $c_\varepsilon << \varepsilon$. Finally, nonlinear models in which the term $\int_\Omega |Du|^2\,dx$ is replaced by a functional of the form $\int_\Omega f(x,Du)\,dx$, have been studied in [1].

2. THE OPTIMIZATION PROBLEM

In this section we study the optimization problem related to (1.5). More precisely, for every function d and every $u \in H^1(\Omega)$ let E(d,u) be the energy associated to problem (1.4), that is

$$E(d,u) = \int_\Omega |Du|^2\,dx \;-\; 2\int_\Omega f(x)u\,dx \;+\; \int_{\partial\Omega} u^2/d\,d\sigma \;.$$

In this way, if u_d denotes the solution of (1.4) and E(d) denotes the minimal energy

$$E(d) = \min\,\{E(d,u)\;:\; u\in H^1(\Omega)\}\,,$$

it is easy to see that

$$E(d) = E(d,u_d) = -\int_\Omega f(x)u_d\,dx \;. \tag{2.1}$$

The optimization criterion we choose is the minimization of E(d) among all functions d with fixed *"total insulation power"*. More precisely, we study the problem

$$\min\,\{E(d)\;:\; d{:}\partial\Omega\to\mathbb{R}^+,\; \int_{\partial\Omega} d\,d\sigma = k\}\,. \tag{2.2}$$

REMARK 2.1. Note that by (2.1) and (2.2), when the heat sources are uniformly distributed in Ω (i.e. $f\equiv c\in\mathbb{R}^+$), our optimization problem (2.2) is equivalent to determine the function d for which the averaged temperature $\int_\Omega u_d\,dx$ is maximum.

Problem (2.2) is just an example of optimization criterion; other criteria should be investigated, such as

$$\min \int_\Omega |u_d - a(x)|^2\,dx \qquad \text{(a(x) is the desired temperature)},$$

or more generally

$$\min\; \int_\Omega g(x,u_d)\,dx \;+\; \int_{\partial\Omega} h(\sigma,d,u_d)\,d\sigma$$

for suitable choices of the functions g and h.

For problem (2.2) the following result holds (see[6]).

THEOREM 2.2. *Let* $f\in L^2(\Omega)$ *be fixed and not identically zero. Then problem (2.2) has a unique solution* d_{opt}; *the corresponding optimal temperature* u_{opt} *can be characterized as the unique solution of the minimum problem*

$$\min\,\{ \int_\Omega |Du|^2\,dx \;-\; 2\int_\Omega f(x)u\,dx \;+\; {}^1\!/_k\big[\int_{\partial\Omega} |u|^2\,d\sigma\big]^2 \;:\; u\in H^1(\Omega)\}\,, \tag{2.3}$$

and we have for a.e.$\sigma \in \partial\Omega$

$$d_{opt}(\sigma) = \frac{k}{\int_{\partial\Omega}|u_{opt}|\,d\sigma}\,|u_{opt}(\sigma)|\,.$$

REMARK 2.3. Note that if $f \equiv 0$, then $u_{opt} \equiv 0$, so that any function $d:\partial\Omega \to \mathbb{R}^+$ with $\int_{\partial\Omega} d\,d\sigma = k$ is a solution of (2.2). Moreover, problem (2.2) does not change if we replace the constraint $\int_{\partial\Omega} d\,d\sigma = k$ by the constraint $\int_{\partial\Omega} d\,d\sigma \le k$. Finally, we remark that the minimum problem (2.3) is equivalent to the differential equation

$$\begin{cases} -\Delta u = f(x) & \text{in } \Omega \\ 0 \in k\,\partial u/\partial\nu + H(u)\int_{\partial\Omega}|u|\,d\sigma & \text{on } \partial\Omega \end{cases}$$

where H(t) is the multimapping

$$H(t) = \begin{cases} \operatorname{sgn} t & \text{if } t \neq 0 \\ [-1,1] & \text{if } t = 0\,. \end{cases}$$

It would be interesting to perform numerical computations for finding approximations of the optimal distribution d_{opt} of insulator, given the shape of the domain Ω and the heat sources f. I conclude by showing a simple example in which this computation can be made (by hand).

In dimension 2 let Ω be the anulus

$$\Omega = \{x \in \mathbb{R}^2 \;:\; 1<|x|<2\}\,,$$

and assume the heat sources are uniformly distributed in Ω (i.e. $f \equiv 1$). The following pictures show the optimal distribution of the insulator around $\partial\Omega$ when the *"total insulation power"* (normalized) $\frac{1}{|\partial\Omega|}\int_{\partial\Omega} d\,d\sigma$ is given. Note that in the third picture the best choice is to put all the insulator on the interior part of $\partial\Omega$.

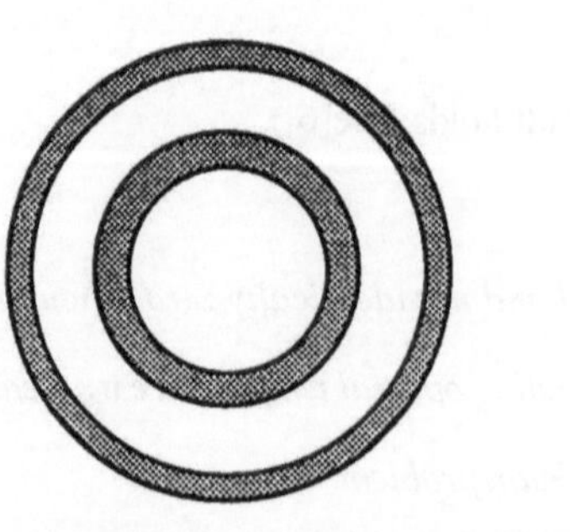

$$\frac{1}{|\partial\Omega|}\int_{\partial\Omega} d\,d\sigma = 0.25$$

$d_{int} = 0.33$

$d_{ext} = 0.21$

$$\frac{1}{|\partial\Omega|}\int_{\partial\Omega} d\, d\sigma = 0.15$$

$$d_{int} = 0.23$$

$$d_{ext} = 0.11$$

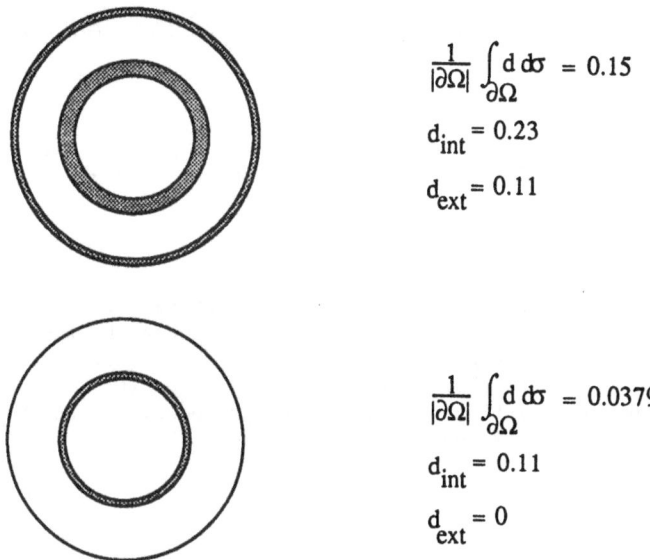

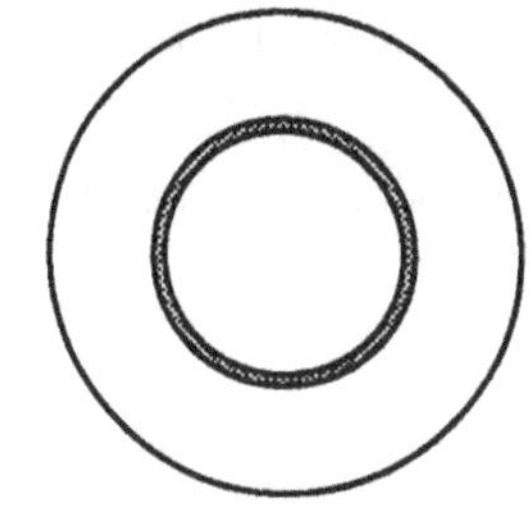

$$\frac{1}{|\partial\Omega|}\int_{\partial\Omega} d\, d\sigma = 0.0379$$

$$d_{int} = 0.11$$

$$d_{ext} = 0$$

REFERENCES

[1] E.ACERBI & G.BUTTAZZO: *Reinforcement problems in the calculus of variations.* Ann.Inst.H.Poincaré Anal.Non Linéaire, (to appear).

[2] E.ACERBI & G.BUTTAZZO: *Limit problems for plates surrounded by soft material.* Arch.Rational Mech.Anal., **92** (1986), 355-370.

[3] H.ATTOUCH: Variational Convergence for Functions and Operators. Pitman, Appl.Math.Ser., Boston (1984).

[4] H.BREZIS & L.CAFFARELLI & A.FRIEDMAN: *Reinforcement problems for elliptic equations and variational inequalities.* Ann.Mat.Pura Appl., **123** (1980), 219-246.

[5] G.BUTTAZZO: *Elliptic problems with a thin insulating layer around the boundary.* Proceedings of "Integral Functionals in Calculus of Variations", Trieste 9-12 September 1985, Springer-Verlag, Lect.Notes in Math., Berlin (to appear).

[6] G.BUTTAZZO: Paper in preparation.

[7] G.BUTTAZZO & G.DAL MASO & U.MOSCO: Paper in preparation.

[8] G.BUTTAZZO & R.V.KOHN: *Reinforcement by a thin layer with oscillating thickness.* Appl.Math.Optim., (to appear).

[9] L.CAFFARELLI & A.FRIEDMAN: *Reinforcement problems in elasto-plasticity.* Rocky Mountain J.Math., **10** (1980), 155-184.

[10] E.SANCHEZ-PALENCIA: Non-Homogeneous Media and Vibration Theory. Springer-Verlag, Lect.Notes in Phys. **127**, Berlin (1980).

SOME EFFECTS OF THE BOUNDARY ROUGHNESS IN A THIN FILM FLOW

M. CHAMBAT(*), G. BAYADA(**), AND J.B. FAURE(**)

Abstract : We study the asymptotic behavior of a Stokes flow between two rough surfaces when the gap η between the two surfaces tends to zero. The roughness is defined by a periodic function whose period ε tends also to zero. In both cases ($\varepsilon \ll \eta$ and $\eta \ll \varepsilon$), the pressure is solution of a Reynolds type equation but the coefficients are truly different.

I. INTRODUCTION

The behavior of a viscous fluid in a narrow gap has long been the subject of intensive studies. Numerous papers are based on the Reynolds equation which is derived from the Stokes equation taking into account the little parameter associated to the film height [1] [9]. However, the surfaces of the gap are not actually smooth and another little parameter appears which is related to the characteristic wavelength of the roughness.

In the mechanical lubrication litterature, two different ways to evaluate the effects of the surfaces roughness can be found [6] [7] [8] [14] [18] :

- If the Stokes equation is considered which is undoubtedly satisfactory from a physical view, all the papers are devoted to small height roughness and no general results are given.
- If the Reynolds equation is considered, computational procedures are proposed to obtain an averaged equation regardless the height of the roughness but the agreement with the preceeding case is not satisfactory.

In both cases, all the results have given rise to many controversies (see for instance [6] [8] [14] and inter comparisons are very difficult due to the various assumptions introduced.

(*) L.A.N., Math., Université LYON I, 69622 Villeurbanne Cedex
(**) Centre de Math., INSA, 69621 Villeurbanne Cedex.
UA 740 CNRS.

For the mathematical aspect of the problem, the situation is less than satisfactory. Most of the papers involve heuristic attempts while some of the others are based on formal asymptotic expansions and no real proof appears in the litterature.

In this paper we consider a deterministic way to describe the surfaces and a periodic roughness is assumed. Section 2 is devoted to the notations, we recall in section 3 the asymptotic behavior of the Stokes equation when the gap height η tends to zero [1], then we study in section 4 the behavior of the previously obtained Reynolds equation when the roughness spacing ε tends to zero, using the theory of the H-convergence [13] ; an homogenized Reynolds equation is found. Section 5 studies what ELROD [7] called the "Stokes roughness", i.e. the roughness wavelength is supposed to be small in front of the gap height, and we are dealt with the homogenization of the Stokes equation in an oscillating domain. In the "fixed part" we are led to a Stokes equation while in the "oscillating part" the velocity tends to zero. Related studies exist for the asymptotic behavior of solutions of partial differential equations with respect to variable domains [3] [15].

2. BASIC EQUATIONS

2.a. Geometrical data and notations :

We shall write $x = (x_1,x_2,x_3)$ for a current point in R^3 and $x = (x_1,x_2)$ for its projection in R^2.

ω is an open set in R^2 with a regular boundary $\partial\omega$.

ε is a small parameter related to the roughness wavelength scale. h is a smooth function, defined for x in $\bar{\omega}$ and y in R^2, periodic with period Y_i in y_i (i=1,2).

We set $Y = [0,Y_1] \times [0,Y_2]$, the periodic cell and we define the average function :

$$\tilde{h}(x) = \frac{1}{mes(Y)} \int_Y h(x,y)dy$$

The real gap between the two surfaces is given by :

$$h_\varepsilon(x) = h(x,x/\varepsilon) \qquad x \in \bar{\omega}$$

We shall give in each section the right hypothesis needed on h.

The three dimensional rescaled domain occupied by the fluid is :

$$\Omega_\varepsilon = \{(x,z) \in R^3,\ x \in \omega,\ o < z < h_\varepsilon(x)\}$$

A smooth lower and upper bound of h_ε is needed in the next chapters. We set :

$$\text{hmin}(x) = \min_{y \in Y} h(x,y) \qquad \text{hmax}(x) = \max_{y \in Y} h(x,y)$$

We suppose that these extrema exist and that :

$$\text{hmax}(x) - \text{hmin}(x) \geqslant \alpha > o \quad \forall x \in \bar{\omega}$$

We introduce several notations (see figure 1)

$$\Omega^- = \{(x,z) \in R^3,\ x \in \omega,\ o < z < \text{hmin}(x)\}$$

$$\Omega^+ = \{(x,z) \in R^3,\ x \in \omega,\ \text{hmin}(x) < z < \text{hmax}(x)\}$$

$$\Sigma^- = \{(x,z) \in R^3,\ x \in \omega,\ z = \text{hmin}(x)\}$$

$$\Sigma^+ = \{(x,z) \in R^3,\ x \in \omega,\ z = \text{hmax}(x)\}$$

$$\Sigma_\varepsilon = \{(x,z) \in R^3,\ x \in \omega,\ z = h_\varepsilon(x)\}$$

$$\Gamma^- = \{(x,z) \in R^3,\ x \in \partial\omega,\ o < z < \text{hmin}(x)\}$$

$$\Gamma^+ = \{(x,z) \in R^3,\ x \in \partial\omega,\ \text{hmin}(x) < z < \text{hmax}(x)\}$$

$$\Gamma_\varepsilon = (x,z) \in R^3,\ x \in \partial\omega,\ o < z < h_\varepsilon(x)\}$$

$$\Gamma_\varepsilon^+ = \Gamma_\varepsilon \setminus \Gamma^-$$

$$\Gamma = \Gamma^+ \cup \Gamma^-$$

2.b. The basic equations :

We are concerned with the thin-film hydrodynamic lubrication of rough surfaces, that is the study of an incompressible viscous fluid flow between two surfaces in motion, as the thickness of the gap is small. To make the model easier to study, we suppose that one of the surface is horizontal and smooth whereas the other one is rough and described by the function h_ε. We take into account the thickness of the fluid film by introducing another small parameter η and a thin domain :

$$\Omega_{\varepsilon\eta} = \{X \in R^3 / x_3 = \eta z,\ (x,z) \in \Omega_\varepsilon\}$$

All the geometrical data introduced in figure 1 have a corresponding one after the change of scale $x_3 = \eta z$. We shall denote them by adding an η subscript. For instance :

$$\partial\Omega_{\varepsilon\eta} = \omega \cup \Gamma_{\varepsilon\eta} \cup \Sigma_{\varepsilon\eta}$$

The basic Stokes system is :

$$\begin{cases} -\Delta U^{\varepsilon\eta} + \nabla . p^{\varepsilon\eta} = o & \quad (2.1) \\ & \text{in } \Omega_{\varepsilon\eta} \\ \operatorname{div} (U^{\varepsilon\eta}) = o & \quad (2.2) \end{cases}$$

where $p^{\varepsilon\eta}$ and $U^{\varepsilon\eta}$ are respectively the pressure and the velocity field while Δ is the Laplace operator and ∇ the gradient vector. For convenience, we shall write with a superscript ^ all the functions defined on the rescaled domain Ω_ε. For instance, we set :

$$\hat{U}^{\varepsilon\eta}(x,z) = U^{\varepsilon\eta}(x,\eta z) \qquad x,z \in \Omega_\varepsilon$$

Boundary conditions are added to solve equations (2.1) (2.2) ; classical operating conditions are Dirichlet ones :

$$U^{\varepsilon\eta} = (k^{\varepsilon\eta},o,o) \text{ on } \partial\Omega_{\varepsilon\eta} \qquad (2.3)$$

where

$$k^{\varepsilon\eta} = \begin{cases} o \text{ on } \Sigma_{\varepsilon\eta} \\ s \text{ on } \omega \end{cases} \qquad (s \in \mathbb{R}^+)$$

$k^{\varepsilon\eta}$ is not easy to evaluate on $\Gamma_{\varepsilon\eta}$ and some examples are given at the end of section 3. We assume here that the function $k^{\varepsilon\eta}$ is a regular connection of these two values, this means that it belongs to the Sobolev space $H^{1/2}(\partial\Omega_{\varepsilon\eta})$, with the additional condition :

$$\int_{\partial\Omega_{\varepsilon\eta}} k^{\varepsilon\eta} \cos(\underline{n},\underline{x}_1) d\sigma = o \qquad (2.4)$$

($\underline{n}$ is the outward unit normal vector on $\partial\Omega_{\varepsilon\eta}$ and $d\sigma$ denotes the surface measure).

2.c. *The variational formulation* :

We introduce the Sobolev spaces $H^1(\Omega_{\varepsilon\eta})$ and $H^{-1}(\Omega_{\varepsilon\eta})$.

We set $L^2_o(\Omega_{\varepsilon\eta}) = \{q \in L^2(\Omega_{\varepsilon\eta}), \int_{\Omega_{\varepsilon\eta}} q dX = o\}$

(,) is the scalar product in $L^2(\Omega_{\varepsilon\eta})$

((,)) is the scalar product in $(L^2(\Omega_{\varepsilon\eta}))^3$

$$a(U,V) = \sum_{i,j=1}^{3} \int_{\Omega_{\varepsilon\eta}} \frac{\partial u_i}{\partial x_j} \frac{\partial v_i}{\partial x_j} dX$$

The weak formulation of the Stokes system can be written :
Find $(U^{\varepsilon\eta}, p^{\varepsilon\eta})$ in $H^1(\Omega_{\varepsilon\eta})^3 \times L_o^2(\Omega_{\varepsilon\eta})$ such that :

$$S_{\varepsilon\eta} \begin{cases} a(U^{\varepsilon\eta},\phi) = \int_{\Omega_{\varepsilon\eta}} p^{\varepsilon\eta} \operatorname{div}(\phi) dX & \forall\phi \in H_o^1(\Omega_{\varepsilon\eta})^3 \quad (2.5) \\ \int_{\Omega\varepsilon\eta} q \operatorname{div}(U^{\varepsilon\eta}) dX = 0 & \forall q \in L_o^2(\Omega_{\varepsilon\eta}) \quad (2.6) \end{cases}$$

We recall the following existence theorem.

THEOREM 2.1 [11]:

Under hypothesis (2.4), there exists one and only one pair $(U^{\varepsilon\eta}, p^{\varepsilon\eta})$ solution of $S_{\varepsilon\eta}$. Moreover it is a solution of the strong formulation of the Stokes system in the following sense :

(2.1) is true in $(H^{-1}(\Omega_{\varepsilon\eta}))^3$

(2.2) is true in $L^2(\Omega_{\varepsilon\eta})$

(2.3) is true in $(H^{1/2}(\partial\Omega_{\varepsilon\eta}))^3$

Remark : all the previous results are still valid if a second member F in $(H^{-1}(\Omega_{\varepsilon\eta}))^3$ is added in (2.1) but in the lubrication theory, the outer forces are generally equal to zero.

3. THE ASYMPTOTIC BEHAVIOR OF THE STOKES EQUATIONS IN THE THIN FILM ASSUMPTIONS

This study falls in the scope of the "Reynolds roughness" when the roughness wavelength is large compared with the gap. We point out that it is not necessary to suppose in the present work that the roughness height is much smaller than the gap height. This problem has already been studied [1] and we give for reader's convenience a review of the results.

We need in this section the following hypothesis on h_ε :

$h_\varepsilon(x) \in C^1(\bar{\omega})$

Denoting by C any constant in ε and η , the following estimates are valid :

THEOREM 3.1 :

$$\|\hat{U}^{\varepsilon\eta}\|_{L^2(\Omega_\varepsilon)^3} \leq C, \quad \|\frac{\partial \hat{u}_i^{\varepsilon\eta}}{\partial x_j}\|_{L^2(\Omega_\varepsilon)} \leq \frac{C}{\eta}, \quad \|\frac{\partial \hat{u}_i^{\varepsilon\eta}}{\partial z}\|_{L^2(\Omega_\varepsilon)} \leq C,$$

$$1 \leq i \leq 3 \quad , \quad 1 \leq j \leq 2$$

and there exists $U^{*\varepsilon}$ in $L^2(\Omega_\varepsilon)^3$ such that :

$$\hat{U}^{\varepsilon\eta} \underset{\eta \to 0}{\longrightarrow} U^{*\varepsilon}, \quad \eta \frac{\partial \hat{U}^{\varepsilon\eta}}{\partial x_j} \underset{\eta \to 0}{\longrightarrow} 0 (j=1,2), \frac{\partial U^{\varepsilon\eta}}{\partial z} \underset{\eta \to 0}{\longrightarrow} \frac{\partial U^{*\varepsilon}}{\partial z} \; L^2(\Omega_\varepsilon)^3\text{-weak}$$

Moreover by a trace theorem on $\Sigma_\varepsilon \cup \omega$, $U^{*\varepsilon} = 0$ on Σ_ε and $\begin{bmatrix} s \\ 0 \\ 0 \end{bmatrix}$ on ω.

We give a similar theorem for the pressure distribution.

THEOREM 3.2 :

$$\|\frac{\partial \hat{p}^{\varepsilon\eta}}{\partial x_i}\|_{H^{-1}(\Omega_\varepsilon)} \leq \frac{C}{\eta^2}, \quad \|\frac{\partial \hat{p}^{\varepsilon\eta}}{\partial z}\|_{H^{-1}(\Omega_\varepsilon)} \leq \frac{C}{\eta}$$

and there exists $p^{*\varepsilon}$ in $L_0^2(\Omega)$ such that :

$$\eta^2 \hat{p}^{\varepsilon\eta} \longrightarrow p^{*\varepsilon} \; L^2(\Omega_\varepsilon)\text{-weak}$$

and $\frac{\partial p^{*\varepsilon}}{\partial z} = 0.$

Now by assuming supplementary assumptions on the boundary condition $k^{\varepsilon\eta}$, we are able to obtain the limit equation when $\eta \to 0$. A first kind of condition which is quite natural with regard to the change of scale $z = x_3/\eta$ will be sufficient to ensure the existence and uniqueness of the limit.

We suppose up to now that there exists $\hat{k}^\varepsilon$ in $H^{1/2}(\Gamma_{\varepsilon\eta})$ which does not depend on η such that :

$$k^{\varepsilon\eta}(x, x_3) = \hat{k}^\varepsilon(x, x_3/\eta) \tag{3.1}$$

We have the following result :

THEOREM 3.3 :

The limit $p^{*\varepsilon}$ belongs to $H^1(\omega)$ and it is the unique solution of the following Neuman problem in ω :

$$\operatorname{div}(h_\varepsilon^3 \nabla p^{*\varepsilon}) = 6s \frac{\partial h_\varepsilon}{\partial x_1} \tag{3.2}$$

$$h_\varepsilon^3 \frac{\partial p^{*\varepsilon}}{\partial n} = 6s\, h_\varepsilon \cos(\underline{n}, \underline{x}_1) - 1^\varepsilon \tag{3.3}$$

$$\int_\omega p^{*\varepsilon}(x) h_\varepsilon(x)\, dx = o \tag{3.4}$$

with $1^\varepsilon = \lim\limits_{\eta \to o} \left(\int_o^{h_\varepsilon(x)} \hat{k}^\varepsilon(x,z)\, dz\right).\underline{n}$

Proof : see [1] theorem 8.

2.d. The choice of the boundary condition on Γ :

We shall focus on the choice of $k^{\varepsilon\eta}$ on the part $\Gamma_{\varepsilon\eta}$ of the boundary. It is difficult to evaluate it by physical measures. Some authors assert that this value is not of importance because η is tending to zero. But it has been shown [1] that the choice of a quadratic or linear link between 0 and s prevents a boundary layer to appear on Γ_ε when η tends to zero. We set :

$$k^{\varepsilon\eta}(X) = 3s\left(\frac{x_3}{\eta h_\varepsilon(x)}\right)^2 - 4s\,\frac{x_3}{\eta h_\varepsilon(x)} + s \tag{3.5}$$

The condition (2.4) is easy to check with this particular choice and then the Stokes equation is well posed. Moreover (3.1) is fulfilled with $1^\varepsilon = o$.

We recall, as already mentionned, that with any other choice than (3.5) for $k^{\varepsilon\eta}$ there will be a boundary layer on Γ, but this phenomenon will have no influence on the convergence result for the pressure.

4. EFFECT OF THE ROUGHNESS : THE HOMOGENIZATION PROCESS FOR $P^{*\varepsilon}$

We are dealt now with the asymptotic behavior of $p^{*\varepsilon}$ when ε tends to zero. Recalling that h_ε is a function of two variables -

the macro one x related to the nominal mean geometry and the micro one $y = \frac{x}{\varepsilon}$, Y periodic in y and representative of the roughness, we observe [3] [15] that the homogenization of (3.2)-(3.4) is not a classical problem due to the fact that the right hand side of (3.2) does not converge in $H^{-1}(\omega)$ when ε tends to zero.

The homogenization of such a kind of equation has been studied in [2] for Dirichlet boundary conditions on $\partial\omega$. The particular form of the right hand side of the equation prevents us from inducing immediatly the results of Dirichlet - to the study of Neuman - boundary conditions. The H-convergence theory [13] is needed to solve the problem. For all definitions and related proofs we address the reader to [2] [10] and we just recall here the few results that are needed throughout.

Firstly, we introduce by a translation a new function p^{ε} instead of $p^{*\varepsilon}$ so that the basic problem (3.2)-(3.4) can be written in the space

$$H^1_m = \{\phi \in H^1(\omega), \int_\omega \phi\, dx = o\}$$

The basic problem becomes :

$$\text{Find } p^{\varepsilon} \in H^1_m, \int_\omega h^3 \nabla p^{\varepsilon} \nabla\phi dx = 6 \int_\omega sh \frac{\partial\phi}{\partial x_1} dx, \forall\phi \in H^1(\omega) \quad (4.1)$$

(We still denote by ∇ the gradient in R^2).

We need to introduce now some further notations.

A^{ε} and β^{ε} are respectively the matrix and the vector defined by :

$$A^{\varepsilon} = \begin{bmatrix} h^3_{\varepsilon} & o \\ o & h^3_{\varepsilon} \end{bmatrix} \qquad \beta^{\varepsilon} = \begin{bmatrix} 6sh_{\varepsilon} \\ o \end{bmatrix}$$

so that the problem (3.4) can be written :

$$\text{Find } p^{\varepsilon} \in H^1_m, \int_\omega A^{\varepsilon} \nabla p^{\varepsilon} . \nabla\phi dx = \int_\omega \beta^{\varepsilon} . \nabla\phi dx \qquad \forall\phi \in H^1(\omega)$$

Let us suppose now that there exists a regular domain Θ such that $\omega \subset\subset \Theta$ and such that the function $h_{\varepsilon}(x)$ has a regular extension in Θ also denoted by h_{ε}.

More precisely, we suppose that :

$$h_\varepsilon(x) = h(x,y)/_{y=\frac{x}{\varepsilon}} \quad \forall x \in \bar{\Theta}$$

$$h(x,y) \in C_o\,(\bar{\Theta}\,,\,\mathcal{L}_p^\infty(R^2)) \quad o \leqslant \alpha \leqslant h(x,y) \tag{4.2}$$

Then, it is proved in [2] (lemma 8) that the matrix A^ε H-converges to the homogenized matrix $A^\circ = (a^\circ_{ij})$ defined below :

Let us introduce the periodic Hilbert space :

$$H^1_p(Y) = \{\phi \in H^1(Y),\ \phi \text{ is Y periodic }\}$$

For each $x \in \omega$, we introduce w_i, solutions of the auxiliary problems in $H^1_p(Y)$:

$$\forall\phi \in H^1_p(Y),\ \int_Y h^3(x,y)\nabla_y w_i \nabla_y \phi dy = - \int_Y h^3(x,y)\, \frac{\partial\phi}{\partial y_i}\, dy$$

It is well known that this problem has a unique solution, up to an additive constant, so we impose to w_i the supplementary condition :

$$\int_Y w_i \, dy = o$$

The homogenized coefficients are given by :

$$a^\circ_{ii}(x) = \frac{1}{\text{mes}(Y)} \left[\int_Y (h^3(x,y) - w_i \frac{\partial h^3}{\partial x_i}) dy \right]$$

$$a^\circ_{ij}(x) = a^\circ_{ji}(x) = \frac{1}{\text{mes}(Y)} \left[\int_Y w_j \frac{\partial h^3}{\partial x_i} dy \right] \quad i \neq j$$

The H-convergence of A^ε is also valid for any subdomain of θ and especially for the initial one ω. We shall use the two following theorems :

THEOREM 4.1 : [10]

Let A^ε be a sequence of matrix which H-converges towards A°. If $v_\varepsilon \in H^1(\omega)$ and $f_\varepsilon \in H^{-1}(\omega)$ are such that :

(i) $\quad \text{div}(A^\varepsilon \nabla v_\varepsilon) = f_\varepsilon \quad$ in $H^{-1}(\omega)$

(ii) $\quad v_\varepsilon$ weakly converges to v° in $H^1(\omega)$

(iii) $\quad f_\varepsilon$ strongly converges to f in $H^{-1}(\omega)$

Then $A^{\varepsilon}\nabla v_{\varepsilon}$ weakly converges to $A^{\circ}\nabla v^{\circ}$ in $(L^2(\omega))^2$, and A° is associated with a continuous and coercive bilinear form on H^1_m.

THEOREM 4.2 : [10]

With the assumption of theorem 3.4, there exists a matrix P^{ε} and a vector Z^{ε} in $L^2(\omega)^2$ such that :

$$Z^{\varepsilon} = \nabla v_{\varepsilon} - P_{\varepsilon}\nabla v^{\circ} \longrightarrow o \quad L^1_{loc}(\omega)^2 \text{ strong}$$

Moreover, if $\int_{\omega} A^{\varepsilon}\nabla v_{\varepsilon}\nabla v_{\varepsilon}dx \to \int_{\omega} A^{\circ}\nabla v^{\circ}\nabla v^{\circ}dx$ then :

$$Z^{\varepsilon} \longrightarrow o \qquad L^1(\omega)^2 \text{ strong}$$

To study the asymptotic behavior of β^{ε} as ε tends to 0, a supplementary assumption is needed to identify its limit. Let us suppose that :

$$h(x,y) \in C^1(\bar{\omega} \times Y) \tag{4.3}$$

This assumption allows us to give the supplementary result :

THEOREM 4.3 :

The vector ${}^tP^{\varepsilon}\beta^{\varepsilon}$ weakly converges to β° in $L^2(\omega)$ and β° lies in $C^1(\bar{\omega})$.

We have now :

THEOREM 4.4 :

The solution p^{ε} of the equation (4.1) weakly converges in $H^1(\omega)$ to the solution p° of the so-called "homogenized problem" :

$$\text{Find } p^{\circ} \in H^1(\omega), \quad \int_{\omega} p^{\circ}dx = o \quad \int_{\omega} A^{\circ}\nabla p^{\circ}.\nabla\phi dx = \int_{\omega}\beta^{\circ}.\nabla\phi dx, \quad \forall\phi \in H^1(\omega).$$

Proof : Setting $\phi = p^{\varepsilon}$ in (4.1) we find :

$$\|\nabla p^{\varepsilon}\|_{L^2(\omega)} \leqslant K$$

But as $p^{\varepsilon} \in H^1_m$, the inequality

$$\|\phi\|^2_{H^1(\omega)} \leqslant K\left[\int_{\omega} \nabla\phi.\nabla\phi dx + \left(\int_{\omega} \phi dx\right)^2\right]$$

implies :

$$\|p^\varepsilon\|_{H^1(\omega)} \leqslant K$$

and as H^1_m is closed in H^1, we deduce that there exists p° in H^1_m such that :

$$p^\varepsilon \rightharpoonup p^\circ \qquad H^1(\omega)\text{-weak} \tag{4.4}$$

For convenience we set $V = H^1(\omega)$, V' its dual and $<,>$ the duality between V and V'. We consider now an auxiliary problem :

$$\text{Find } v_\varepsilon \in H^1_m, \int_\omega A^\varepsilon \nabla v_\varepsilon . \nabla\phi dx = <f,\phi>, \forall\phi \in V, \tag{4.5}$$

where f is any element of V' such that

$$<f,1> = o \tag{4.6}$$

It well known [16] that the condition (4.6) implies the existence and uniqueness of v_ε by the Fredholm alternative ; moreover we have, exactly as for p^ε :

$$\|v_\varepsilon\|_{H^1} \leqslant K$$

and there exists $v^\circ \in H^1_m$ such that

$$v_\varepsilon \rightharpoonup v^\circ \qquad V\text{-weak} \tag{4.7}$$

We can now use the theorem 4.1 with $f_\varepsilon = f'$, the restriction of f to $H^{-1}(\omega)$. f' is a constant in ε , so :

$$A^\varepsilon \nabla v_\varepsilon \rightharpoonup A^\circ \nabla v^\circ \qquad (L^2(\omega))^2\text{- weak} \tag{4.8}$$

and using (4.5) and (4.8), we obtain :

$$\int_\omega A^\circ \nabla v^\circ \nabla\phi dx = <f,\phi> \qquad \forall\phi,\ \phi \in H^1(\omega) \tag{4.9}$$

For peculiar $\phi = v^\circ$;

$$\int_\omega A^\circ \nabla v^\circ \nabla v^\circ dx = <f,v^\circ> \tag{4.10}$$

But from (4.5) with $\phi = v_\varepsilon$

$$\int_\omega A^\varepsilon \nabla v_\varepsilon v_\varepsilon dx = <f,v_\varepsilon> \tag{4.11}$$

Letting now ε tend to zero in (4.11) and using (4.7), we obtain :

$$\int_\omega A^\varepsilon \nabla v_\varepsilon . \nabla v_\varepsilon dx \longrightarrow \langle f, v^\circ \rangle = \int_\omega A^\circ \nabla v^\circ \nabla v^\circ dx$$

Recalling now (4.8), we use the strong convergence of theorem (4.2). There exist P^ε and Z^ε such that :

$$Z^\varepsilon = \nabla v_\varepsilon - P^\varepsilon \nabla v^\circ \qquad (4.12)$$

$$\text{and } Z^\varepsilon \longrightarrow o \text{ in } L^1(\omega) \text{ strong} \qquad (4.13)$$

We use now v_ε as a test function in (3.4) ; and substituting (4.13) in the right hand side, we obtain :

$$I_\varepsilon = \int_\omega A^\varepsilon \nabla p^\varepsilon \nabla v_\varepsilon dx = \int_\omega (\beta^\varepsilon . Z^\varepsilon + \beta^\varepsilon . P^\varepsilon \nabla v^\circ) dx$$

$$= \int_\omega \beta^\varepsilon . Z^\varepsilon dx + \int_\omega \nabla v^\circ . {}^t P^\varepsilon \beta^\varepsilon dx$$

Using now (4.12) and theorem 4.3, we deduce that β^ε tends to $(\tilde{h}, o)$ in $\mathcal{L}^\infty(\omega)^2$ weak-star :

$$I_\varepsilon \longrightarrow \int_\omega \beta^\circ . \nabla v^\circ dx$$

But I_ε is nothing else than $\langle f, p^\varepsilon \rangle$ which tends to $\langle f, p^\circ \rangle$, so by (4.9) with $\phi = p^\circ$, we obtain :

$$\langle f, p^\circ \rangle = \int_\omega A^\circ \nabla v^\circ \nabla p^\circ dx = \int_\omega \beta^\circ . \nabla v^\circ dx \qquad (4.14)$$

To conclude the proof, it suffices to show that for any v° in $H^1(\omega)$, we can find an f in V' such that v° is the solution of (4.9) and becomes a test function for (4.14).

The bilinear form $a(v^\circ, v) = \int_\omega A^\circ \nabla v^\circ \nabla v dx$ defines for each v° in $H^1(\omega)$ a linear continuous mapping from V into R. So it is an element f of V' such that :

$$a(v^\circ, v) = \langle f, v \rangle \qquad \forall v \in H^1(\omega)$$

which ends the proof. □

The last point to show is that the translated unknown $p^{*\varepsilon}$ also converges. This is pointed out in the following theorem :

THEOREM 4.5 :

The solution $p^{*\varepsilon}$ of problem (3.2)-(3.4) weakly converges in $H^1(\omega)$ towards the solution p^{*1} of :

$$\operatorname{div}(A^\circ \nabla p^{*1}) = \operatorname{div}(\beta^\circ) \tag{4.15}$$

$$\nabla p^{*1} . \underline{n}_o = \beta^\circ . \underline{n} \tag{4.16}$$

$$\int_\omega \tilde{h} p^{*1} dx = o \tag{4.17}$$

where $\underline{n}$ is the outward unit normal vector on $\partial\omega$, $\underline{n}_o = A^\circ \underline{n}$.

Proof : Let us introduce the constant c_ε such that $p^\varepsilon = p^{*\varepsilon} + c_\varepsilon$.

From (3.4) we have : $c_\varepsilon \int_\omega h_\varepsilon dx = \int_\omega h_\varepsilon p^\varepsilon dx$

As p^ε strongly converges in $L^1(\omega)$ and h_ε converges in $L^\infty(\omega)$ weak star, we deduce that :

$$c_\varepsilon \rightarrow c_o = \frac{\int_\omega \tilde{h} p^\circ dx}{\int_\omega \tilde{h} dx}$$

Now $p^{*\varepsilon}$ converges in $H^1(\omega)$ weak to $p^{*1} = p^\circ - c_o$ with

$\int_\omega \tilde{h} p^{*1} dx = o$ which is (4.17) □

5. HOMOGENIZATION OF THE STOKES EQUATION IN A HIGHLY VARYING DOMAIN

This study can be regarded as the homogenization of the Stokes equation with respect of the domain. In this section, the only hypothesis needed on h(x,y) is that : $h(x,y) \in \mathscr{C}^\circ(\bar{\omega}, \mathcal{L}^\infty_p(R^2))$. A related study can be found for a uniform homogenization (h does not depend on x but only on y) for the Laplacian operator in [3] . For any function v defined on $\Omega_{\varepsilon\eta}$ we denote by $\bar{v}$ the function equal to v on $\Omega_{\varepsilon\eta}$ and extended by zero to the whole Ω_η. The following theorems are concerned with the asymptotic behaviour of the velocity field and of the pressure. Presently we can make a rigorous study in Ω_η for the velocity field but only in Ω^-_η for the pressure.

We need a formal expansion to be able to describe the limit behavior of p in Ω_η^+.

THE ASYMPTOTIC BEHAVIOR IN Ω_η^-

We introduce the following Stokes problem S_η in Ω_η^- :

Find $(U^{*\eta},p^{*\eta})$ in $H^1(\Omega_\eta^-)^3 \times L_o^2(\Omega_\eta^-)$ such that :

$$S_\eta \begin{cases} -\Delta U^{*\eta} + \nabla p^{*\eta} = o & (5.1) \\ \operatorname{div}(U^{*\eta}) = o & (5.2) \\ U^{*\eta} = (k^\eta,o,o) \text{ on } \partial\bar{\Omega}_\eta, & (5.3) \end{cases}$$

where $k^\eta = k^{\varepsilon\eta}$ on $\Gamma_\eta^- U\omega$ and 0 on Σ_η^-

To make no restriction on the kind of roughness of the surfaces, longitudinal as well as transverse striations must be taken into account. By the way the function h_ε must have the possibility to oscillate on the edge of $\Sigma_{\varepsilon\eta}$ and $k^{\varepsilon\eta}$ becomes then a rapidly varying function. That provides a further difficulty when ε tends to zero. Thereby, we shall keep the boundary condition (2.7) for the only case where h_ε does not depend on ε on $\partial\omega$. In every other cases, we choose :

$$k^{\varepsilon\eta}(x) = \begin{cases} 3s\left(\dfrac{x_3}{\eta h\min(x)}\right)^2 - 4s\,\dfrac{x_3}{\eta h\min(x)} + s & o \leqslant x_3 \leqslant h\min(x) \quad (5.4) \\ 0 & h\min(x) \leqslant x_3 \leqslant h_\varepsilon(x) \end{cases}$$

Conditions (2.4) and (3.1) are still valid for this choice of $k^{\varepsilon\eta}$ and the Stokes equation is well posed in $\Omega_{\varepsilon\eta}$. Moreover we remark that choice (5.4) implies that $k^{\varepsilon\eta}$ does not depend on ε and problem S_η has one and only one solution in Ω_η^- .

Now we give the convergence theorems :

THEOREM 5.1:Let $(U^{\varepsilon\eta}, p^{\varepsilon\eta})$ be the solution of the Stokes problem $S_{\varepsilon\eta}$, then :

$\bar{U}^{\varepsilon\eta}$ weakly converges to $U^{*\eta}$ in $H^1(\Omega^-)^3$

$\underset{\sim}{p}^{\varepsilon\eta}$ weakly converges to $p^{*\eta}$ in $L_o^2(\Omega_\eta^-)$,

where $(U^{*\eta},p^{*\eta})$ is the solution of S_η and $\underset{\sim}{p}^{\varepsilon\eta}$ the translated pressure :

$$\underset{\sim}{p}^{\varepsilon\eta} = p^{\varepsilon\eta} - \left(\int_{\Omega_\eta^-} p^{\varepsilon\eta} dx \right) / \text{mes}\,(\Omega_\eta^-) \tag{5.5}$$

Proof : Let $(U^{\varepsilon\eta}, p^{\varepsilon\eta})$ be the solution given by theorem 2.1. We choose a special test function in (2.5)$\phi = U^{\varepsilon\eta} - k^\eta$ where k^η is a lift of $H^1(\Omega_\eta^-)$ of $k^{\varepsilon\eta}$ extended by zero to $\Omega_{\varepsilon\eta}$ and we put $q = p^{\varepsilon\eta}$ in (2.6) :

$$a(U^{\varepsilon\eta}, U^{\varepsilon\eta}) = a(U^{\varepsilon\eta}, K^\eta)$$

Because of the zero condition on $\Sigma_{\varepsilon\eta}$, we can extend all the functions by zero on Ω_η and we take the Poincaré norm :

$$\|\bar{U}\|_\eta = \left(\sum_{i,j=1}^{3} \int_{\Omega_\eta} \left(\frac{\partial \bar{u}_i}{\partial x_j}\right)^2 dX \right)^{1/2}$$

Because of the choice of k^η, which is equal to $U^{\varepsilon\eta}$ on $\partial\Omega_{\varepsilon\eta}$ and which is independent of ε, we find :

$$\|\bar{U}^{\varepsilon\eta}\|_\eta \leqslant \|\bar{k}^\eta\|_\eta \tag{5.6}$$

The right side is a constant in ε and there exists $U^{*\eta}$ in $H^1(\Omega_\eta)^3$ such that a subsequence of $\bar{U}^{\varepsilon\eta}$ weakly converges to $U^{*\eta}$.

To obtain an a priori estimate on the pressure in $L_o^2(\Omega_\eta)$, we have to find estimates of the derivatives in $H^{-1}(\Omega_\eta)$ and this is not possible with test functions in $H_o^1(\Omega_{\varepsilon\eta})^3$. This is why we have to consider $\underset{\sim}{p}^{\varepsilon\eta}$(see 5.5) instead of $p^{\varepsilon\eta}$ and we estimate its derivatives in $H^{-1}(\Omega_\eta^-)$.

For any φ_1 in $H_o^1(\Omega_\eta^-)$, $\bar{\varphi}_1$ belongs to $H_o^1(\Omega_{\varepsilon\eta})$. We set $\phi = (\bar{\varphi}_1, o, o)$ in (2.5)

$$\sum_{i=1}^{3} \int_{\Omega_\eta^-} \frac{\partial u_1^{\varepsilon\eta}}{\partial x_i} \frac{\partial \varphi_1}{\partial x_i} dX = \int_{\Omega_\eta^-} p^{\varepsilon\eta} \frac{\partial \varphi_1}{\partial x_1} dX$$

$$\left\langle \frac{\partial p^{\varepsilon\eta}}{\partial x_1}, \varphi_1 \right\rangle_{H^{-1}(\Omega_\eta^-), H_o^1(\Omega_\eta^-)} \leqslant \|\bar{U}^{\varepsilon\eta}\|_\eta \, \|\varphi_1\|_{H^1(\Omega_\eta^-)}$$

From (5.6) : $\left\| \dfrac{\partial \underset{\sim}{p}^{\varepsilon\eta}}{\partial x_1} \right\|_{H^{-1}(\Omega_\eta^-)} \leqslant C^{te}/\varepsilon$

The domain Ω_η^- does not depend on ε, so :

$$\|p^{\varepsilon\eta}\|_{L_o^2(\Omega_\eta^-)} \leq C^{te}/\varepsilon, \|\nabla p^{\varepsilon\eta}\|_{H^{-1}(\Omega_\eta^-)} \leq C^{te}/\varepsilon$$

which shows the second point of the theorem.

Taking now any function ϕ in $H_o^1(\Omega_\eta^-)^3$ extended by zero to $\Omega_{\varepsilon\eta}$, we find the limiting equations :

$$\sum_{i,j=1}^{3} \int_{\Omega_\eta^-} \frac{\partial u_i^{*\eta}}{\partial x_j} \frac{\partial \varphi_i}{\partial x_j} dX = \int_{\Omega_\eta^-} p^{*\eta} \mathrm{div}\phi dX$$

$$\int_{\Omega_\eta^-} q \,\mathrm{div}\, U^{*\eta} dX = 0, \qquad \forall\phi \in H_o^1(\Omega_\eta^-)^3, \forall q \in L_o^2(\Omega_\eta^-)$$

The boundary conditions (5.3) are obvious because k^η does not depend on ε and because of the trace on Ω_η^-, which ends the proof.

□

The next point to look at is the behavior of the velocity $\bar{U}^{\varepsilon\eta}$ on Ω_η^+.

THEOREM 5.2 : the velocity field $\bar{U}^{\varepsilon\eta}$ weakly converges to 0 in $H^1(\Omega_\eta^+)^3$.

Proof : We have already shown, see (5.6) that $U^{-\varepsilon\eta}$ converges in $H^1(\Omega_\eta^+)^3$- weak. Setting :

$$\chi^\varepsilon = \chi(\Omega_{\varepsilon\eta}^+),$$

It is clear that :

$\chi^\varepsilon(x) = 1_B(\frac{x}{\varepsilon})$, where 1_B is the indicatrix function of the set :

$$B(x) = \{y \in R^2, \eta\, \mathrm{hmin}(x) < x_3 < \eta h(x,y)\} \qquad (5.7)$$

It is a periodic function in y of period Y which is not uniform in x and where x occurs as a parameter. We have to take the limit of χ^ε. We consider the change of variables :

$$\xi = x,\ t = (x_3 - \eta \mathrm{hmin}(x))/(\eta\, \mathrm{hmax}(x) - \eta \mathrm{hmin}(x))$$

$$(x,x_3) \in \Omega_\eta^+ \Leftrightarrow (\xi,t) \in \Theta = \omega \times]0,1[$$

Taking any ψ in $L^1(\Omega_\eta^+)$, we calculate :

$$I_\varepsilon = \int_{\Omega_\eta^+} \chi^\varepsilon (X)\psi(X)dX = \int_\Theta \chi_\Theta^\varepsilon (\xi,t)j(\xi)\psi_\Theta(\xi,t)d\xi dt \qquad (5.8)$$

where j is the Jacobian. But $\chi_\Theta^\varepsilon(\xi,t) = 1_{B(\xi,t)}(\xi/\varepsilon)$. Recalling that if $v(\xi,y)$ belongs to $\mathscr{C}^\circ(\omega;L_p^\infty R2)$:

$$v(\xi,\xi/\varepsilon) \underset{\varepsilon\to 0}{\rightharpoonup} \frac{1}{mes(Y)} \int_Y v(\xi,y)dy \qquad L^\infty(\omega)\text{-weak}^*$$

for each t, we apply this result to χ_θ^ε and using Fubini and Lebesgue theorem, we let ε tend to zero in (5.8), so :

$$\chi^\varepsilon \rightharpoonup \frac{1}{mes(Y)} \int_Y 1_B(y)dy = \tilde{1}_B \qquad L^\infty(\Omega_\eta^+)\text{-weak}^*$$

We already know that $\bar{U}^{\varepsilon\eta}$ strongly converges to $U^{*\eta}$ in $L^2(\Omega_\eta^+)^3$ so :

$$\chi_\varepsilon \bar{U}^{\varepsilon\eta} \rightarrow 1_B \; U^{*\eta} \qquad \text{in } \mathscr{D}'(\Omega_\eta^+)$$

On the other hand, $\chi_\varepsilon \bar{U}^{\varepsilon\eta} = \bar{U}^{\varepsilon\eta}$, we obtain

$$\tilde{1}_B \; U^{*\eta} = U^{*\eta} \text{ in } \Omega_\eta^+$$

By the assumption made in 2.a on the rough part of the surfaces we know that the average $\tilde{1}_B$ is strictly less than 1, which proves the theorem. □

THEOREM 5.3 : the convergence of the velocity field in Ω_η is actually a strong one.

Proof : In (2.5), we can take $\phi = \bar{U}^{\varepsilon\eta} - U^{*\eta}$ which lies in $H_0^1(\Omega_{\varepsilon\eta})^3$, so :

$$a(\bar{U}^{\varepsilon\eta} - U^{*\eta}, \bar{U}^{\varepsilon\eta} - U^{*\eta}) = a(U^{*\eta}, U^{*\eta} - \bar{U}^{\varepsilon\eta})$$

which tends to zero from the theorem 5.1. □

For a rigorous approach, the behavior of the pressure in the oscillating part Ω_η^+ of the domain is to be considered. This is still an open question.

To end the study of section 5, it remains to know what happens to $U^{*\eta}$ and $p^{*\eta}$ when η tends to zero. They are solutions of a Stokes problem in Ω_η^- and this leads to an already known step : the transition between the Stokes and the Reynolds equation, see section 3. The limit p^{*2} is the solution of a Reynolds equation with a different gap height :

$$\mathrm{div}\,(hmin^3 \nabla p^{*2}) = 6s\,\frac{\partial hmin}{\partial x_1}$$

$$hmin^3\,\frac{\partial p^{*2}}{\partial n} = 6 s hmin \cos(\underline{n},\underline{x}_1)$$

$$\int_\omega hmin\, p^{*2} dx = o$$

It has been pointed out that the way how the two little parameters ε (the roughness wavelength) and η (the gap height) tend to zero leads to the same kind of limit equation but with truly different coefficients. This situation is to be found in some other two little parameters works [4] [5] [12] [16].

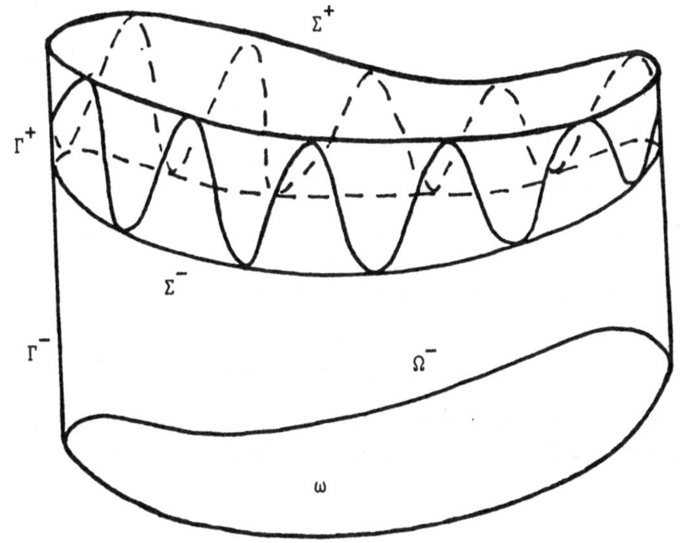

Fig. 1 : The open rescaled domain Ω_ε

References

[1] G. BAYADA and M. CHAMBAT, The transition between the Stokes equation and the Reynolds equation : a mathematical proof, Appl. Math. and Opt., 14, 73-93 1986 .

[2] G. BAYADA AND J.B. FAURE, Application des techniques d'homogénéisation à des phénomènes de rugosité en lubrification hydrodynamique, rapport L.A.N. Lyon St Etienne, UA CNRS 040740, 32, 1-52 1984 .

[3] R. BRIZZI and J.P. CHALOT, Homogénéisation de frontière, thèse, Nice, 1978.

[4] G. BUTTAZO, An optimization problem for thin insolating layers around a conducting medium (this issue).

[5] D. CAILLERIE, Homogénéisation des équations de la diffusion stationnaire dans des domaines cylindriques applatis, R.A.I.R.O. Anal. Num., 15 n°4, 295-319 1981 .

[6] K.K. CHEN and D.H. SUN, On the statistical treatment of rough surface in hydrodynamic lubrication problems, Proc. of the 4th Leeds-Lyon Symp. on "Surface roughness in lubrication", I.M.E., 41-45 1977.

[7] H.G. ELROD, A review of theories for the fluid dynamic effects of roughness on laminar lubricating films, proc. of the 4th Leeds-Lyon Symp. on 'Surface roughness in lubrication', I.M.E. 11-29, 1977.

[8] H.G. ELROD, A general theory for laminar lubrication with Reynolds roughness, Trans. ASME, J. Lub. Tech., 101, n°1, 8-14 1979 .

[9] B. FANTINO, J. FRENE and M. GODET, Conditions d'utilisation de l'équation de Reynolds en mécanique des films minces, C.R. Acad. Sc. Paris, 262 A, 155-165 (1971).

[10] J.B. FAURE, Application des techniques d'homogénéisation à la prise en compte des phénomènes de rugosité en lubrification hydrodynamique, thèse, Lyon 1, 1986.

[11] V. GIRAULT and P.A. RAVIART, Finite Element Approximation of the Navier Stokes Equations, Springer Verlag, Berlin, 1981.

[12] R. KOHN and M. VOGELIUS, A new model for thin plates with rapidly varying thickness, Int. J. Solid & Struct.,20, 333-350 1984 .

[13] F. MURAT, H-Convergence, Séminaire d'Analy. fonct. et num., Alger 1977-78.

[14] N. PHAN-THIEN, On the effects of the Reynolds and Stokes surface roughness in a two-dimensional slider bearing, Proc. R. Soc. Lond., A 377, 349-362 1981 .

[15] O. PIRONNEAU and C. SAGUEZ, Asymptotic behavior with respect to the domain of solutions of partial differential equations, Tech. Rep. INRIA 218 1977 .

[16] J. SAINT JEAN PAULIN, Homogénéisation et perturbation dans un problème lié à l'échauffement d'un cable électrique, Ann. Fac. Sc. Toulouse, Vol. V, 43-59 1983 .

[17] E. SANCHEZ-PALENCIA, Non homogeneous media and vibration theory, lect. Notes in physics n°127, Springer Verlag, Berlin 1978.

[18] D.C. SUN and K.K. CHEN, First effects of the Stokes roughness on hydrodynamic lubrication, Trans. ASME, J. lub. Tech. 99, 2-9 1977 .

FREE BOUNDARY PROBLEMS IN DISSOLUTION-GROWTH PROCESSES

Francis CONRAD (*)(**) and Michel COURNIL (*)

(*) Ecole des Mines, 158 Cours Fauriel 42023 SAINT-ETIENNE (France)

(**) C.M.A. Ecole des Mines de Paris, Sophia-Antipolis 06565 VALBONNE (France)

ABSTRACT

This paper is concerned with the analysis of evolution or stationary free boundary problems arising in growth and dissolution phenomena. The heterogeneous system considered is a grain exchanging mass with a liquid solution. The evolution of the concentration of the liquid solution is described by a parabolic moving boundary problem (5). The associated elliptic stationary free boundary problem may exhibit 0, 1, 2 or 3 solutions. The stability of the equilibrium solutions is studied by means of analytical and numerical techniques. The results obtained are genuinely more complex than those, wellknown , for fixed boundary value problems. Finally, some important open problems are listed.

1 - INTRODUCTION - THE PHYSICAL PROBLEM

The general framework is the following : the physical system is a heterogeneous medium with several phases ; the existence of chemical reactions in the medium (possibly limited at the interfaces) implies creation or consumption of some phases ; thus the geometry of the system (of the interfaces) may evolve. Therefore, such phenomena are an important source of free boundary problems.

More specifically we will consider the following case : small solid grains of a single compound exist in a liquid phase, which is a dilute solution of the compound ; exchange of matter occurs at the interfaces of the solid grain. The evolution of the concentration in the liquid phase is modelled by a diffusion equation plus a non linear source term . The existence of a source term is due, for instance to exchange of matter between the liquid phase and a porous medium (e.g. in geology, see (2, 6)) ; in other cases it may be related to the evolution of another species existing in the solution, but which is not taken into account [11]. On the interfaces, we have standard boundary conditions, plus an equation modelling the growth or dissolution of the grains according to the boundary concentration.

Let us mention some important classes of applications where dissolution-growth phenomena are crucial :

- setting of binders,
- sintering in a liquid phase,
- mineralization in geology,
- primary steps of industrial cristallization.

Remark The dissolution-growth process does not include nucleation ; in other words, creation of grains is not taken into account by the models.

2 - MATHEMATICAL MODEL

Due to the complexity of a complete system of small grains (shape of the grains, possibility of contact) we consider here only one grain of spherical geometry isolated in a liquid solution (Fig. 1). The objective of this first study of a model problem is to analyze the behavior of that system W.R.T. the physical parameters.

Let R(t) denote the radius of the grain and C(X,t), $R < X < L$ denote the concentration of the aqueous medium.

Then we have the following system of equations describing the evolutions of R(t) and C(X,t) (see [7,10,11] for more details) :

$$\begin{cases} \dfrac{\partial C}{\partial t} = D \Delta C + F(C) \ ; \ R(t) < X < L, \ t > 0 & (1) \\ D \dfrac{\partial C}{\partial X} = \left(\dfrac{1}{V} - C\right) \dfrac{dR}{dt} \ ; \ X = R(t), \ t > 0 & (2) \\ C = C_L \ ; \ t > 0 \ ; \ X = L, \ t > 0 & (3) \\ \dfrac{dR}{dt} = K(C - C_o) \ ; \ X = R(t), \ t > 0 & (4) \\ R(0), \ C(X,0), \ R(0) < X < L \text{ are given initial conditions} & (5) \end{cases}$$

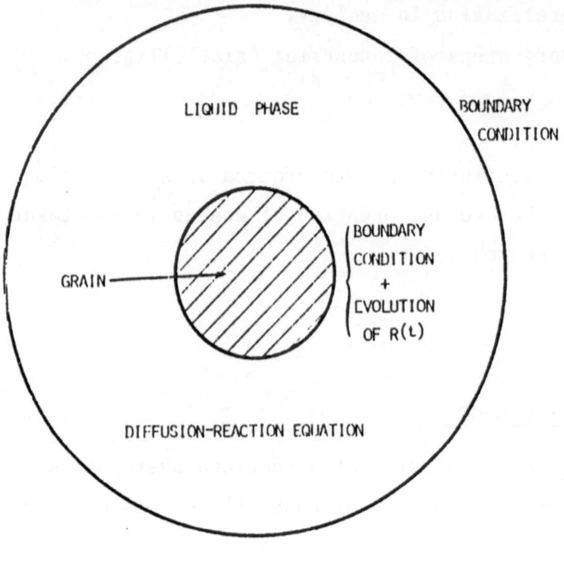

FIGURE 1

Equation (1), of diffusion-reaction type derives from a mass balance in the aqueous medium whereas (2) is issued from a mass balance on the interface ; (3) is a classical Dirichlet boundary condition (but Neumann or mixed conditions are also realistic for some systems (7)) ; (4), a consequence of Nernst law, describes the evolution of the boundary ; in the Gibbs - Kelvin theory:

$$C_o = G(R(t)) = C_* \exp\left(\frac{\gamma}{R(t)}\right)$$

All the constants involved in the model are positive. The nonlinear source term F is a nonnegative, nonincreasing function.

Our purpose in the following : first, study the stationary problem : how many equilibrium states W.R.T. the parameters ? Then study the evolution problem : what are the stable equilibria of the system, and how to characterize them.

3 - STATIONARY SOLUTIONS

We drop the time variable in system (1) - (5) ; setting $r = \frac{X}{L}$, $C_S = \frac{C}{C_L}$, $R_S = \frac{R}{L}$ $\beta = \frac{C_*}{C_L}$, $\delta = \frac{Y}{L}$, $g(y) = \beta \exp(\frac{\delta}{y})$ and $\lambda = \frac{L^2 F(0)}{D\, C_L}$, $f(C_S) = \frac{F(C_L C_S)}{F(0)}$, we get the stationary problem in a nondimensional form (S) :

$$\begin{cases} \Delta C_S + \lambda f(C_S) = 0 \;;\; R_S < r < 1 & (1) \\ C_S'(R_S) = 0 & (2) \\ C_S(1) = 1 & (3) \\ C_S(R_S) = g(R_S) & (4) \end{cases}$$

Though λ is fixed by physical given data, it is interesting to consider (S) as a nonlinear eigenvalue problem W.R.T. λ, that is, the unknowns in (S) are λ, R_S, C_S (we restrict to radial solutions).

Remark If (λ, R_S) is known then C_S is known as a solution of an initial value problem :

$$\begin{cases} C_S'' + \frac{2}{r} C_S' + \lambda f(C_S) = 0 \;;\; R_S < r < 1 \\ C_S(R_S) = g(R_S) \\ C_S'(R_S) = 0 \end{cases}$$

hence the graph (λ, R_S) is an adequate bifurcation diagram as far as multiplicity is concerned. The physical domain is restricted to $\lambda \geq 0$.

3-1 EXISTENCE AND MULTIPLICITY

LEMMA Let $R \in]0,1[$ be fixed and w be a solution of

$\Delta w + \lambda f(w) = 0$ in $]R,1[$ with any two of the following boundary conditions :

$$w(R) = g(R) \; ; \; w'(R) = 0 \; ; \; w(1) = 0$$

Then (λ, R, w) is a solution of (S) iff

$$g(R) - 1 = \lambda \int_R^1 f(w)\, r\, (1-r)\, dr$$

Proof Multiply equation $\Delta w + \lambda f(w) = 0$ by $r(1-r)$ and integrate par parts over $[R,1]$:

$$-R\,(1-R)\, w'(R) + w(1) - w(R) + \lambda \int_R^1 f(w)\, r\, (1-r)\, dr = 0 \qquad \square$$

The preceeding Lemma provides the basis of an existence result and qualitative behaviour of the bifurcation diagram.

More precisely, let (λ, R, C) be a solution of (S). Then either $f(C) \equiv 0$ on $(R,1)$ and $C(r) \equiv 1$ which is only possible if $g(R) = 1$ and $f(1) = 0$; λ is then arbitrary.

Or $f(C) \not\equiv 0$ and (S) $\Leftrightarrow$

$$\begin{cases} \Delta C + \dfrac{g(R) - 1}{\displaystyle\int_R^1 f(C)\, r\, (1-r)\, dr}\, f(C) = 0 \text{ in }]R,1[\\ C(R) = g(R) \\ C(1) = 1 \end{cases}$$

Let Φ be harmonic, $\Phi(R) = g(R)$, $\Phi(1) = 1$; let G be the Green operator for $-\Delta$ on $]R,1[$ with homogeneous Dirichlet boundary conditions.

Then (S) is clearly equivalent to the fixed point formulation :

$$C = T(C) = \Phi + \frac{g(R) - 1}{\int_R^1 f(C)\, r\, (1-r)\, dr}\, G\, f(C)$$

THEOREM 1 Suppose that f > 0

a) for any fixed $R \in]0,1[$ there exists $(\lambda, C) \in \mathbb{R} \times C^2([R,1])$ such that (λ, R, C) is solution of (S)

b) if $\lambda \geq 0$ (hence $g(R) \geq 1$) (λ, C) is unique.

Proof Let $C_o = \{v \in C([R,1]) \ / \ v(R) = v(1) = 0\}$ and B be the closed convex subset of $C([R,1])$:

$$B = \{v \in \Phi + C_o \ / \ |v|_\infty \leq \max(1, g(R))\}$$

First, we prove that $T(B) \subset B$; in fact, if $v \in B$, then $w = T(v)$ satisfies :

$$\begin{cases} -\Delta w = \dfrac{g(R) - 1}{\int_R^1 f(v)\, r\, (1-r)\, dr}\, f(v) \text{ in }]R,1[\\ w(R) = g(R) \\ w(1) = 1 \end{cases}$$

Notice that the integral is nonzero since $v(1) = 1$ and f > 0 ; hence $(r^2 w')'$ has the sign of $1 - g(R)$. Since $w'(R) = 0$, w is decreasing when $g(R) > 1$ and w is increasing when $g(R) < 1$.

In any case $|w|_\infty = \max(w(R), w(1)) = \max(1, g(R))$; thus $T(B) \subset B$.

Second, T is obviously compact on $\mathcal{C}([R,1])$ thanks to the regularity of the Green operator. So the existence follows from Schauder's fixed point theorem.

For the uniqueness when $\lambda \geq 0$, let (λ_1, C_1) (λ_2, C_2) be two solutions of (S) ; $Z = C_1 - C_2$ satisfies :

$$\begin{cases} -\Delta Z - \lambda_1 f'(\xi) Z = (\lambda_2 - \lambda_1) f(C_2) \\ Z(R) = 0 \ ; \ Z(1) = 0 \end{cases}$$

where $|\xi(X)| \leq \max\{|C_1(X)|, |C_2(X)|\}$.

Consequentely, $-\Delta Z - \lambda_1 f'(\xi) Z \geq 0$ or ≤ 0 wether $\lambda_2 \geq \lambda_1$ or $\lambda_2 \leq \lambda_1$.

If $Z \not\equiv 0$ then, by the strong maximum principle of Hopf, the normal derivative of Z at R must be nonzero (recall that $-f'(\xi) \geq 0$) which contradicts the boundary condition $w'(R) = 0$; hence $C_1 = C_2$ and, by the integral formula, $\lambda_1 = \lambda_2$. □

Remark

Part b) of Theorem 1 indicates that R is an adequate parameter in the domain $(\lambda \geq 0)$.

THEOREM 2 Suppose $f > 0$ and $R \in]0,1[$ be fixed. Let (λ, R, C) be a solution of (S)

a) if $g(R) \leq 1$, C increases from $g(R)$ to 1
if $g(R) \geq 1$, C decreases from $g(R)$ to 1

b) as $R \to 0+$, $\lambda \to +\infty$

c) as $R \to 1-$ $\lambda \to \begin{cases} -\infty & \text{if } g(1) < 1 \\ +\infty & \text{if } g(1) \geq 1 \end{cases}$

Proof Part a) is achieved in the same way as the monotony of w in the preceeding theorem.

To prove parts b) and c) we use the integral formula :

$$\lambda = \frac{g(R) - 1}{\int_R^1 f(C)\, r\, (1-r)\, dr}$$

Since $\int_R^1 f(C)\, r\, (1-r) dr \leqq \int_0^1 f(C)\, r\, (1-r) dr \leqq \frac{|f|_\infty}{6}$

we get $\lim_{R \to 0+} \lambda = +\infty$

On the other hand, if $g(1) \neq 1$, as $R \to 1-$:

$$\lambda = \frac{g(R) - 1}{(1-R)\, f\, (C(\xi))\, \xi\, (1-\xi)} \longrightarrow \begin{cases} +\infty & \text{if } g(1) > 1 \\ -\infty & \text{if } g(1) < 1 \end{cases}$$

If $g(1) = 1$, then $\lambda = - \dfrac{g'(\eta)}{f\, (C(\xi))\, \xi\, (1-\xi)}$ $\quad (R < \xi < 1)$

Since $g(1) = \beta\, e^{\delta} = 1$, $g'(1) = -\beta\, \delta\, e^{\delta} \neq 0$ and $\lambda \to +\infty$ as $R \to 1_-$. □

Remarks Suppose $f \neq 0$

1) If $g(1) \geqslant 1$, (S) has at least two solutions for fixed λ

2) If $g(1) < 1$, multiple solutions are also possible in the physical domain (for instance with $f \equiv 1$ see Fig. 4).

When $f \not\equiv 0$ vanishes, the situation is more complex.

THEOREM 3 Suppose $\tau = \inf\{t > 0, f(t) = 0\} \leq 1$

a) if $g(1) \geq 1$, (S) has no solution

b) if $g(1) < 1$, solutions can only exist if $g(R) = 1$ or $g(R) < \tau \leq 1$

c) $R = g^{-1}(1)$, $C \equiv 1$ is a solution for any λ

d) if (λ, C) exists for any $R \in \,]g^{-1}(\tau), 1[$ then
$\lim \lambda = -\infty$ as $R \to 1-$,
$\lim \lambda = -\infty$ as $R \to g^{-1}(\tau)+$ provided $\tau < 1$.

Proof a) if (λ, R, C) is a solution of (S) then $C \geq \min(g(R), 1) = 1$, hence $f(C) \equiv 0$ and $C \equiv 1$, which contradicts $C(R) = g(R) > 1$

b) if $\tau \leq g(R)$, then $C \geq \tau$, hence $f(C) \equiv 0$ and $C \equiv 1$, then necessarily $g(R) = 1$

c) is obvious

d) the limit of λ as $R \to 1-$ is obtained as in THEOREM 2. When $R \to g^{-1}(\tau)+$ we write : $C \geq \min(g(R), 1) = g(R)$ hence

$$\lambda = \frac{g(R) - 1}{\int_R^1 f(C)\ r\ (1-r)\ dr} \leq \frac{6(g(R) - 1)}{f(g(R))\ (1-R)^2\ (1+2R)} \longrightarrow -\infty$$

as $g(R) \to \tau < 1$

The existence part in b) is more difficult to establish since the integral $\int_R^1 f(C)\ r\ (1-r)\ dr$ is not bounded away from 0 ; we cannot prove that the mapping T of THEOREM 1 is compact.

Remarks

1) The case $\tau > 1$ has also been considered. We get existence for all R such that $g(R) < \tau$ and non existence if $g(R) > \tau$. If $g(1) \geq \tau$, (S) has no solution.

2) We see that existence for all $R \in (0,1)$ fails. Moreover if $f \equiv 0$, solutions exist only for $g(R) = 1$ (and $C \equiv 1$, λ is arbitrary) and uniqueness for fixed R fails too.

From THEOREM 1 to 3, we can depict the bifurcation diagrams (λ, R) according to the different cases (Figs. 2, 3).

3-2 NUMERICAL RESULTS

a) Principle of the method

Tne numerical procedure has been performed by V. Katossky, of Ecole des Mines de Saint-Etienne (10). We recall a sketch of the procedure :

1) first, write (S) on a fixed domain ; set $r = R + x(1-R)$, $x \in (0,1)$ and $C(r) = u(x)$; then (λ,R,u) is solution of the following system :

$$\begin{cases} u''(x) + \dfrac{2(1-R)}{R + x(1-R)} u'(x) + \lambda (1-R)^2 f(u) = 0 \quad \text{in } (0,1) \\ u(0) = g(R) \\ u'(0) = 0 \\ u(1) = 1 \end{cases}$$

In other words, we have now a nonlinear equation $G(\lambda,R,u) = 0$ where u is in a fixed functional space. This is a standard framework for the use of continuation methods.

2) second, apply a continuation method on an augmented problem, as described for instance in (8) :

$$\begin{cases} G(\lambda,R,u) = 0 \\ N(\lambda,R,u,s) = 0 \end{cases}$$

where s is a new adequate parameter (of course, the differential equation has to be discretized). As expected (see Figs 2-3) λ is not always an adequate parameter ; however, in the cases tested numerically, s = R is always possible ; in other words, R is not only an adequate parameter for a global representation of the diagram (λ,R), but allows also the use of the implicit function theorem on the augmented system to get λ(R), u(R), for any R ∈ (0,1) (see (10) for the details of the procedure).

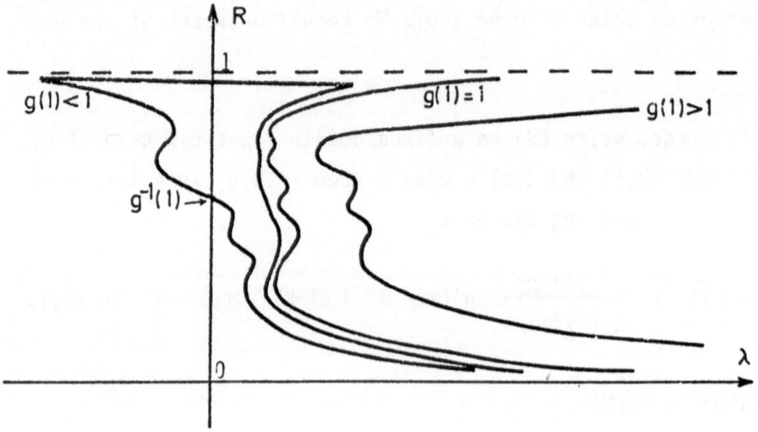

FIGURE 2 :CASE f > 0

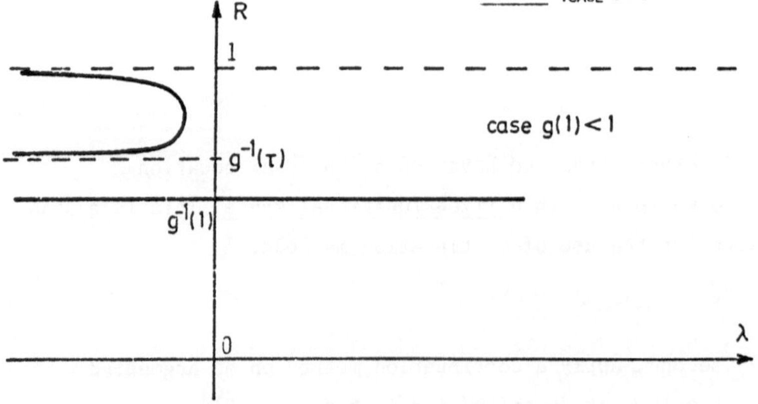

FIGURE 3 : CASE f(τ) = 0

b) Numerical results

The parameter δ is kept constant, β varies in such a way that $g(1) = \beta\, e^{\delta}$ crosses the value 1 ($f(1)$ is always positive). Three realistic source terms have been tested :

$$f(C) \equiv 1 \quad \text{(Fig. 4)}$$

$$f(C) = \quad \left(1 - \frac{\alpha C}{\beta}\right)^{+} \quad \text{(Fig. 5)} \quad (\alpha = 0.5)$$

$$f(C) = \quad \left(1 - \frac{\alpha^2 C^2}{\beta^2}\right)^{+} \quad \text{(Fig. 6)} \quad (\alpha = 0.5)$$

All these results illustrate the theoretical conclusion of subsection 3.1. Moreover we see that multiple solutions are possible even with $f \equiv 1$; in fact, the precise form of the source term does not appear to be very important.

As a conclusion for this section, we can say that S - shaped or U - shaped bifurcation diagrams are very easy to obtain with our class of models. That is, (S) admits 0, 1, 2 or 3 solutions according to the physical parameters. A natural question is the following : what are the physically observable - i.e. the stable - equilibria of (S) ?

FIGURE 4

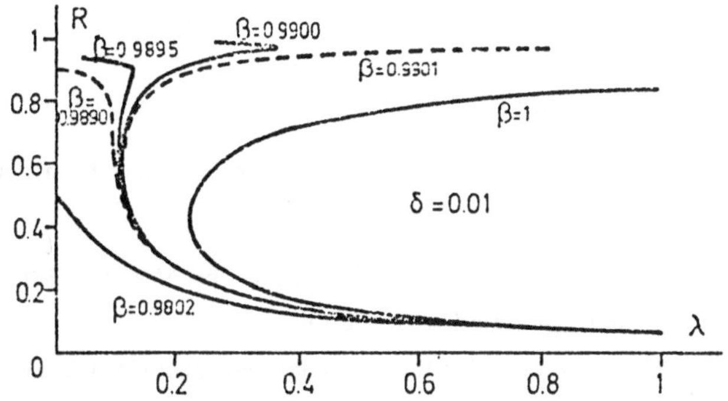

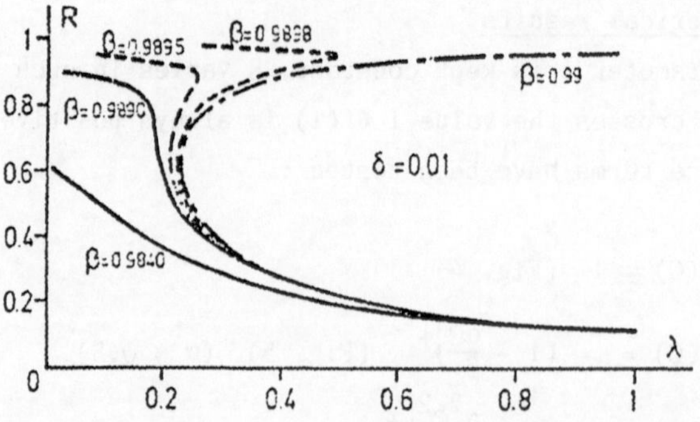

FIGURE 5

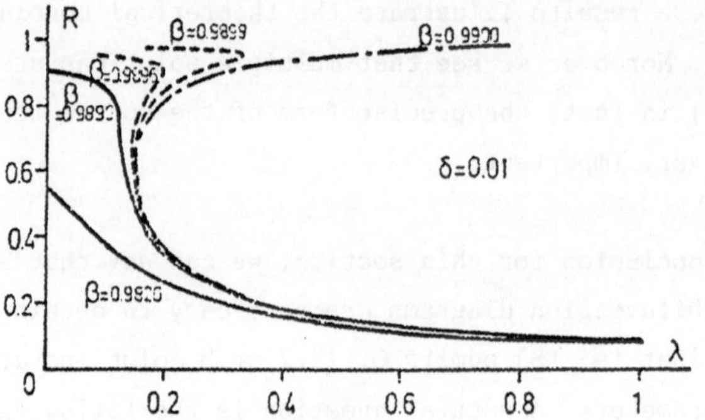

FIGURE 6

4 - STABILITY ANALYSIS

This study, based on the evolution problem, has been set up by two methods :

- numerically by solving the evolution problem for initial conditions near equilibrium solutions,
- analytically by defining a linearized stability criterion, in an adequate sense.

4-1 NUMERICAL METHOD

The evolution problem (1) - (5) :

$$\begin{cases} \dfrac{\partial C}{\partial t} = D \, \Delta C + F(C) \; ; \; R(t) < X < L, \, t > 0 & (1) \\ D \dfrac{\partial C}{\partial X} = \left(\dfrac{1}{V} - C\right) \dfrac{dR}{dt} \; ; \; X = R(t), \; t > 0 & (2) \\ C = C_L \; ; \; X = L, \; t > 0 & (3) \\ \dfrac{dR}{dt} = K \left(C - G(R(t))\right) \; ; \; X = R(t), \; t > 0 & (4) \\ + \text{ I.C.} & (5) \end{cases}$$

has been solved numerically by the following procedure :

a) first transform system (1) - (5) into a system on a fixed domain, thanks to the transform :

$$y = \frac{X - R(t)}{L - R(t)} \; , \quad 0 < y < 1, \; t > 0, \text{ setting :}$$

$$C(X,t) = C(R + (L-R) \, y, \, t) = U(y,t) \; ,$$

we get :

$$\begin{cases} \dfrac{\partial U}{\partial t} = \dfrac{D}{(L-R)^2} \dfrac{\partial^2 U}{\partial y^2} + \left(\dfrac{2D}{X(L-R)} - \dfrac{\dot{R}(X-L)}{(L-R)^2} \right) \dfrac{\partial U}{\partial y} + F(U) \; ; & \\ 0 < y < 1, \; t > 0 & (1) \\ \dfrac{D}{L-R} \dfrac{\partial U}{\partial y} (0,t) = \left(\dfrac{1}{V} - U(0,t)\right) \dot{R} & (2) \\ U(1,t) = C_L & (3) \\ \dot{R}(t) = K \left(U(0,t) - G(R(t))\right) & (4) \\ U(y,0) = C(X,0) \; ; \; 0 < y < 1 \; ; \; R(0) = R_0 & (5) \end{cases}$$

b) second, discretize in time and space this coupled system and set up a decoupling procedure between R(t) and C(X,t) ; an alternative method is to discretize in space only and solve the resulting O.D.E system by a standard method ; but, as the time scales on C(X,t) and R(t) are different we have to take care on this stiff problem. The first method revealed quite satisfactory and has been used, in an implicit form since the asymptotic behaviour is required (see (3) for details).

<u>Results</u> As initial conditions we choose for R(0) a perturbation of a stationary solution, and for C(X,0) a constant state not too far from an equilibrium solution C(r) ; F is a constant ; three types of bifurcation diagrams have been tested, according to the generic shapes (see Figs 7, 8, 9).

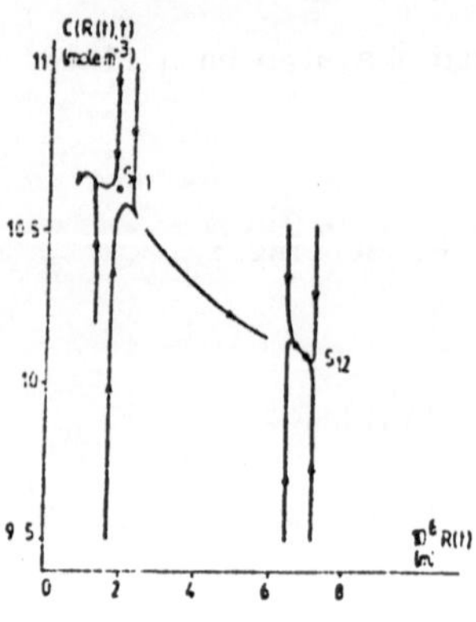

Figure 7 ($\beta = 1; \lambda = 0.4$)

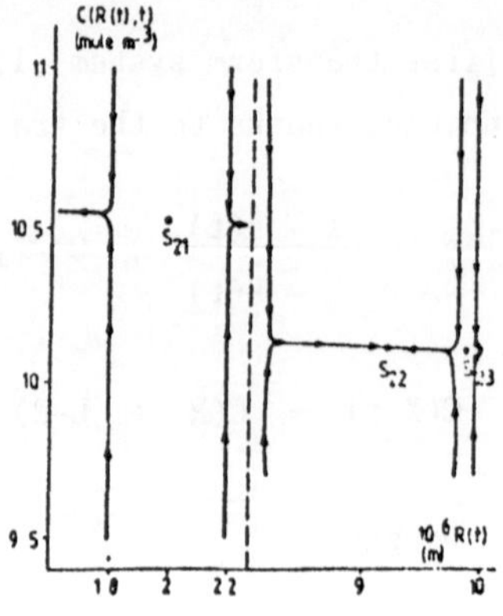

Figure 8 ($\beta = 0.99$; $\lambda = 0.2323$)

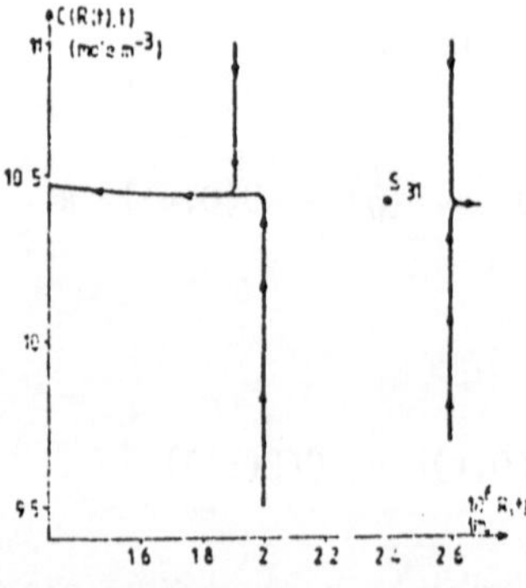

Figure 9 ($\beta = 0.9802$; $\lambda = 0.153$)

From these numerical experiments, we conjecture that the generic situations concerning stability are those which appear on Fig. 10. Thus the dynamical behaviour of our system is not the standard one observed for parabolic equations on fixed domains. Recall that in that case, uniqueness implies in general stability and in case of three solutions, 2 are stable, and the intermediate one is unstable. However, a little more has to be done numerically. Also occurence of Hopf bifurcation type has to be tested. In that case we know that uniqueness of a stationary state does not imply its stability.

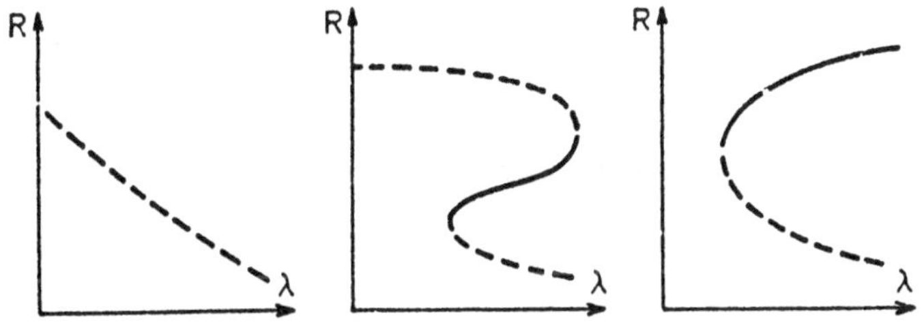

FIGURE 10

4-2 LINEARIZED STABILITY ANALYSIS

a) A theory for the general case

Consider the evolution problem (1) - (5) written on a fixed domain $(0,1)$ and set $U = U^* + V$, $R = R^* + S$, where (U^*, R^*) is a solution of the stationary problem, on $(0,1)$:

$$\begin{cases} \dfrac{D}{(L-R^*)^2}\dfrac{d^2U^*}{dy^2} + \dfrac{2D}{X(L-R^*)}\dfrac{dU^*}{dy} + F(U^*) = 0 \text{ in } (0,1) \\ \dfrac{dU^*}{dy}(0) = 0 \\ U^*(1) = C_L \\ U^*(0) = G(R^*) \end{cases}$$

By a formal linearization of (1) - (5) we obtain, up to the first order in (V,S) the following linear system on the fixed domain :

$$\frac{\partial V}{\partial t} = \frac{D}{(L-R^*)^2}\frac{\partial^2 V}{\partial y^2} + \frac{2D}{X(L-R^*)}\frac{\partial V}{\partial y} + F'(U^*)V$$

$$- \frac{2S}{L-R^*}F(U^*) - \frac{2DSL}{X^2(L-R^*)^2}\frac{dU^*}{dy} + \frac{\dot{S}(1-y)}{L-R^*}\frac{dU^*}{dy} \qquad (1)_L$$

$$\frac{D}{L-R^*}\frac{\partial V}{\partial y}(0,t) = \left(\frac{1}{V} - G(R^*)\right)\dot{S} \qquad (2)_L$$

$$V(1,t) = 0 \qquad (3)_L$$

$$\dot{S} = K\left(V(0,t) - G'(R^*)S\right) \qquad (4)_L$$

$$V(y,0) \ ; \ S(0) \ \text{small} \qquad (5)_L$$

or, if we come back to the moving domain :

$$\left\{\begin{array}{ll} \frac{\partial V}{\partial t} = D\,\Delta V + F'(U^*)V - \frac{2S}{L-R^*}F(U^*) - \frac{2DSL}{X^2(L-R^*)}\frac{dU^*}{dX} & \\ R^* < X < L \ , \ t > 0 & (1)_L \\ D\frac{\partial V}{\partial X}(R^*,t) = \left(\frac{1}{V} - G(R^*)\right)\dot{S} \ ; \ t > 0 & (2)_L \\ V(L,t) = 0 \ ; \ t > 0 & (3)_L \\ \dot{S} = K\left(V(R^*,t) - G'(R^*)S\right) \ ; \ t > 0 & (4)_L \\ V(X,0) \ , \ S(0) \ \text{are given (small)} & (5)_L \end{array}\right.$$

In the first equation, the two first terms are the usual ones due to linearization ; the other extra terms are due to the existence of a moving boundary.

Remark If, instead of transforming the moving boundary problem into a problem on $]0,1[$, we use an affine transform to get a problem on $]a,b[$ the linearized system $(1)_L - (5)_L$ written on the moving domain is exactly the same (but not the one written on $]a,b[$, of course). This will lead to a partially intrinsec definition of linearized stability which is of course necessary. Such an invariance property has to be established for more general transforms.

In the general case, it seems hard to perfom some analysis from Equations $(1)_L - (5)_L$. For further investigation, we limit the study to a special case.

b) Suppose U* is a constant

Hence $U^* = C_L$ and $F(U^*) = 0$ with $R^* = G^{-1}(C_L)$ (this case occurs if $F(C_L) = 0$, $G(L) < C_L$ for instance).

Then the linearized system is the following :

$$\begin{cases} \dfrac{\partial V}{\partial t} = D \, \Delta \, V + F'(C_L) \, V & (1)_L \\ (2)_L - (5)_L \text{ are unchanged} \end{cases}$$

Classically we set $V(X,t) = v(t)\, w(X)$; then $(1)_L \Rightarrow v(t) \sim \exp(\lambda t)$, $(2)_L \Rightarrow S(t) \sim \exp(\lambda t)$ whereas w satisfies the nonstandard eigenvalue problem (E.V.P.) :

$$\begin{cases} \Delta w + \dfrac{F'(C_L)}{D} w = \mu w & \text{in }]R^*,L[\\ w'(R^*) = \alpha \dfrac{\mu}{\mu+\beta} w(R^*) \\ w(L) = 0 \;;\; w \not\equiv 0 \end{cases}$$

where $\mu = \dfrac{\lambda}{D}$; $\alpha = \dfrac{K}{D}\left(\dfrac{1}{V} - C_L\right)$; $\beta = \dfrac{G'(R^*)\, K}{D}$

By the special form of v(t), we shall say that (R*,U*) is a linearly stable equilibrium if all solutions μ of E.V.P. have negative real part.

This, however, is formal since it has not been proved that the set of eigenfunctions w of E.V.P. is complete.

We set $w(r) = \frac{q(r)}{r}$, then (E.V.P.) is equivalent to :

$$\begin{cases} q'' = (\mu-\gamma)\, q \qquad \gamma = \dfrac{F'(C_L)}{D} \\ q'(R^*) - \dfrac{q(R^*)}{R^*} = \alpha \dfrac{\mu}{\mu+\beta}\, q(R^*) \\ q(L) = 0 \end{cases} \qquad \text{(E.V.P.)}$$

If, moreover, we suppose now $C_L > \inf \{z/F(z) = 0\}$ then $\gamma = 0$. Setting $\mu = \eta^2$ in (E.V.P.), and solving the second order equation, we are led to the following equation for η :

$$\text{th}\ \eta(L-R^*) = - \frac{R^*\,\eta}{1 + \dfrac{\alpha\,\eta^2}{\eta^2+\beta} R^*}$$

Elementary analysis shows that this equation has at least a positive root η (see (3) for all the details).

Thus (E.V.P.) admits at least one positive real eigenvalue ; a stationary state (U*,R*) with U* constant cannot be stable. When $F \equiv 0$, this result is consistent with the numerical experiments, that is, in case of a unique solution, we have instability. (see also (4), (7) for similar results).

5 - OPEN PROBLEMS AND FURTHER DEVELOPPEMENT

The modelization of dissolution-growth processes leads to interesting mathematical studies. In particular, the stationary free boundary problem may admit 0, 1, 2 or 3 solutions.

The stability of the solution branches does not follow the classical rules known for fixed domains (1,9); however, more has to be done numerically to enforce the conjecture.

Also the concept of linearized stability has to be precised; in particular its complete intrinsec character W.R.T. change of coordinates into fixed domain problems has to be established.

The evolution problem has to be studied mathematically. N. Yebari, of Ecole des Mines has obtained a local existence theorem for system (1) - (5). He also studies the asymptotic behaviour from a mathematical point of view.

Of course, the present study is only a first step before investigating more complex and more realistic systems, in particular with several grains.

Finally, we have to mention that modelization of dissolution-growth processes leads also to interesting theoretical homogeneization problems, which will be investigated in the future.

REFERENCES

(1) Aris R., 1975, The mathematical theory of diffusion and reaction in permeable catalysts, Clarendon Press, Oxford.

(2) Conrad F., Guy B. and Cournil M., 1983, Bilan et "condition" d'entropie dans la métasomatose de percolation, C.R. Acad. Sci., Paris 296, 1965.

(3) Conrad F., Cournil M., Multiplicity and stability analysis in a free boundary problem arising from a dissolution-growth process (submitted for publication).

(4) Cournil M., 1983, Stabilité d'un système hétérogène constitué d'un solide pulvérulent et de sa solution aqueuse, C.R. Acad. Sci., Paris 297(II), 463.

(5) Friedman A., 1982, Variational principles and free boundary problems, J. Wiley.

(6) Guy B., Conrad F., Cournil M. and Kalaydjian F., 1984, Chemical instabilities and "shocks" in a non-linear convection problem issued from geology, in Nicolis G. and Baras F. (editors), R. Reidel, Dordrecht, 341.

(7) Kalaydjian F. and Cournil M., 1986, Stability of steady-states in some solid-liquid systems, React. Sol. (accepted).

(8) Keller H.B., 1977, Numerical solution of bifurcation and non - linear eigenvalue problems, Rabinowitz Editor, Acad. Press.

(9) Sattinger D.H., 1973, Topics in stability and bifurcation theory, Lect. Notes in Math. 309, Springer, Berlin.

(10) Treguer-Katossky V., 1984, Thesis, Saint-Etienne, France.

(11) Treguer-Katossky V. and Cournil M., 1986, Study of a free boundary problem arising in dissolution-growth phenomena ; multiple solutions in the stationary case (submitted for publication) .

SHAPE OPTIMIZATION AND CONTINUATION METHOD

Chung,S., Deng,S.M., Kernevez,J.P., Liu,Y. and Wang,Z.

UTC, B.P. 233, 60206, COMPIEGNE, FRANCE.

Abstract. A first part describes a simple method for interactive optimization of the shape of a thin plate. The aim is to minimize the stresses in a given region of the plate. A second part shows how the size of a 2-dimensional domain Ω can influence the number and stability of the steady states in a reaction-diffusion system defined on Ω and describes some results on the optimal control of this system in the presence of bifurcations.

I Optimum Design of a thin plate.

There are, in Journals dealing with numerical methods in Engineering, many examples of structural shape optimization problems. See [1] and [2] among many others. A system being governed by P.D.E.s on a 2 or 3 dimensional spatial domain Ω with boundary conditions on $\Gamma=\partial\Omega$, the problem is to act on Γ or a part of it in order to minimize some "cost function" depending upon Γ via the solution of the P.D.E.s. A difficulty is to parametrize Γ in order for Ω to have an admissible shape. On the other hand the Engineer is not practically faced to this original problem, but rather to an aprroximation to it, generally by the Finite Element Method. Then the domain Ω is replaced by juxtaposed elements and the boundary of this new domain Ω_h is defined by a finite number of nodes. It is generally admitted that choosing, as control variables, the coordinates of some of these nodes, without any constraint, is not a good way to deal with the optimization problem. One of the reasons is that the aspect of elements whose some nodes move can become very bad, when too much freedom is let to the moving nodes.

However it is this method that we adopted for thin plates. Our method consists in performing only a few optimization steps, in order for the nodes to move just a little, then remesh the new domain, perform a few optimization steps again, etc..., the user seeing the change of shape and its effect on the system state, and deciding to continue, to stop or to come back to earlier steps, with the possibility to change himself the position of some nodes. We exploit the possibilities of interactive processing, thus avoiding the cumbersome programming of constraints and the long computer time for an optimization program to treat these constraints.

Our goal is to optimize the shape of a thin plate Ω. The deflection w is governed by the equation

$$\Delta^2 w = f \quad \text{in } \Omega \tag{1}$$

For example the plate is clamped along a part Γ_o of the boundary Γ of Ω and free elsewhere. Our aim is to minimize the stresses near Γ_o by acting on the shape of the boundary.

We have worked on a Finite Element model of the plate and chosen, as control parameters, the coordinates of boundary nodes. This choice is a priori one of the simplest. It is well known that it presents advantages and drawbacks :

(i) the number of control variables may be large if the boundary on which these control points lie is long or if the mesh is fine. However we did not encounter difficulties of computer time for the configurations we studied.

(ii) a possible drawback is the appearance of irregularities on the boundaries, necessitating the introduction of constraints. We never encountered such a situation and did work without optimization constraints.

(iii) integrity of the mesh during the optimization process : the characteristic of our optimization algorithm is to make only a few iterations (7 or 8) with a given mesh. During these iterations, only those boundary elements with moving nodes are going to vary. Then, with the new boundary obtained after 7 or 8 iterations we re-mesh the domain (this is done iteratively and quickly) and we can run again the program for a few more optimization iterations, etc...

(iv) an advantage is that, since there is no a priori constraint on the shape, it is possible to obtain any shape minimizing the stresses, and the finite elements automatically adapt to it. Otherwise, constraints could prevent the shape to attain an optimal form.

This method constitutes a Computer Aided Design tool, enabling the user :

(i) to generate the initial shape of the plate

(ii) to mesh

(iii) to choose the control nodes and the region where he wishes to minimize the stresses

(iv) to run the program for some optimization iterations

(v) to visualize the stresses

(vi) to go back to (ii) until satisfaction.

Optimization being without any constraint, the computations are fast.

Kirchoff - Poisson model for thin plates and position of the problem

A thin plate Ω is submitted to a force in a region Φ, $\Omega \supset \Phi$. It is clamped on a part Γ_0 of the boundary Γ and free on the remaining part Γ_1o. Important stresses appear near Γ_0, in a region $\mathbb{C}$. Our aim is to determine the shape of the plate in order to minimize the stresses in $\mathbb{C}$.

State of the system

We are led [3] to the variational formulation

$$\begin{aligned} & a(w,v) = (f,v) \qquad \forall v \in V \\ & w \in V \end{aligned} \tag{2}$$

where

$$a(w,v) = D \int_\Omega \{(w_{xx} + w_{yy})(v_{xx} + v_{yy}) + (1-\mu)\,[2\, w_{xy} v_{xy} - w_{xx} v_{yy} - w_{yy} v_{xx}\,]\}\, dx\, dy$$

$$(f,v) = \int_\Omega f\, v \;dx\, dy$$

$$V = \{ v \in H^2(\Omega),\ v = v_n = 0 \ \text{on}\ \Gamma_0 \}$$

$$D = E h^3 / (12(1-\mu^2))$$

w is the deflection, E the Young modulus, μ the Poisson ratio, h the thickness, v_n the normal derivative of v.

This corresponds to the minimization of $E_b - E_e$ where E_b is the bending energy

$$E_b = 0.5 \int_\Omega K^T D_b K \; dx\, dy ,$$

$$K^T = [\, \partial\beta_x/\partial x ,\ \partial\beta_y/\partial y ,\ \partial\beta_y/\partial x + \partial\beta_x/\partial y \,] ,$$

$$\beta_x = -w_x , \quad \beta_y = -w_y$$

$$D_b = D \begin{bmatrix} 1 & \mu & 0 \\ \mu & 1 & 0 \\ 0 & 0 & (1-\mu)/2 \end{bmatrix}$$

and E_e the potential energy

$$E_e = \int_\Omega f w \; dx\, dy$$

It can be shown that (2) implies (1).

<u>Cost function</u>

Suppose that the boundary Γ is determined by some control varaibale α. Then we define the cost function

$$J(\alpha) = 0.5 \int_{C(\alpha)} | D_b K(\alpha) |^2 \, d\Omega \; / \int_{C(\alpha)} 1 \, d\Omega$$

where $K(\alpha)$ and $D_b K(\alpha)$ are the "strain" and "stress" corresponding to the boundary defined by the control variable α.

<u>Finite Element Approximation using a 12 degrees of freedom quadrilateral element [4]</u>

The degrees of freedom at each node are w, θ_x and θ_y where $\theta_x = w_y$ and $\theta_y = - w_x$. On each element $\Omega^{(m)}$, K and w are approximated by

$$w = N^{(m)}(x,y)\, U^{(m)} \quad \text{and} \quad K = B^{(m)}(x,y)\, U^{(m)}$$

where

$$U^{(m)} = [\, w_1, \theta_{x1}, \theta_{y1}, w_2, \theta_{x2}, \theta_{y2}, \ldots, w_4, \theta_{x4}, \theta_{y4} \,]^T \in \mathbb{R}^{12}$$

and $N^{(m)}$ and $B^{(m)}$ are respectively 1 x 12 and 3 x 12 matrices.

The approximate solution U of the state equation is given by

$$K(\alpha)\, U = F$$

where K and F are obtained by the assembly of element stiffness matrices K_m and force vectors F_m:

$$K_m = \int_{\Omega m} B^{(m)T} D_b\, B^{(m)}\, dx\, dy \quad ,$$

$$F_m = \int_{\Omega m} N^{(m)T} f\, dx\, dy$$

The cost function is approximated by

$$J(\alpha) = 0.5\; U(\alpha)^T Q(\alpha)\, U(\alpha) / A(\alpha)$$

where

$$Q = \sum_{m} Q_m , \quad Q_m = \int_{\Omega m} B^{(m)T} D_b^T D_b\, B^{(m)}\, dx\, dy$$
$$\supset\!\!\!\!\subset\Omega_m$$

$$A = \sum_{\substack{m \\ \mathbb{C} \supset \Omega_m}} A_m, \qquad A_m = \int_{\Omega m} dx\, dy$$

(We suppose the region $\mathbb{C}$ to be the union of some elements Ω_m).

Calculation of the gradient J '(α) in order to apply a gradient method

We define the Lagrangian function :

$$L(U, \alpha, P) = J^*(U, \alpha) - (P, K(\alpha) U - F)$$

where $P \in \mathbb{R}^n$ (n being the dimension of U and F) and :

$$J^*(U, \alpha) = 0.5\ U^T Q(\alpha)\ U / A(\alpha)$$

Then

$$J(\alpha) = L(U(\alpha), \alpha, P)$$

and

$$\partial J / \partial \alpha_i = \{ 0.5\, U^T \partial Q / \partial \alpha_i\ U / A(\alpha) \} - 0.5/A^2\, U^T Q U\ \partial A / \partial \alpha_i - (P, \partial K / \partial \alpha_i\, U)$$

provided the adjoint variable P satisfies : $\partial L / \partial U = 0$, i.e.

$$K(\alpha) P = Q(\alpha) U / A(\alpha)$$

The difficulty of course is the calculation of $\partial K/\partial \alpha_i$, $\partial Q/\partial \alpha_i$ and $\partial A/\partial \alpha_i$. For an element Ω_m which has no node on the variable boundary $\partial K_m/\partial \alpha_i = 0$. For an element not in $\mathbb{C}$ $\partial Q_m/\partial \alpha_i = 0$. We used finite differencing, i.e. $\partial K_m/\partial \alpha_i \approx 0.5/\varepsilon\ [K_m(\alpha_i+\varepsilon) - K_m(\alpha_i-\varepsilon)]$, etc...

Numerical results

1st case : We start from a rectangular plate clamped along the side

Γ_0, force acting on Φ

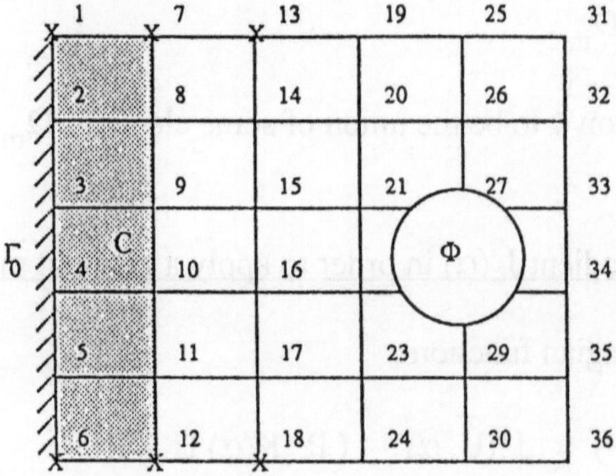

X indicates the nodes which can move

The stress $|\sigma| = (\sigma^2_x + \sigma^2_y + \tau^2_{xy})^{0.5}$ ranges from 0 to the maximun value 560. After 9 iterations we find the shape in figure 1(a) and the maximum value 420. Note the symmetry of the optimized plate.

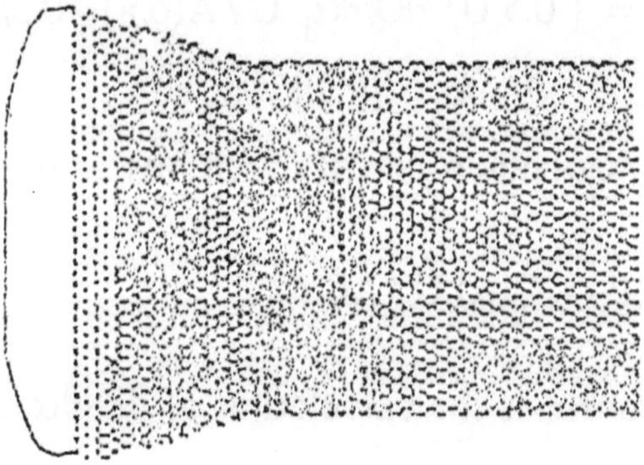

Fig. 1(a)

<u>2nd case</u> : With the same rectangular starting shape, we take different boundary conditions : side Γ_0 clamped, nodes 33 and 34 simply supported, force acting on 2 disks Φ.

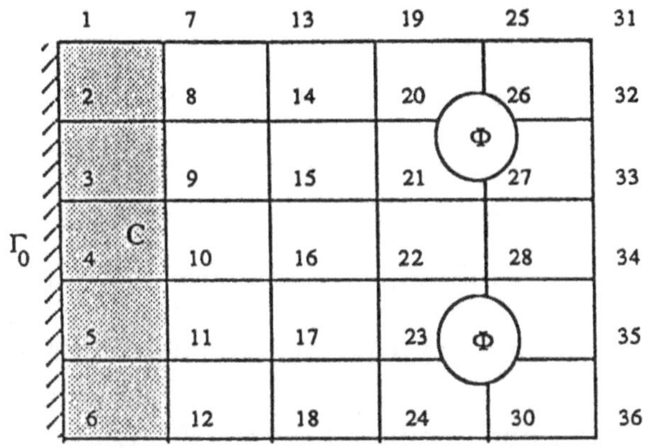

Before any optimization the stress σ_x ranges form -130 to +180.

After 6 iterations we have $-93 \leq \sigma_x \leq +140$ and the plate has the shape depicted in Fig. 1(b)

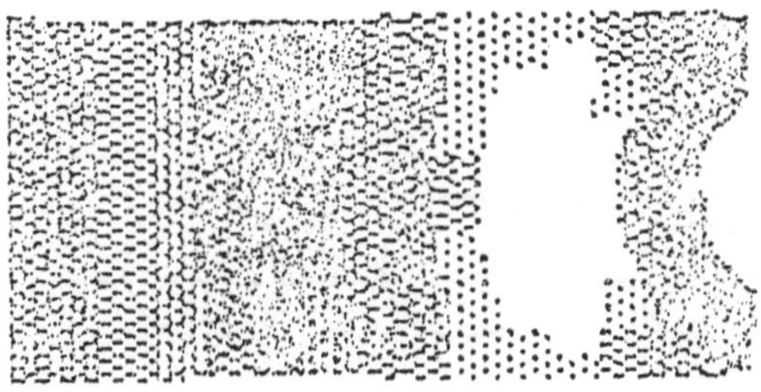

Fig. 1(b)

<u>3rd case</u> : Starting from the above shape Fig. 1(b), we now change the boundary conditions : plate clamped along the left-hand side Γ_0 and simply supported at nodes 31,32,35 and 36. Starting with values of σ_x between -130 and +170, we find, after 6 iterations, σ_x between -110 and +150, with a shape only slightly changed.

II Morphogenesis in an enzyme reaction diffusion system

The aim of this part is to give an example of a system where, as its size varies, a steady state may lose its stability at the benefit of bifurcating solutions. This system is intended to be a model of someaspects of a growing embryo. We suppose that the actual spatial domain Ω_a (the "embryo") grows and remains homothetic to a reference domain Ω (an open bounded set of $\mathbb{R}^2$), with measure $(\Omega_a) = \lambda$ measure (Ω). Thus λ is the size of the "embryo".

The state of this system as a function of its size λ is represented in figure 2(a) where some norm of the solution is plotted against λ, and the full (resp. dashed) lines represent stable (resp. unstable) steady states.

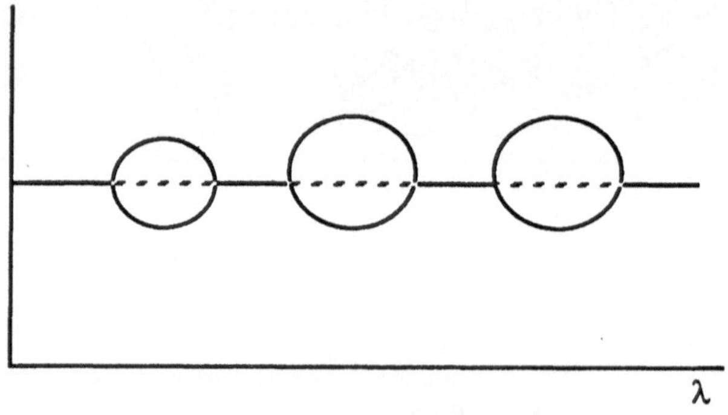

Figure 2(a)

One sees a "trivial" solution represented by a straight line and which loses stability at the benefit of closed loops of solutions.

The system is governed by the following P.D.E.s

$$-\Delta s \quad + \lambda [\, \rho \, R\,(s,a) - \quad (s_0 - s)\,] = 0$$

$$-\beta\Delta a + \lambda[\rho R(s,a) - \alpha(a_0 - a)] = 0 \qquad (1)$$

in an open bounded set Ω of $\mathbb{R}^2$, together with the B.C.s

$$\partial s/\partial n = 0 \quad \text{and} \quad \partial a/\partial n \quad \text{on } \Gamma = \partial\Omega \qquad (2)$$

where s_0, a_0, ρ, α, β and λ are positive parameters and

$$R(s,a) = a\,s/(1+s+k\,s^2)\ , \qquad k>0 \qquad (3)$$

For obtaining such results take for example Ω as in Fig. 2(b) and

$$k=0.1,\ \ s_0=102.5,\ \ a_0=92.8,\ \ \rho = 13,\ \ \alpha=1.2\,,\ \beta=5. \qquad (4)$$

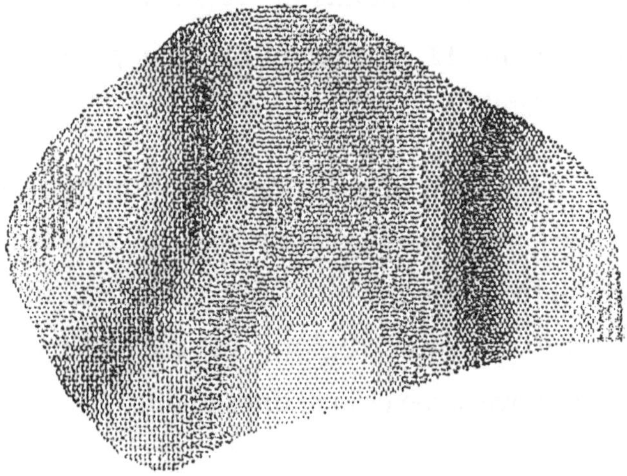

Fig. 2(b)

Equations (1) express the interaction between 3 phenomena

(i) Transport of 2 substrates S and A from a reservoir, where they are at concentration s_0 and a_0, to the membrane Ω, where they are at concentrations s and a (whence terms s_0-s and $\alpha(a_0$-a)).

(ii) Diffusion of S and A within Ω (whence the terms $-\Delta s$ and $-\beta\Delta a$)

(iii) Reaction of consumption of S and A under the catalytic action of an enzyme (whence the term $R(s,a)$)

The parameters α and β are the ratios of A and S diffusion coefficients, and ρ and λ are ratios of characteristic times

$$\rho = \theta_T/\theta_R \quad \text{and} \quad \lambda = \theta_D/\theta_T$$

where θ_T, θ_R and θ_D are the characteristic times respectively for transport of S from the reservoir to the membrane across an inactive layer, reaction of S and A in the membrane and diffusion of S within the membrane. Since $\theta_D = L^2/D_s$ where L is the diameter of Ω_a (the largest distance between 2 points of Ω_a) and D_s is the diffusion coefficient of S within Ω_a , one sees that λ is proportional to the actual size of the "embryo" Ω_a.

The substrates S and A are what Turing called "morphogens" [5], i.e. chemical species whose presence or absence induces some cell differentiation in the embryo. For example some cell differentiation will occur only in those regions of the embryo where S concentration is above some critical threshold.

Our model (1), (2), (3) shows sequential such "structurations in space" as λ grows, the other parameters being hold fixed : the "trivial" family of solutions represented by a straight line in Fig. 2(a) corresponds to spatially uniform solutions of (1), (2), (3), i.e. for every point (x,y) on $\Omega \cup \Gamma$, $s(x,y) = \tilde{s}$ and $a(x,y) = \tilde{a}$ where $\tilde{s}$ and $\tilde{a}$ are such that the brackets in (1) be zero

$$\rho\, R\,(s^{\sim},a^{\sim}) \;=\; (s_0 - s^{\sim}) \;=\; \alpha\,(a_0 - a^{\sim}) \qquad (5)$$

On the contrary the bifurcated solutions on the clossed loops correspond to fields of concentrations which are no more spatially uniform, but present with $s^{\sim}$ and $a^{\sim}$ deviations which are roughly proportional to the ith non trivial eigenfunction of the Laplacian operator with no-flux B.C.s for the ith loops

$$s(x,y,\lambda) \approx s^{\sim} + k(\lambda)\, w_i(x,y) \quad \text{and}$$

$$a(x,y,\lambda) \approx a^{\sim} + l(\lambda)\, w_i(x,y) \quad , 2 \leq i \qquad (6)$$

$$-\Delta w_i = \mu_i\, w_i \quad \text{in } \Omega, \quad \partial w_i/\partial n = 0 \text{ on } \Gamma,$$

$$\mu_1 = 0 < \mu_2 < \mu_3 < \ldots < \mu_i < \ldots \qquad (7)$$

<u>In conclusion</u> we have a model for cell differentiation in embryos as they grow and we address now the problem to maximize the concentration gradients by acting not only on λ but also on such environmental variables as s_0 and a_0 or a parameter like ρ which is proportional to the concentration of enzyme in the membrane.

Because of the importance of large concentration gradients in this model of cell differentiation it is natural to see whether they can be increased by acting on such parameters as λ,ρ,s_0,a_0. Whence the optimal control problem : the state (s,a) and the control (λ,ρ,s_0,a_0) being related by equations (1),minimize the cost function

$$J = -0.5 \int_\Omega (|\nabla s|^2 + |\nabla a|^2)\, d\Omega \quad +$$

penalty terms on λ, ρ, s_0, a_0 (these parameters cannot be "too large")

It has been shown in [6] that fairly close approximations to s and a are obtained by the Galerkin method

$$s \approx s^* = \sum_{i=1}^{N} s_i w_i(x,y) \quad \text{and} \quad a \approx a^* = \sum_{i=1}^{N} a_i w_i(x,y)$$

where the w_i's are defined in [7] and the coefficients s_i and a_i satisfy

$$\int_\Omega \{ -\Delta s^* + \lambda [\rho R(s^*,a^*) - (s_0 - s^*)] \} w_i \, d\Omega = 0$$

$$\int_\Omega \{ -\beta \Delta a^* + \lambda [\rho R(s^*,a^*) - \alpha (s_0 - s^*)] \} w_i \, d\Omega = 0 \qquad (8)$$

$$1 \le i \le N$$

It can be shown [6] that, as λ varies, the family of Galerkin approximations admit the same branch of trivial solutions $s^* = \tilde{s}$, and $a^* = \tilde{a}$ as in the exact case, and there are N-1 loops of non trivial solutions, bifurcating from the trivial branch for the same values of λ as in the exact case.

Whence the problem : the state (s^*,a^*) in a 2N-dimensional space and the control variables $(\lambda,\rho) \in \mathbb{R}^2$ being related by (8), minimize the cost function

$$J = -0.5 \sum_{i=2}^{N} \mu_i (s_i^2 + a_i^2) + m(\lambda^2 + \rho^2), \quad m=0.1$$

We now give some numerical results for this problem obtained by a generalized gradient method already described in [8]. This method enables to optimize in the presence of bifurcations. For more results, on the application of this one and other methods, such as penalty methods, to this problem, we refer to [3].

The parameter values being $k=0.1, \alpha=1.2, \beta=5, s_0=102.5, a_0=92.8$ and using N=5 eigenfunctions we started from points on the bifurcating loops close to the trivial branch. After several bifurcations we found local extrema with impressive gradients of concentrations. For example

First loop : starting with values of s ranging from 7.91 to 8.11 for $\lambda=5.33$ and $\rho=13$, we arrive at a point, for $\lambda=8.46$ and $\rho=10.82$, for which $3.61 \leq s \leq 22.03$.

Second loop : starting with values of $s \in [8.07, 8.21]$ for $\lambda=11.65$ and $\rho=13$, we arrive at a point, for $\lambda=18.77$ and $\rho=10.85$, for which $2.81 \leq s \leq 23.19$.

Although the detail is rather complicated (there are sequential bifurcations) however it can be seen that each optimum can be attained by several ways.

REFERENCES

[1] Bhavikatti, S.S., and Ramakrishnan, C.V. Optimum Shape Design of Rotating Disks. Computers and Structures 11 (1980) :

397 -401.

[2] Krzysztof, D., and Zenon, M. Multiparameter Structural Shape Optimization by the Finite Element Method. International Journal for Numerical Methods in Engineering 13 (1978) : 247 - 263.

[3] Deng, S.M. Optimisation de quelques systèmes distribués. Thesis, Compiègne, 1985.

[4] Batoz, J.L., and Ben Tahar, H. Evaluation of a new quadrilateral thin bending element. International Journal for Numerical Methods in Engineering 18 (1982) :1655 - 1677.

[5] Turing,A.M. The chemical basis of morphogenesis, Phil. Trans. Roy. Soc. London, Vol. B 237 (1952) 37-72.

[6] Joly,G., Kernevez,J.P., and Sharan,M. Continuation of Galerkin approximations for reaction-diffusion problems, Acta Applicandae Mathematicae,1,263-270,1983.

[7] Kernevez,J.P. Enzyme Mathematics, North Holland, Amsterdam, 1980 1-262 pages.

[8] Kernevez,J.P., Joly,G. and Sharan,M. Control of Systems with Multiple Steady-States, p.635-649 in Computing Methods in Applied Sciences and Engineering,North Holland, Amsterdam, 1982.

FURTHER DEVELOPMENT IN SHAPE SENSITIVITY ANALYSIS VIA PENALIZATION METHOD

M.C. DELFOUR
C R M
Univ. de Montréal
CANADA

J.P. ZOLESIO
C N R S
Univ. de Montpellier
FRANCE

1. INTRODUCTION

The object of this paper is the development of a penalization technique to compute the shape derivative of cost functionals where the state is the solution of a non-linear equation and/or a linear variational inequality. This type of problem is frequently encountered in Shape Sensitivity Analysis.

For partial differential equations where the state is the minimizing element of a quadratic energy functional over a linear subspace of a Hilbert space, the shape derivative can be computed by differentiating a MinMax problem with respect to an appropriate vector field (cf. DELFOUR and ZOLESIO [1,2,3]). This approach readily lends itself to some class of non-differentiable cost functions, but difficulties are encountered when the energy functional is non-linear or when the state is given as the minimizing element over a closed convex set which is not linear.

2. STATEMENT OF THE PROBLEM AND ORIENTATION

Let $E : \mathbb{R}^+ \times K \to \mathbb{R}$ be an *energy functional* defined over a closed convex subset K of a Banach space B. Assume that for each t in $\mathbb{R}^+$, the map

$$\varphi \mapsto E(t,\varphi) \tag{1}$$

is convex and continuous on K and that there exists on unique solution $y = y(t) \in K$ to the minimization problem

$$E(t,y) = \operatorname*{Inf}_{\varphi \in K} E(t,\varphi) \overset{\Delta}{=} e(t) \tag{2}$$

In particular y is completely characterized by the variational inequality

$$y \in K, \quad dE(t,y\ ;\ 0,\ \varphi-y) \geq 0 \quad , \forall\ \varphi \in K\ , \tag{3}$$

where for each ψ in B

$$dE(t,y \; ; \; 0,\psi) = \lim_{s \searrow 0} \frac{E(t,y+s\psi) - E(t,y)}{s} \qquad (4)$$

Associate with the above problem a cost functions

$$J(t) = F(t,y(t)) \qquad (5)$$

for some functional

$$F : \mathbb{R}^+ \times K \to \mathbb{R} . \qquad (6)$$

Assume that for all t in a neighborhood of 0 the map

$$\varphi \mapsto F(t,\varphi) \qquad (7)$$

is convex and continuous on K for some topology T_B weaker than the norm topology of B.

Our objective is to investigate the existence the Right-Gateaux differential of J at 0

$$dJ(0) = \lim_{s \searrow 0} \frac{J(S) - J(0)}{S} \qquad (8)$$

and to characterize it.

2.1 Construction of a MinSup problem : the Lagrangian approach

In many cases the above problem can be reformulated with the help of a Lagrangian of the form

$$L(t,\varphi \; ; \; \psi) = F(t,\varphi) + dE(t,\varphi \; ; \; 0,\psi). \qquad (9)$$

When $K = B$

$$J(t) = \mathop{\text{Inf}}_{\varphi \in B} \; \mathop{\text{Sup}}_{\psi \in B} \; L(t,\varphi \; ; \; \psi) \qquad (10)$$

If in addition L is convex and lower semi continuous in φ and concave and upper semi continuous in ψ the Lagrangian has saddle points (φ_t, ψ_t) which are completely characterized by the following system of equations (we assume F and E are sufficiently differentiable in φ)

$$dF(t,\varphi_t \; ; \; 0,\varphi) + d^2E(t,\varphi_t \; ; \; 0,\psi_t \; ; \; 0,\varphi) = 0 \; \forall \varphi \in B \qquad (11)$$

$$dE(t,\varphi_t \; ; 0,\psi) = 0 \quad , \forall \psi \in B. \tag{12}$$

For non-linear energy functionals $E(t,\varphi)$ the convexity of the Lagrangian with respect to φ is usually lost as can be seen from the thermal radiator problem (cf. DELFOUR, PAYRE and ZOLESIO [1]) where

$$E(\varphi) = \frac{1}{2} \int_\Omega |\nabla\varphi|^2 dx + \int_{\Sigma_3} (\frac{1}{5}|\varphi|^5 - q_s\varphi)d\sigma - \int_{\Sigma_1} q_{in} \varphi \, d\delta \tag{13}$$

where $q_s > 0$, $q_{in} > 0$, Ω is a volume of revolution with boundary $\Sigma = \Sigma_1 \cup \Sigma_2 \cup \Sigma_3$, Σ_1 is the interface between the radiator and the heat source, Σ_3 is the radiating surface and Σ_2 is the lateral adiatatic surface

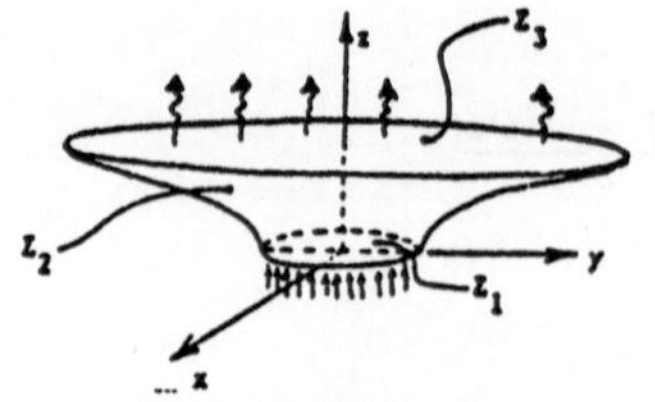

Figure 1

Volume Ω and its boundary

$\Sigma = \Sigma_1 \cup \Sigma_2 \cup \Sigma_3$

It is readily seen that the underlying space B is

$$B = \{\varphi \in H^1(\Omega) : \varphi|_{\Sigma_3} \in L^5(\Sigma_3)\} \tag{14}$$

which is a reflexive Banach space. However

$$dE(\varphi;\psi) = \int_\Omega \nabla\varphi.\nabla\psi \, dx + \int_{\Sigma_3} [|\varphi|^3 \varphi\psi - q_s\psi]d\delta - \int_{\Sigma_1} q_{im} \varphi \, d\sigma \tag{15}$$

and when ψ is negative on a subset of non-zero measure of Σ_3, $dE(\varphi;\psi)$ is no longer convex with respect to φ.

Another interesting and difficult case is the one where K is no longer a linear subspace of B, but a closed convex set of B. There the InfSup formulation (10) could be modified as follows

$$J(t) = \operatorname*{Inf}_{\varphi \in K} \; \operatorname*{Sup}_{\mu \geq 0} \; \operatorname*{Sup}_{\psi \in K} \{F(t,\varphi) - \mu \, dE(t,\varphi \; ; 0,\psi-\varphi)\} . \tag{16}$$

but in general, we again loose the convexity with respect to φ. For the characterization of optimal controls in this context the reader is referred to SHI SHUZHONG [1]. It is also interesting to note that the Right-Hand-Side of (16) can also be written in the following form

$$J(t) = \operatorname*{Inf}_{\varphi \in K} \operatorname*{Sup}_{\phi \in T_K(\varphi)} \{F(t,\varphi) - dE(t,\varphi \; ; \; 0,\phi\} \tag{16a}$$

where $T_K(\varphi)$ is the closure of the cone $\mathbb{R}^+(K-\varphi)$ in B. Notice also that the following two variational inequalities are equivalent for a closed convex set K

$$\exists\, y \in K,\ \forall\, \phi \in K,\quad dE(y\; ;\; \phi-y) \geq 0$$

and

$$\exists\, y \in K,\ \forall\, \phi \in T_K(y),\quad dE(y;\phi) \geq 0.$$

2.2 Construction of a non-Lagrangian formulation

To get around the above difficulties we propose to replace the Lagrangian by the following functional

$$G(t,\varphi,\mu) = F(t,\varphi) + \mu[E(t,\varphi) - e(t)] \tag{17}$$

where $\mu \in \mathbb{R}^+$ and

$$C(t) = \operatorname*{Inf}_{\varphi \in K} E(t,\varphi) = E(t, y_t)\ . \tag{18}$$

It is readily seen that

$$J(t) = \operatorname*{Inf}_{\varphi \in K} \operatorname*{Sup}_{\mu \geq 0} G(t,\varphi,\mu)\ . \tag{19}$$

When $F(t,\varphi)$ and $E(t,\varphi)$ are convex with respect to φ, the functional G is convex in φ for all $\mu \geq 0$ and linear (hence concave) in μ for all φ.

In this case, the InfSup problem (19) is equivalent to the InfSup problem (14)

$$\operatorname*{Inf}_{\varphi \in K} \operatorname*{Sup}_{\mu \geq 0, \phi \in K} \{F(t,\varphi) - \mu\, dE(t,\varphi\; ;\; 0,\ \phi-\varphi)\}$$

Indeed if $\qquad \exists\, \varphi \in K\ ,\ E(\varphi) = e$

then φ is completely characterized by

$$dE(\varphi,\psi-\varphi) \geq 0,\ \forall\ \psi \in K \iff \sup_{\psi \in K} - dE(\varphi,\psi-\varphi) = 0.$$

Conversely if

$$\exists \varphi \in K,\ \sup_{\psi \in K} - dE(\varphi\ ;\ \psi-\varphi) = 0,$$

then

$$E(\varphi) - \inf_{\psi \in K} E(\psi) = \sup_{\psi \in K} \{E(\varphi) - E(\psi)\}$$

$$\leq \sup_{\psi \in K} - dE(\varphi\ ;\ \psi-\varphi) = 0$$

which implies

$$\exists \varphi \in K,\ E(\varphi) \leq \inf_{\psi \in K} E(\psi) \implies \exists\ \varphi \in K,\ E(\varphi) = e.$$

The inequalities characterizing a saddle point $(\psi_t,\ \mu_t) \in K \times \mathbb{R}^+$ (if it exists) of (19) would be

$$dF(t,\varphi_t\ ;\ 0,\varphi-\varphi_t) + \mu_t\ dE(t,\varphi_t\ ;\ 0,\varphi-\varphi_t) \geq 0,\ \forall\ \varphi \in K \tag{20}$$

$$(\mu-\mu_t)\ [E(t,\varphi_t) - e(t)] \leq 0,\quad \forall\ \mu \geq 0. \tag{21}$$

The last inequality (21) is equivalent to

$$\left.\begin{array}{l} \mu_t\ [E(t,\varphi_t) - e(t)] = 0 \\ \mu_t \geq 0\ ,\ \ [E(t,\varphi_t) - e(t)] \leq 0 \end{array}\right\} \tag{22}$$

So either

$$E(t,\varphi_t) - e(t) = 0 \implies \varphi_t - y_t \text{ and } \mu_t \geq 0 \text{ arbitrary}$$

or

$$\mu_t = 0 \implies E(t,\varphi_t) - e(t)\ )\ 0 \implies \varphi_t = y_t$$

If $\mu_t \geq 0$ is finite, equation (20) reduces to

$$dF(t,y_t\ ;\ 0,\varphi-y_t) \geq 0,\ \forall\ \varphi \in K$$

which is equivalent to say that

$$F(t,y_t) = \inf_{\varphi \in K} F(t,\varphi). \tag{23}$$

This implies that the solution y_t of (18) also minimizes $F(t,\varphi)$ over all φ in K. This is a special case. In all other cases $\mu_t = +\infty$ which makes it difficult to extract any information from (20).

At this stage the existence of saddle points is questionable and we have seemingly lost the adjoint state which quite naturally comes out of a Lagrangian formulation. To get around this difficulty we study the following family of problems indexed by $\varepsilon > 0$

$$J_\varepsilon(t) = \operatorname*{Inf}_{\varphi \in K} G_\varepsilon(t,\varphi) \tag{24}$$

where

$$G_\varepsilon(t,\varphi) = G(t,\varphi,\frac{1}{\varepsilon}) = F(t,\varphi) + \frac{1}{\varepsilon}[E(t,\varphi) - e(t)] \tag{25}$$

Under appropriate hypothesis the minimizing elements φ_ε^t would be characterized by

$$dF(t,\varphi_\varepsilon^t ; 0,\varphi-\varphi_\varepsilon^t) + \frac{1}{\varepsilon} dE(t,\varphi_\varepsilon^t ; 0,\varphi-\varphi_\varepsilon^t) \geq 0, \ \forall \varphi \in K \tag{26}$$

So the steps are now clear. We must introduce appropriate hypothesis so that

$$\lim_{S\searrow 0} J_\varepsilon(t) = J(t) .$$

In the process we shall construct the variable

$$p_\varepsilon^t = (\varphi_\varepsilon^t - \varphi_o^t) / \varepsilon$$

which will converge in an appropriate sense to the usual *adjoint state* variable p which is typical of a Lagrangian approach. Thus we shall recover everything without the afore mentioned limitation of a Lagrangian method.

3. THE FAMILY OF PROBLEMS INDEXED BY t

In this section a more precise problem formulation is given and specific hypothesis are introduced in order to make sense of the constructions outlined in the previous section.

3.1 Problem formulation and hypotheses

Let $E : \mathbb{R}^+ \times K \to \mathbb{R}$ be an energy functional defined over a closed convex subset K of a Banach space B. Assume that the following hypothesis is verified.

For each t in $[0,T]$ the map

H1 $$\varphi \mapsto E(t,\varphi) \tag{1}$$

is convex and continuous on K and there exists a unique solution $y = y(t) \in K$ to the minimization problem

$$E(t,y) = \text{Inf}\,\{E(t,\varphi) : \varphi \in K\} \overset{\text{def}}{=} e(t). \tag{2}$$

In particular y is completely characterized by the variational inequality

$$y \in K, \quad dE(t,y\ ;\ 0,\varphi-y) \geq 0, \ \forall\, \varphi \in K \tag{3}$$

where for each ϕ in B

$$dE(t,y\ ;\ 0,\phi) = \lim_{S \searrow 0} \ (E(t,y+S\phi) - E(t,y))/S \tag{4}$$

Associate with the above problem a cost function :

$$J(t) = F(t,\ y(t)) \tag{5}$$

for some functional

$$F : \mathbb{R}^+ \times K \to \mathbb{R}\ . \tag{6}$$

For the moment, assume that the map $\varphi \mapsto F(t,\varphi)$ is convex and lower semi continuous on B.

Our main objective is to short that, under appropriate hypothesis, the cost function $J(t)$ can be expressed in the form

$$J(t) = J(0) + \int_0^t f(s)\, ds \tag{7}$$

for some function f in $L^\infty(0,T)$ which will be characterized in terms of the state $y(t)$ and the solution $p(t)$ to an appropriate adjoint unilatéral problem for each t. Under an additional hypothesis we shall also show that f belongs to $C^0(0,T)$, that is J belongs to $C^1(0,T)$ and $dJ(0) = f(0)$.

3.2 Penalized problems

Instead of tackling the problem directly we introduce a family of penalized problem indexed by $\varepsilon > 0$:

$$J_\varepsilon(t) = \inf_{\varphi \in K} \{F(t,\varphi) + \frac{1}{\varepsilon}\,[E(t,\varphi) - e(t)]\}. \tag{8}$$

H2 (i) There exist $T > 0$ and $\bar{\varepsilon} > 0$ such that for all t in $[0,T]$ and ε in $[0,\bar{\varepsilon}]$ there exists a unique minimizing element y_ε^t in K of the functional

$$G_\varepsilon(t,\varphi) = F(t,\varphi) + \frac{1}{\varepsilon}\,[E(t,\varphi) - e(t)] \tag{9}$$

over all φ in K.

(ii) For all in $[0,T]$

$$y_\varepsilon^t \to y_0^t \text{ in } B. \tag{10}$$

Hypothesis H2 contains hypothesis H1 and $y(t) = y_0^\varepsilon$. Existence and uniqueness of solution y_ε^t in a neighborhood of $(t,\varepsilon) = (0,0)$ may result from a positivity hypothesis on $F(t,.)$ on K or from a growth property of $F(t,\varepsilon)$ as $||\varphi||$ goes to infinity. In the sequel we shall denote by y the solution $y(0) = y_0^0$.

To make sense of the adjoint state we need the following additional hypotheses in a neighborhood N of y in B.

H3 The map $\varphi \mapsto E(t,\varphi)$ is twice Gateaux differentiable in N : that is for all φ in N and ϕ and ξ in B the following limit exist

$$dE(t,\varphi\ ;\ 0,\phi) = \lim_{s \searrow 0} [E(t,\varphi+s\phi) - E(t,\varphi)]/s$$

$$d^2e(t,\varphi\ ;\ 0,\phi\ ;\ 0,\xi) = \lim_{s \searrow 0} [dE(t,\varphi+s\xi;0,\phi) - dE(t,\varphi;0,\phi)]/s$$

H4 There exists a Hilbert space V, $B \subset V$, with continuous embedding such that the map

$$\phi \mapsto F(t,\phi)$$

is convex and V-continuous. Moreover for all φ in $N \cap K$

the maps

$$\psi \mapsto dE(t,\varphi\ ;\ 0,\psi)\ ,\ (\psi,\xi) \mapsto d^2E(t,\varphi\ ;\ 0,\psi\ ;\ 0,\xi)$$

extende continuously to V and $V\times V$, respectively and

$$\exists \alpha > 0 \text{ such that } \forall \psi \in V,\ d^2E(t,\varphi\ ;\ 0,\psi\ ;\ 0,\psi) \geq \alpha ||\psi||_V^2.$$

H5 Given convergent sequences $\varphi_n \to y_o^t$ in B, $\psi_n \to \psi$ in V (strong) and $\xi_n \to \xi$ in V (weak), there exists a subsequence $\{\varphi_{n_k}\}$ such that

$$d^2E(t,\varphi_{n_k}\ ;\ 0,\psi_{n_k}\ ;\ 0,\xi_{n_k}) \to d^2E(t-y_o^t\ ;\ 0,\psi\ ;\ 0,\xi)$$

As mentionned in section 2 we shall introduce the approximate adjoint state

$$p_\varepsilon^t = (y_\varepsilon^t - y_o^t)/\varepsilon \in B$$

and study its behaviour as ε goes to zero. This will require the following additional hypotheses.

H6 Given any two sequences $\{\varphi_n\}$ in $N \cap K$ and $\{\psi_n\}$ in V such that $\varphi_n \to y_o^t$ in B and $\psi_n \to \psi$ weakly in V for some ψ in V, there exist subsequences (still denoted $\{\varphi_n,\psi_n\}$) such that

$$\liminf_{n \to \infty}\ d^2E(t,\varphi_n\ ;\ 0,\psi_n\ ;\ 0,\psi_n) \geq d^2E(t,y_o^t\ ;\ 0,\psi\ ;\ 0,\psi).$$

3.3 A priori estimates for the penalized problems

Lemma 1. Assume that hypothesis H2 to H4 are verified. There exist a constant $c(t) > 0$ such that

$$|E(t,\ y^t) - E(t,y_o^t)| < \varepsilon\ C(t)\ ||y_\varepsilon^t - y_o^t||_V \tag{11}$$

$$||y_\varepsilon^t - y_o^t||_V \leq \varepsilon\ C(t)/\alpha \tag{12}$$

$$||p_\varepsilon^t||_V \leq C(t)/\alpha\ . \tag{13}$$

Proof. By definition of the minimizing element y_ε^t we have

$$F(t,y_\varepsilon^t) + \frac{1}{\varepsilon}[E(t,y_\varepsilon^t) - E(t,y_o^t)] \leq F(t,y_o^t) \tag{14}$$

By V-continuity and convexity of $\varphi \mapsto F(t,\varphi)$, there exists a support functional to $F(t,\varphi)$ at $\varphi = y_o^t$, that is

$$\exists\, x_o^* \in V',\ \forall\, \varphi \in V,\ F(t,\varphi) \geq F(t,y_o^t) + \langle x_o^*, \varphi - y_o^t\rangle \ .$$

Hence

$$F(t,\varphi) \geq F(t,y_o^t) - C(t)\, \|\varphi - y_o^t\|_V \ ,\ \forall\, \varphi \in V\ , \tag{15}$$

with $C(t) = \|x_o^*\|_{V'}$. From (14) we have

$$|E(t,y_\varepsilon^t) - E(t-y_o^t) \leq \varepsilon\ |F(t,y_\varepsilon^t) - F(t,y_o^t)|.$$

But from (15)

$$0 \leq F(t,y_o^t) - F(t,y_\varepsilon^t) \leq C(t)\ \|y_o^t - y_\varepsilon^t\| \ .$$

and hence (11). By hypothesis H3, there exists $\theta \in\ ]0,1[$ such that (use the variational inequality (3))

$$E(t,y_\varepsilon^t) - E(t,y_o^t) \geq d^2E(t,y_o^t + \theta(y_\varepsilon^t - y_o^t)\ ;\ 0,y_\varepsilon^t - y_o^t; 0,y_\varepsilon^t - y_o^t)$$

and by hypothesis H4

$$E(t,y_\varepsilon^t) - E(t,y_o^t) \geq \alpha\ \|y_\varepsilon^t - y_o^t\|_V^2 \ .$$

Combining the last inequality with (11) we obtain (12) and (13). ∎

Lemma 2. Under hypothesis H2 to H4, for all t in $[0,T]$

$$J_\varepsilon(t) \rightarrow J_o(t) \quad \text{as} \quad \varepsilon \rightarrow 0 \tag{16}$$

and for each $\varepsilon > 0$ there exists $\theta \in\ ]0,1[$ such that

$$\begin{aligned} 0 \leq \frac{1}{\varepsilon}\, dE(t,y_o^t;0,p_\varepsilon^t) + d^2E(t,y_o^t + \theta(y_\varepsilon^t - y_o^t)\ ;\ 0,p_\varepsilon^t\ ;\ 0,p_\varepsilon^t) \\ \leq -\, dF(t,y_o^t\ ;\ 0,p_\varepsilon^t) \end{aligned} \tag{17}$$

But

$$-\, dF(t,y_o^t\ ;\ 0,p_\varepsilon^t) \leq C(t)^2/\alpha \tag{18}$$

and

$$0 \leq \frac{1}{\varepsilon} dE(t,y_o^t \; ; \; 0,p_\varepsilon^t) \leq C(t)^2/\alpha \tag{19}$$

$$\begin{aligned} 0 &\leq \limsup \frac{1}{\varepsilon} dE(t,y_o^t \; ; \; 0,p_\varepsilon^t) \\ &\leq \liminf \{dF(t,y_o^t \; ; \; 0,p_\varepsilon^t) + d^2E(t,y_o^t+\theta(y_\varepsilon^t-y_o^t) \; ; \; 0,p_\varepsilon^t \; ; \; 0,p_\varepsilon^t)\} \end{aligned} \tag{20}$$

Proof. By definition $J_\varepsilon(t) \leq J_o(t)$ from (14). But

$$E(t,y_\varepsilon^t) \geq E(t,y_o^t) \Rightarrow F(t,y_\varepsilon^t) \leq J_\varepsilon(t)$$

and necessarily

$$F(t,y_\varepsilon^t) \leq J_\varepsilon(t) \leq J_o(t) = F(t,y_o^t).$$

By hypothesis H4 and estimate (12) in Lemma 1, we obtain (16). We know by hypothesis H2 to H4 that

$$E(t,y_\varepsilon^t) - E(t,y_o^t) \geq dE(t,y_o^t \; ; \; 0,y_\varepsilon^t - y_o^t) \geq 0$$

Combining this with (11) and (12) we obtain (19). Now by hypothesis H3, there exists $\theta \in]0,1[$ such that

$$E(t,y_\varepsilon^t) - E(t,y_o^t) = dE(t,y_o^t \; ; \; 0,y_\varepsilon^t-y_o^t) + d^2E(t,y_o^t+\theta(y_\varepsilon^t-y_o^t);0,y_\varepsilon^t-y_o^t). \tag{21}$$

But y_ε^t verifies the variational inequality

$$dF(t,y_\varepsilon^t \; ; \; 0,\varphi-y_\varepsilon^t) + \frac{1}{\varepsilon} dE(t,y_\varepsilon^t \; ; \; 0,\varphi-y_\varepsilon^t) \geq 0, \; \forall \varphi \in K \tag{22}$$

and

$$dF(t,y_\varepsilon^t \; ; \; 0,\varphi-y_\varepsilon^t) + dF(t,\varphi \; ; \; 0,y_\varepsilon^t-\varphi) \leq 0 \; . \tag{23}$$

By setting $\varphi = y_o^t$ in the above inequalities we obtain (17) and (20). ∎

Remark 1. For $0 \leq \varepsilon_1 \leq \varepsilon_2$

$$E(y_o) \leq E(y_{\varepsilon_1}) \leq E(y_{\varepsilon_2}) \; , \; F(y_{\varepsilon_2}) \leq F(y_{\varepsilon_1}) \leq F(y_o)$$

$$J_o = F(y_o) \geq J_{\varepsilon_1} \geq J_{\varepsilon_2} \quad , \quad 0 \geq \frac{J_{\varepsilon_1} - J_o}{\varepsilon_1} \geq \frac{J_{\varepsilon_2} - J_o}{\varepsilon_2}$$

$$0 \geq \frac{F(y_{\varepsilon_1}) - F(y_o)}{\varepsilon_1} \geq \frac{F(y_{\varepsilon_2}) - F(y_o)}{\varepsilon_2}$$

But

$$0 \leq dE(y_o \; ; \; p_\varepsilon) \leq \frac{1}{\varepsilon} [E(y_\varepsilon) - E(y_o)] \leq F(y_o) - F(y_\varepsilon) \leq -dF(y_o ; y_\varepsilon - y_o)$$

and

$$\lim_{\varepsilon \searrow 0} dE(y_o \; ; \; p_\varepsilon) = 0 \quad , \quad \limsup_{\varepsilon \searrow 0} dF(y_o \; ; \; p_\varepsilon) \leq 0.$$

Hence

$$0 \geq dF_o = \lim_{\varepsilon \searrow 0} \frac{F(y_\varepsilon) - F(y_o)}{\varepsilon} \geq \frac{F(y_{\varepsilon_1}) - F(y_o)}{\varepsilon_1} \geq \frac{F(y_{\varepsilon_2}) - F(y_o)}{\varepsilon_2}$$

and

$$\lim_{\varepsilon \searrow 0} \frac{1}{\varepsilon^2} [E(y_\varepsilon) - E(y_o)] = dJ_o - dF_o \geq 0$$

Also

$$0 \geq dF_o \geq \limsup_{\varepsilon \searrow 0} dF(y_o \; ; \; p_\varepsilon)$$

$$0 \geq dJ_o \geq \limsup_{\varepsilon \searrow 0} dF(y_o \; ; \; p_\varepsilon) + \limsup_{\varepsilon \searrow 0} \frac{1}{\varepsilon} dE(y_o \; ; \; p_\varepsilon)$$

and

$$0 \leq \limsup_{\varepsilon \searrow 0} \frac{1}{\varepsilon} dE(y_o \; ; \; p_\varepsilon) \leq dJ_o - dF_o \; . \qquad \blacksquare$$

3.4 Limiting behaviour of p_ε^t as ε goes to zero

In lemma 1 we have seen that the elements p_ε^t are bounded in V. So by construction they have weak limit points in the tangent convex cone

$$T_K(y_o^t) = V\text{-closure } \{\lambda(\varphi - y_o^t) : \varphi \in K, \lambda \geq 0\}. \quad (24)$$

Lemma 3. Assume that hypotheses H1 to H4 and H6 are verified and that p in V is a weak limit point of $\{p_\varepsilon^t : \varepsilon > 0\}$. Then

$$dE(t, y_o^t ; 0,p) = 0 \ , \ p \in T_K(y_o^t) \tag{25}$$

$$0 \leq \liminf \frac{1}{\varepsilon} \ dE(t, y_o^t ; 0, p_\varepsilon^t) \tag{26}$$

$$0 \leq \limsup \frac{1}{\varepsilon} dE(t,y_o^t ; 0,p_\varepsilon^t) \leq -[dF(t,y_o^t;0,p)+d^2E(t,y_o^t;0,p;0,p)] \tag{27}$$

Proof. Identity (25) is a direct consequence of inequalities (19). As for (26) it follows from (3) by setting $\varphi = y_o^t$ and dividing by $\varepsilon > 0$. Finally (27) follows from (20) and is a consequence of the weak lower semicontinuity of $\psi \mapsto dF(t,y_o^t ; 0,\psi)$ and hypothesis H6. ∎

So the weak limit points of $\{p_\varepsilon^t ; \varepsilon > 0\}$ belong to the closed convex cone

$$S(t) = T_K(y_o^t) \cap \nabla E(t,y_o^t)^\perp \ , \tag{28}$$

where

$$\nabla E(t,y_o^t)^\perp = V\text{-closure } \{\psi \in B \mid dE(t,y_o^t ; 0,\psi) = 0\}. \tag{29}$$

In fact they belong to a smaller set for which the condition

$$0 \leq \limsup \frac{1}{\varepsilon} dE(t,y_o^t ; 0,p_\varepsilon^t) \leq c(t)^2/\alpha \ ,$$

holds, but that set is hard to characterize.

3.5 Variational inequality for the limit points

We now construct a cone $A(t)$ and a variational inequality for the limit points of $\{p_\varepsilon^t\} = \{p_\varepsilon^t : \varepsilon > 0\}$. Let

$$A(t) = \left\{ \psi \in V \ \middle| \ \begin{array}{l} \exists \, \{\varphi_\varepsilon : \varepsilon > 0\} \ K, \ \psi_\varepsilon = (\varphi_\varepsilon - y_o^t)/\varepsilon \text{ such that} \\ \varphi_\varepsilon \to \psi \text{ in } V \text{ (weak) as } \varepsilon > 0 \to 0 \text{ and} \\ \lim\limits_{\varepsilon \searrow 0} \frac{1}{\varepsilon} dE(t,y_o^t ; 0,\psi_\varepsilon) = 0. \end{array} \right. \tag{30}$$

Lemma 4. (i) The set $A(t)$ is a cone with vertex at 0 in V.

Moreover

$$\mathbb{R}^+(K-y_o^t) \cap \nabla E(t,y_o^t)^{\perp} \subset A(t) \subset T_K(y_o^t) \cap \nabla E(t,y_o^t)^{\perp}. \tag{31}$$

(ii) If

$$\lim_{\varepsilon \searrow 0} \frac{1}{\varepsilon} \, dE(t,y_o^t \,;\, 0,p_\varepsilon^t) = 0 \,, \tag{32}$$

then all weak points of $\{p_\varepsilon^t\}$ is in $\overline{co}\, A(t)$.

Proof. (i) To show that $0 \in A(t)$, choose $\varphi_\varepsilon = y_o^t$, $\forall\, \varepsilon > 0$. Given $\lambda > 0$ and $\psi \in A(t)$

$$\exists\, \{\varphi_\varepsilon\} \subset K,\ \psi_\varepsilon = (\varphi_\varepsilon - y_o^t)/\varepsilon \ \rightarrow\ \psi \ \text{ in } V(\text{weak}) \text{ as } \varepsilon \rightarrow 0.$$

and

$$\lim_{\varepsilon \searrow 0} \frac{1}{\varepsilon} \, dE(t,y_o^t \,;\, 0,\psi_\varepsilon) = 0.$$

Then choose

$$\overline{\varphi}_\varepsilon = \varphi_{\varepsilon\lambda} \quad , \quad \overline{\psi}_\varepsilon = (\overline{\varphi}_\varepsilon - y_o^t)/\varepsilon$$

and notice that

$$\overline{\psi}_\varepsilon = \lambda \, \frac{\varphi_{\varepsilon\lambda} - y_o^t}{\varepsilon\lambda} = \lambda\, \psi_{\varepsilon\lambda} \ \rightarrow\ \lambda\psi \ \text{ in } V(\text{weak}) \text{ as } \ \varepsilon \rightarrow 0.$$

Moreover

$$\frac{1}{\varepsilon} \, dE(t,y_o^t \,;\overline{\psi}_\varepsilon) = \lambda \, \frac{1}{\varepsilon\lambda} \, dE(t,y_o^t \,;\, 0,\psi_{\varepsilon\lambda}) \ \rightarrow\ \lambda.0 = 0.$$

So we have shown that $A(t)$ is a cone with vertex at 0.

The next step is to show that any element

$$\psi \in \mathbb{R}^+(K-y_o^t) \cap \nabla E(t,y_o^t)^{\perp}$$

belongs to $A(t)$. This is equivalent to show that $\forall\, \lambda \geq 0$ and $\varphi \in K$ such that

$$dE(t,y_o^t \,;\, 0,\varphi - y_o^t) = 0.$$

Then

$$\psi = \lambda(\varphi - y_o^t) \in A(t).$$

To see that choose for ε such that $\varepsilon\lambda \geq 1$

$$\varphi_\varepsilon = (1-\varepsilon\lambda)y_o^t + \varepsilon\lambda\varphi \in K.$$

Then

$$\phi_\varepsilon = (\varphi_\varepsilon - y_o^t)/\varepsilon \quad = \lambda(\varphi - y_o^t),\ \frac{1}{\varepsilon}\ dE(t,y_o^t\ ;\ 0,\phi_\varepsilon) = 0.$$

and $\phi \in A(t)$. This proves the first part of (31). For the second one, it is clear that

$$\phi_\varepsilon \in \mathbb{R}^+(K-y_o^t) \ \Rightarrow\ \phi \in T_K(y_o^t).$$

Moreover there exists $\varepsilon_o > 0$ such that for any $\varepsilon \leq \varepsilon_o$

$$0 \leq dE(t,y_o^t\ ;\ 0,\phi_\varepsilon) \leq \varepsilon.$$

But ε goes to zero and necessarily

$$0 \leq dE(t,y_o^t\ ;\ 0,\phi) \leq 0.$$

(ii) is a corollary to (ii). ■

Remark 2. If

$$dE(t,y_o^t\ ;\ 0,\varphi) = 0,\quad \forall\, \varphi \in B \tag{33}$$

then

$$\bar{A}(t) = T_K(y_o^t) = \overline{co}\ A(t) \tag{34}$$

and all limit points of $\{p_\varepsilon^t\}$ belong to $\bar{A}(t)$. ■

Remark 3. If (32) is true and y_o^t minimizes $F(t,\varphi)$ over K, then $p_\varepsilon^t \to 0$ in V(strong). To see this use (20) and hypothesis H6. ■

Theorem 1. (i) Under hypothesis H1 to H6 any limit point p of $\{p_\varepsilon^t\}$ in V(weak) belongs to

$$S(t) = T_K(y_o^t) \cap \nabla E(t,y_o^t)^{\perp} \tag{35}$$

and verifies the variational inequality

$$dF(t,y_o^t\ ;\ 0,\phi) + d^2E(t,y_o^t\ ;\ 0,\phi\ ;\ 0,p) \geq 0 \quad \forall\, \phi \in A(t) \tag{36}$$

and the inequality

$$dF(t,y_o^t ; 0,p) + d^2E(t,y_o^t ; 0,p ; 0,p) \leq 0. \tag{37}$$

(ii) If

H7 $\overline{co}\ A(t) = S(t)$

and the map

$$\phi \mapsto dF(t,y_o^t ; 0,\phi) \tag{38}$$

is linear, then $p_\varepsilon^t \to p$ in V(weak), where p is the unique solution in $S(t)$ of the variational inequality

$$\left.\begin{aligned} & p \in S(t), \forall \phi \in S(t) \\ & dF(t,y_o^t ; 0,\phi-p) + d^2E(t,y_o^t ; 0,\phi-p ; 0,p) \geq 0. \ \blacksquare \end{aligned}\right\} \tag{39}$$

<u>Remark 4</u>. Hypothesis H7 is weaker than the classical hypothesis

$$V\text{-closure } \{\mathbb{R}^+(K-y_o^t) \cap \nabla E(t,y_o^t)^\perp\} = S(t) \tag{40}$$

(cf. F. MIGNOT [1], J. SOKOLOWSKI [1]). ■

<u>Proof of Theorem 1.</u> Since the parameter t is fixed, we shall drop it everywhere in the proof. (i) We already know that the weak limit points of $\{p_\varepsilon^t\}$ belongs to $S(t)$ and that (37) is verified. To established (36) we fix a weak limit point p of $\{p_\varepsilon\}$ and the associated sequence $\{\varepsilon_k > 0\}$, $\varepsilon_k \to 0$ such that

$$p_k = p_{\varepsilon_k} \to p \text{ in } V \text{ (weak)}.$$

Consider an arbitrary element ϕ in $A(t)$ and its associated $\{\varphi_\varepsilon : \varepsilon > 0\} \subset K$ such that

$$\phi = \text{weak} \lim_{\varepsilon \searrow 0} \phi_\varepsilon , \quad \phi_\varepsilon = (\varphi_\varepsilon - \varphi_o)/\varepsilon$$

$$\text{and} \quad \lim_{\varepsilon \searrow 0} \frac{1}{\varepsilon}\, dE(y_o ; \phi_\varepsilon) = 0 .$$

The above properties remain true with ε_k in place of ε .

We now turn to the variational equation for $y_k = y_{\varepsilon_k}$

$$dF(y_k\ ; \varphi - y_k) + \frac{1}{\varepsilon_k} dE(y_k\ ; \varphi - y_k) \geq 0,\ \forall\ \varphi \in K. \tag{41}$$

Let $\varphi = \varphi_k = \varphi_{\varepsilon_k}$ in (41). By hypothesis H4, there exists $\theta_k \in]0,1[$ such that

$$dE(y_k\ ; \varphi_k - y_k) = dE(y_o\ ; \varphi_k - y_k)$$
$$+ d^2E(y_o + \theta_k(y_k - y_o)\ ; \varphi_k - y_k\ ; y_k - y_o).$$

So (41) yields

$$0 \leq dF(y_k\ ; \frac{\varphi_k - y_o}{\varepsilon_k}) + \frac{1}{\varepsilon_k} dE(y_o\ ; \frac{\varphi_k - y_o}{\varepsilon_k}) + d^2E(y_o + \theta_k(y_k - y_o)\ ; \frac{\varphi_k - y_o}{\varepsilon_k}; p_k)$$
$$- dF(y_o\ ; p_k) - \frac{1}{\varepsilon_k} dE(y_o\ ; p_k) - d^2E(y_o + \theta_k(y_k - y_o)\ ; p_k\ ; p_k) \tag{42}$$

where we have used the fact that

$$dF(y_o\ ; y_k - y_o) + dF(y_k\ ; y_o - y_k) \leq 0.$$

Multiply (42) by λ_k^n and sum over k from n to N_n :

$$0 \leq - \Sigma \lambda_k^n \frac{1}{\varepsilon_k} dE(y_o\ ; p_k) - \Sigma\ \lambda_k^n\ [dF(y_o\ ; p_k) + d^2E(y_o + \theta_k(y_k - y_o); p_k; p_k)]$$
$$+ \Sigma\ \lambda_k^n \frac{1}{\varepsilon_k} dE(y_o\ ; \psi_k) + \Sigma\ \lambda_k^n\ dF(y_k\ ; \psi_k) \tag{43}$$
$$+ \Sigma\ \lambda_k^n\ d^2E(y_o + \theta_k(y_k - y_o)\ ; \psi_k\ ; p_k)\ ;$$

where

$$\sum_{k=n}^{N_n} \lambda_k^n = 1\ ,\ \lambda_k^n \geq 0\ ,\ \psi_k = (\varphi_k - y_o)/\varepsilon_k\ . \tag{44}$$

The first term on the first line of (43) is negative. Take the lim sup of the remaining terms on the Right-Hand-Side of (43) on use the following result : given a sequence $\{f_k\}$ of real numbers such that $f_k \to f$ in $\mathbb{R}$, then

$$\bar{f}_n = \sum_{k=n}^{N_n} \lambda_k^n\ f_k \to f\ ,\ \sum_{k=n}^{N_n} \lambda_k^n = 1,\ \lambda_k^n \geq 0. \tag{45}$$

By lemma 3

$$\liminf_{\varepsilon \searrow 0} [dF(y_o ; p_\varepsilon) + d^2E(y_o+\theta_\varepsilon(y_\varepsilon-y_o) ; p_\varepsilon ; p_\varepsilon] = a \leq 0 \qquad (46)$$

exists and is negative (cf. (20) in lemma 2).

So using (45) and (46), the second term in the first line of (43) is less than $-a$ as k goes to ∞. By definition of ψ we know that

$$\lim_{k \to \infty} \frac{1}{\varepsilon_k} dE(y_o ; \psi_k) = 0$$

and by using (45), the first term in the second line of (43) goes to 0 as k goes to ∞. By hypothesis H5 there exists a subsequence of $\{\varepsilon_k\}$, still denoted $\{\varepsilon_k\}$ such that

$$d^2E(y_o+\theta_k(y_k-y_o) ; \psi_k ; p_k) \to d^2E(y_o ; \psi ; p)$$

and by using (45) again the term in the last line of (43) goes to $d^2E(y_o ; \psi ; p)$. The only term left is

$$g_n = \sum_{k=n}^{N^n} \lambda_k^n \, dF(y_k ; \psi_k) . \qquad (47)$$

Recall that for a convex continuous function F, the map

$$\psi \mapsto dF(\varphi ; \psi) : V(\text{strong}) \to \mathbb{R}$$

is convex and locally Lipschitz continuous and that

$$(\varphi,\psi) \mapsto dF(\varphi ; \psi) : V(\text{strong}) \times V(\text{strong}) \to \mathbb{R}$$

is upper semicontinuous. As a result

$$\left.\begin{aligned} g_n - dF(y_o ; \psi) &= \Sigma \lambda_k^n [dF(y_k ; \psi_k) - dF(y_o ; \psi)] \\ &= \Sigma \lambda_k^n [dF(y_k ; \psi_k) - dF(y_k ; \psi)] \\ &+ \Sigma \lambda_k^n [dF(y_k ; \psi) - dF(y_o ; \psi)] \end{aligned}\right\} \qquad (48)$$

By local Lypschitz continuity, there exists a neighborhood N of y_o and a constant $c > 0$ such that

$$\forall\, y \in N,\ \forall\, \phi_1, \phi_2 \in V,\ |dF(y\ ;\ \phi_2) - dF(y\ ;\ \phi_1)| \leq C||\phi_2 - \phi_1||_V \tag{49}$$

As a result the first term on the Right-Hand-Side of (48) is bounded by

$$\Sigma\, \lambda_k^n\, C\, ||\phi_k - \phi||_V \;=\; C||\Sigma\, \lambda_k^n\, \phi_k - \phi||_V \to 0 \tag{50}$$

As for the second term denote by

$$\ell = \limsup_{k \to \infty}\ dF(y_k\ ;\ \phi) \leq dF(y_o\ ;\ \phi) \tag{51}$$

Then always by (45)

$$\limsup_{n \to \infty}\ \Sigma\, \lambda_k^n\, dF(y_k\ ;\ \phi) \;=\; \limsup_{k \to \infty}\ dF(y_k\ ;\ \phi) \tag{52}$$

and the second term is negative.

In conclusion we have shown the following inequality for all ϕ in $A(t)$

$$0 \leq -a + dF(y_o\ ;\ \phi) + d^2E(y_o\ ;\ \phi\ ;\ p)$$

But in view of lemma 3, we know that

$$a = dF(y_o\ ;\ p) + d^2E(y_o\ ;\ p\ ;\ p) \leq 0\ .$$

Recall that the set $A(t)$ in a cone; So for any ϕ in $A(t)$ and $\lambda > 0$

$$dF(y_o\ ;\ \lambda\phi) + d^2E(y_o\ ;\ \lambda\phi\ ;\ p) \geq a$$

and

$$dF(y_o\ ;\ \phi) + d^2E(y_o\ ;\ \phi\ ;\ p) \geq \operatorname{Inf}\{a/\lambda\ :\ \lambda > 0\} = 0\ .$$

(ii) When (38) is linear, inequality (36) holds for all ϕ in $\overline{\text{co}}\, A(t)$ and by combining it with (37)

$$\left.\begin{array}{l} p \in S(t)\ ,\ \forall\, \phi \in \overline{\text{co}}\, A(t) \\[1em] dF(y_o\ ;\ \phi - p) + d^2E(y_o\ ;\ \phi - p\ ;\ p) \geq 0\ . \end{array}\right\} \tag{53}$$

So when hypothesis H7 is true, (39) has a unique solution which necessarily coincides with all weak limit points of $\{p_\varepsilon\}$.

This yields the uniqueness of the weak limit point and its complete characterization. ■

Remark 4. Another interesting cone with vertex at U for which inequality (36) holds is

$$B(t) = \left\{ \phi \in \nabla E(t,y_o^t)^\perp \;\middle|\; \begin{array}{l} \exists\{\lambda_\varepsilon > 0\},\ \exists\{\varphi_\varepsilon\} \subset K,\ \ \phi_\varepsilon = \lambda_\varepsilon(\varphi_\varepsilon - y_\varepsilon^t)/\varepsilon \\ \text{such that } \phi_\varepsilon \to \phi \text{ in } V(\text{strong}) \text{ as } \varepsilon \to 0 \\ \text{and } \limsup_{\varepsilon \to 0} \frac{1}{\varepsilon}\, dE(t,y_o^t\, ;\, 0,\phi_\varepsilon) \le 0 \end{array} \right\} \tag{54}$$

By definition, it is easy to check that

$$\overline{co}\, A(t) - p \subset \overline{co}\, B(t)$$

for all limit points p of $\{p_\varepsilon^t\}$ in V(weak). It is easy to show that

$$\mathbb{R}^+(K-y_o^t) \cap \nabla E(t,y_o^t)^\perp \subset B(t) \subset T_k(y_o^t) \cap \nabla E(t,y_o^t)^\perp . \tag{55}$$

So condition H7 could be further weakened to

H7 $\quad \overline{co}\, \{C(t),\, B(t)\} = S(t).$ ■

Remark 5. If inequality (36) is to be verified only on $\mathbb{R}^+(K-y_o^t) \cap \nabla E(t,y_o^t)^\perp$, then hypothesis H5 can be weakened to

H5' There exists a dense subspace D of V such that

$\quad \forall \varphi \in N \cap K,\ \forall \phi \in D,\ (\varphi,\xi) \mapsto d^2E(t,\varphi\ ;\ 0,\xi\ ;\ 0,\phi)$

$\quad$ is continuous from $B \times V$(weak) into $\mathbb{R}$.

4. LIMITING BEHAVIOUR OF $J_\varepsilon(t)$ AS A FUNCTION OF t AND DERIVATIVE OF $J_o(t)$

The object of this section is to determine conditions under which $J_o \in W^{(1)}(0,T)$ and study the limit of $dJ_o(t)$ as t goes to zero.

4.1 Differentiability of $J_\varepsilon(t)$ with respect to t

We first compute the derivative of $J_\varepsilon(t)$, $t \in [0,T]$ from the right

$$dJ_\varepsilon(t) = \lim_{s \searrow 0} [J_\varepsilon(t+s) - J_\varepsilon(t)]/s \tag{1}$$

where J_ε is defined by (8) as

$$J_\varepsilon(t) = \mathrm{Min}\,\{G_\varepsilon(t,\varphi) \mid \varphi \in K\} \tag{2}$$

with

$$G_\varepsilon(t,\varphi) = F(t,\varphi) + \frac{1}{\varepsilon}[E(t,\varphi) - e(t)]. \tag{3}$$

Introduce the sets

$$A_\varepsilon(t) = \{\phi \in K \mid G(t,\phi) = J_\varepsilon(t)\}. \tag{4}$$

We first need an intermediate result from J.P. ZOLESIO [3] which will be applied to $e(t)$ and $J_\varepsilon(t)$.

Theorem 1. Let $G : \mathbb{R} \times B \to \mathbb{R}$ be a functional defined on a reflexive Banach space B and be A a subset of B. Let

$$J(t) = \mathrm{Inf}\{G(t,\varphi) \mid \varphi \in A\},\ A(t) = \{\phi \in A \mid J(t) = G(t,\phi)\} \tag{5}$$

with the following hypothesis : there exists $T > 0$ such that :

HH1 $A(t) \neq \emptyset$, $0 \leq t \leq T$

HH2 $\forall y^o \in A(0)$, $\forall y^t \in A(t)$, the functions $s \mapsto G(s,y^o)$ and $s \mapsto G(s,y^t)$ are differentiable in a neighborhood of zero

HH3 $\forall y^o \in A(0)$, $s \mapsto \partial_s G(s,y^o)$ is upper semi-continuous

HH4 $\exists$ a topology $\mathcal{T}$ on B and a compact subset ϕ of B such that $A(t) \cap Q \neq \emptyset$, $0 \leq t \leq T$

HH5 The map $(s,\varphi) \mapsto \partial_s G(s,\varphi)$ is lower semi-continuous on $\mathbb{R} \times B(\mathcal{T})$

HH6 (i) $\forall \varphi \in B$, $t \mapsto G(t,\varphi)$ is upper semi-continuous at $t = 0$

(ii) $\exists$ a topology $\tilde{\mathcal{T}}$ on B such that the map $t,\varphi \mapsto G(t,\varphi)$ is lower semi-continuous.

Then the Right-Hand-Side derivative of J is given by

$$dJ(t) = \text{Inf}\{\partial_t G(0,\varphi) \mid \varphi \in A(0)\}. \qquad \blacksquare \qquad (6)$$

We now proceed in two steps. First we use Theorem 1 to show that under appropriate hypotheses, $e(t)$ is continuously differentiable on $[0,T[$. Then using that result and Theorem 1 once more, we obtain the differentiability of J_ε in $[0,T]$.

Lemma 1. Assume that hypothesis H1 is verified and that

H7 $\forall\, \varphi \in N$, $t \mapsto E(t,\varphi) : [0,T] \to \mathbb{R}$
is of class C^1 and the map
$t,\varphi \mapsto E(t,\varphi)$ and $(t,\varphi) \mapsto dE(t,\varphi\,;\,1,0)$
are weakly lower semi-continuous on $[0,T] \times B$

Then the function $e(t)$ is of class C^1 on $[0,T]$ and

$$e'(t) = de(t;1) = dE(t,y_o^t\,;\,1,0), \quad 0 \leq t \leq T. \qquad (7)$$

Proof. By direct application of Theorem 1, we obtain the R.H.S. derivative de $(t;1)$ given by (7). But since the set $A(0)$ is reduced to the single element y_o^t, then

$$de(t;1) = -\,de(t;-1) = e'(t)$$

is the usual derivative at t. $\blacksquare$

Lemma 2. Assume that hypothesis H1, H2 and H7 are verified and that

H8 for each $\varepsilon \geq 0$, the function $t \mapsto y_\varepsilon^t : [0,T] \to B$ is continuous

H9 $\forall\, \varphi \in N$ the functions $t \mapsto F(t,\varphi) : [0,T] \to \mathbb{R}$ is of class C^1
and the maps
$(t,\varphi) \mapsto F(t,\varphi)$, $(t,\varphi) \mapsto dF(t,\varphi\,;\,1,0)$
are weakly lower semi-continuous on $[0,T] \times B$.

Then for each $\varepsilon > 0$ and $0 \leq t \leq T$,

$$dJ_\varepsilon(t) = dF(t,y_\varepsilon^t\,;\,1,0) + \frac{1}{\varepsilon}\,[dE(t,y_\varepsilon^t\,;\,1,0) - dE(t,y_o^t\,;\,1,0)] \quad (8)$$

Proof. Direct application of Theorem 1. ■

4.2 Absolute continuity of J_o

We first construct the pointwise limit $f(t)$ of $dJ_\varepsilon(t)$ as goes to zero. Then we use a boundedness hypothesis to get the absolute continuity of the limit function $J_o(t)$ on $[0,T]$.

H10 The map

$$\varphi \mapsto dF(t,\varphi\ ;\ 1,0) : V \to \mathbb{R}$$

is continuous in N.

H11 For all ψ in B and t in $[0,T]$, the limit

$$d^2(t,\varphi;1,0;0,\psi) = \lim_{s\searrow 0}[dE(t+s,\varphi;1,0;0,\psi) - dE(t,\varphi;1,;0,\psi)]/s$$

exists for all φ in N.

H12 For all t in $[0,T]$, the map

$$\varphi,\psi \mapsto d^2E(t,\varphi\ ;\ 1,0\ ;\ 0,\psi)$$

is continuous on $N \times V(\text{weak})$.

Lemma 3. Assume that hypotheses H1 to H12 are verified and that the map (3.38) is linear, then

$$\forall\, t \in [0,T],\quad dJ_\varepsilon(t) \to f(t) \quad \text{as} \quad \varepsilon \to 0$$

where

$$f(t) = dF(t,y_o^t\ ;\ 1,0) + d^2E(t,y_o^t\ ;\ 1,0\ ;\ 0,p_o^t)\ . \tag{9}$$

Proof. From H11, there exists θ, $0 < \theta < 1$, such that

$$[dE(t,y_\varepsilon^t\ ;\ 1,0) - dE(t,y_o^t\ ;\ 1,0)]/\varepsilon$$
$$= d^2E(t,y_o^t + \theta(y_\theta^t - y_o^t)\ ;\ 1,0\ ;\ 0,p_\varepsilon^t).$$

By H12, the R.H.S. of the above expression goes to

$$d^2E(t,y_o^t\ ;\ 1,0\ ;\ 0,p_o^t)\ .$$

Similarily by H10

$$dF(t,y_\varepsilon^t \; ; \; 1,0) \to dF(t,y_o^t \; ; \; 1,0).$$

Then (9) is obtained by going to the limit in (8) as ε goes to zero.

We now introduce the boundedness hypothesis to apply Lebesgue Dominated Convergence Theorem and

$$J_o(t) = \lim_{\varepsilon \searrow 0} J_\varepsilon(t) = J_o(0) + \lim_{\varepsilon \searrow 0} \int_0^t dJ_\varepsilon(s) \; ds = J_o(0) + \int_0^t f(s)ds.$$

Recall from Remark 3.1 that

$$J_\varepsilon(t) \nearrow J_o(t) \quad \text{as} \quad \varepsilon \to 0.$$

The boundedness hypothesis is

H13 $\quad \exists \; M > 0 \quad$ such that, $\forall \; t \in [0,T], \; \forall \; \varphi \in N, \; \forall \; \psi \in V$

$$|d^2E(t,\varphi \; ; \; 1,0 \; ; \; 0,\psi)| \le M||\psi||_V$$

and the map

$$t,\varphi \mapsto dF(t,\varphi \; ; \; 1,0)$$

is bounded in $[0,T] \times N$.

<u>Theorem 2</u>. Under hypotheses H1 to H13, the linearity of the map (3.28) and the density hypothesis H7 for all t in $[0,T]$, the function J_o is absolutely continuous. Its derivative coincides almost everywhere with the function f in $L^\infty(0,T)$ and hence J_o belongs to $W^{1,\infty}(0,T)$:

$$dJ_o(t) = dF(t,y_o^t \; ; \; 1,0) + d^2E(t,y_o^t \; ; \; 1,0 \; ; \; 0,p_o^t) \qquad (10)$$

where p_o^t is the unique solution in S_t of the variational inequality : for all ψ in S_t

$$dF(t,y_o^t \; ; \; 0,\psi-p_o^t) + d^2E(t,y_o^t \; ; \; 0, \; \psi-p_o^t \; ; \; 0,p_o^t) \ge 0. \quad \blacksquare$$

<u>Remark 1</u>. Hypothesis H8 requires the continuity of the function $t \mapsto y_\varepsilon^t$, $\varepsilon \ge 0$, in the B-norm. It is clear that the technique of lemma 3.1 would only give the continuity in V. Thus a stronger

result is required which can be obtained in each case depending on the structure of E and F. $\blacksquare$

4.3 Differentiability of $J_o(t)$ at $t = 0$.

As this juncture Theorem 2 seems to be the most reasonable result when K is not a subspace of V. The delicate point is the continuity of p_o^t as a function of t at 0 in V(weak). It is crucially related to the limiting behaviour of the sets

$$S_t = T_K(y_o^t) \cap \nabla E(t,y_o^t)^{\perp} \tag{12}$$

This point is readily explained in the following are dimensional example.

<u>Example</u> $\quad K = \{\varphi \in \mathbb{R} : \varphi \geq 0\}$

$$E(u,\varphi) = \frac{1}{2}\varphi^2 + u\varphi \quad , \quad F(u,\varphi) = \frac{1}{2}(\varphi-1)^2 \tag{13}$$

It is easy to verify that

$$y_u = \begin{cases} 0 \ , & \text{if } u \geq 0 \\ -u \ , & \text{otherwise} \end{cases}$$

and that

$$J(u) = \frac{1}{2}(y_u-1)^2 = \begin{cases} 1/2 \ , & u \geq 0 \\ (u+1)^2/2 \ , & \text{otherwise} \end{cases}$$

For $t = 0$ as a function of u the function $J(u)$ is represented in Figure 1.

The directional derivative at u in the direction v is

$$dJ(u\ ;\ v) = \begin{cases} 0 & , \quad u \geq 0 \\ (u+1)v & , \quad \text{for } u < 0 \end{cases} \tag{14}$$

So J is differentiable everywhere except at $u = 0$

$$dJ(0\ ;\ v) = \min\ \{0,v\}\ .$$

Now fix u, v and $t \geq 0$

$$\tilde{E}(t,\varphi) = E(u+tv,\varphi)\ ,\ \tilde{F}(t,\varphi) = F(u+tv,\varphi)$$

$$y_t = y_{u+tv}\quad ,\quad \tilde{J}(t) = J(u+tv)$$

Choose $u = 0$. Then for $t \geq 0$

$$y_t = \begin{cases} 0 & , \quad \text{if } v \geq 0 \\ -tv & , \quad \text{if } v < 0 \end{cases}$$

for $v = 1$ and $t > 0$

$$y_\varepsilon^t = \begin{cases} (\varepsilon-t)/(\varepsilon+t) & , \ 0 \leq t \leq \varepsilon \\ 0 & , \ \varepsilon < t \end{cases}\ ,\ y_o^t = 0$$

$$p_\varepsilon^t = \begin{cases} (\varepsilon-t)/\varepsilon(\varepsilon+1), & 0 \leq t \leq \varepsilon \\ 0 & , \ \varepsilon < t \end{cases}\ ,\ p_o^t = 0$$

But for $t = 0$

$$y_\varepsilon^o = \varepsilon/(\varepsilon+1)\quad ,\quad y_o^o = 0\quad ,\quad p_\varepsilon^o = 1/(\varepsilon+1)\quad ,\quad p_o^o = 1$$

As a result

$$\lim_{t \searrow 0} p_o^t = 0 \neq 1 = p_o^o\ .$$

For $v = -1$ and $t \geq 0$

$$y_\varepsilon^t = \frac{\varepsilon+t}{\varepsilon+1}\quad ,\quad p_\varepsilon^t = \frac{1-t}{\varepsilon+1}\quad ,\quad p_o^t = 1-t \rightarrow p_o^o = 1\ .$$

Finally

$$dJ_\varepsilon(t) = p_\varepsilon^t \tag{15}$$

and in each case we recover the results on the begining. ■

<u>Proposition 1.</u> (i) Assume that hypothesis H1 to H13 (H7 for all t in $[0,T]$) hold, that the map (3.38) is linear and that

H14 $\quad t,\phi \mapsto d^2E(t,y_o^t \,;\, 1,0 \,;\, 0,\phi) : [0,T] \times V(\text{weak})$ is continuous

H15 $\quad t \mapsto dF(t,y_o^t \,;\, 1,0)$ is continuous at $t = 0$

H16 $\quad p_o^t \to p$ (unique) in $V(\text{weak})$.

Then

$$dJ_o(0) = dF(0,y_o^o \,;\, 1,0) + d^2E(0,y_o^o \,;\, 1,0 \,;\, 0,p). \tag{16}$$

(ii) If, in addition, $p = p_o^o$, then p is completely characterized by (11) with $t = 0$. ■

When the cones $S(t)$ have an appropriate behaviour as t goes to 0, it is possible to obtain a variational equation for the limit point p of p_o^t as t goes to zero.

<u>Proposition 2.</u> Assume that the hypothesis of Proposition 1 (i) hold and that

H17 $\quad \lim_{t \searrow 0} dF(t,y_o^t \,;\, 0,\phi) = dF(0,y_o^o \,;\, 0,\phi), \ \forall \phi \in V$

$\quad \liminf_{t \searrow 0} dF(t,y_o^t \,;\, 0,p_o^t) \geq dF(0,y_o^o \,;\, 0,p)$

H18 $\quad \lim_{t \searrow 0} d^2E(t,y_o^t \,;\, 0,\phi \,;\, 0,p_o^t) = d^2E(0,y_o^o \,;\, 0,\phi \,;\, 0,p)$

$\quad \liminf_{t \searrow 0} d^2E(t,y_o^t \,;\, 0,p_o^t \,;\, 0,p_o^t) \geq d^2E(0,y_o^o \,;\, 0,p \,;\, 0,p)$

H19 $\quad \exists\, T > 0$ such that

$$\forall\, 0 < t_1 \leq t_2 \leq T, \quad S(t_1) \subset S(t_2). \tag{17}$$

Then p is the unique solution in the closed convex cone

$$S = \bigcap_{0<t\leq T} S(t) \tag{18}$$

of the variational inequality

$$\left.\begin{aligned} & p \in S,\ \forall\, \phi \in S \\ & dF(0,y_o\ ;\ 0,\phi-p) + d^2E(0,y_o\ ;\ 0,\phi-p\ ;\ 0,p) \geq 0. \ \blacksquare \end{aligned}\right\} \quad (19)$$

5. SHAPE DERIVATIVE FOR THE RADIATOR PROBLEM

Let $V = V(t,x_1,x_2,z)$ be a velocity field, $V \in C^o([0,T],\ C^1(\mathbb{R}^3;\mathbb{R}^3))$ such that

$$V(t,\ x_1,\ x_2,\ 0) = 0\ . \tag{1}$$

since Σ_1 is invariant in the deformation of the domain. If the field V is written as $V = (V_{x_1},\ V_{x_2},\ V_z)$, then the condition

$$V_z(t,\ x_1,\ x_2,\ L) = 0 \tag{2}$$

implies that Σ_3 remains linear in the deformation.
Denote by $T_t = T_t(V)$, the transformation associated to V

$$\frac{dT}{dt} = (V)X = V(t,\ T_t(V)X)\ ,\ T_o(V)X = X,\ t \geq 0.$$

Consider the matrix

$$A(t) = J(t)\ \ (DT_t)^{-1}.\ {}^*(DT_t)^{-1}$$

where $J(t) = \det(DT_t)$. On Σ_3 $J(t)$ is to be understood as $\det(D\tilde{T}_t)$ where $\tilde{T}_t$ is a m apping from $\mathbb{R}^2$ in $\mathbb{R}^2$, namely

$$\tilde{T}_t(x_1,\ x_2) = T_t(x_1,x_2,L)$$

and $D\tilde{T}_t$ is the 2×2 matrix. In fact $\tilde{T}_t$ is the transformation associated with the velocity field

$$\tilde{V}(t,x_1,x_2) = (V_{x_1}(t,x_1,x_2,L),\ V_{x_2}(t,x_1,x_2,L))$$

For $t \in [0,T]$ and any matrix norm we have

$$\left.\begin{array}{ll} ||A(t,x)|| \leq c\ ||T_t||_{W^{1,\infty}(\Omega)} , & \forall x \in \bar{\Omega} \\ |J(t)| \leq c\ ||T_t||_{W^{1,\infty}(\Omega)} , & \forall x \in \bar{\Omega} \end{array}\right\} \quad (3)$$

Now the norm $||T^t||_{W^{1,\infty}(\Omega)}$ is continuous in t and a fortiori bounded in $[0,T]$. We also recall (from J.P. ZOLESIO [1]) that the following continuity properties

$$\left.\begin{array}{ll} ||A(t) - A(s)||_{L^\infty(\Omega)} \rightarrow 0 & \text{when } s \rightarrow t \\ |J(t) - J(t)|_{L^\infty(\Omega)} \rightarrow 0 & \text{when } s \rightarrow t \end{array}\right\} \quad (4)$$

Denote by Ω_t the perturbated domain with its boundary $\Omega_t = T_t(V)\ (\Omega)$ in three pieces :

$$\Sigma_i^t = T_t(V)\ (\Sigma_i)\ ,\ 1 \leq i \leq 3$$

But from (1) $\Sigma_1^t = \Sigma_1$.

For each t in $[0,t]$ we consider the Banach space

$$B_t = B(\Omega_t) = \{\varphi \in H^1(\Omega_t) : \varphi|_{\Sigma_3^t} \in L^5(\Sigma_3^t)\}$$

and the energy functional defined on this space :

$$E_t(\varphi) = \int_{\Omega_t} \frac{1}{2}\ |\nabla\varphi|^2\ dx + \int_{\Sigma_3^t} (\frac{1}{5}\ |\varphi|^5 - \varphi d_s) d\Sigma - \int_{\Sigma_1} q_1 \varphi\ d\Sigma$$

E_t is convex, lower semi-continuous on B_t and there exists a unique element $y_t \in B_t$ which minimizes E_t on B_t (see M.C. DELFOUR, G. PAYRE, J.P. ZOLESIO [1,2]).

The cost function associated to the radiator problem is $F_t : B_t \rightarrow \mathbb{R}^+$ defined by

$$F_t(\varphi) = \int_{\Omega_t} [(\varphi - T_1)^+]^2\ dx$$

Then a unique element $y_{\varepsilon,t} \in B_t$ minimizes on B_t the penalized energy :

$$\varepsilon > 0 \ , \quad E_t(y_{\varepsilon,t}) + \varepsilon\, F_t(y_{\varepsilon,t}) \leq E_t(\varphi) + \varepsilon\, F_t(\varphi), \ \forall\, \varphi \in B_t \tag{5}$$

Consider the function $\hat{y} = \max\,(y_{\varphi,t},\ q_s^{\frac{1}{4}})$ and assume that

$$T_1 > q_s^{\frac{1}{4}} \tag{6}$$

then

$$(\hat{y} - T_1)^+ = (y - T_1)^+$$

and $F_t(\hat{y}) = F_t(y_{\varepsilon,t})$; from M.C. DELFOUR, G. PAYRE, J.P. ZOLESIO [1,2] we then know that

$$E_t(\hat{y}) + \varepsilon\, F_t(\hat{y}) \leq E_t(y_{\varepsilon,t}) + \varepsilon\, F_t(y_{\varepsilon,t})$$

By uniqueness of the minimum in (5) we get $\hat{y} = y_{\varepsilon,t}$ that is :

$$y_{\varepsilon,t} \geq q_s^{\frac{1}{4}} \quad \text{on} \quad \Omega_t \tag{7}$$

It is immediate that

$$\Delta\, y_{\varepsilon,t} = \varepsilon\, (y_{\varepsilon,t} - T_1)^+ \quad \text{in} \quad \Omega_t \tag{8}$$

Then, $\Delta\, y_{\varepsilon,t}$ being in $L^2(\Omega_t)$, $\frac{\partial}{\partial n}\, y_{\varepsilon,t}$ is defined on $H^{-\frac{1}{2}}\,(\partial\Omega_t)$ and

$$\frac{\partial}{\partial n}\, y_{\varepsilon,t} = 0 \quad \text{on } \Sigma_2^t \tag{9}$$

$$\frac{\partial}{\partial n}\, y_{\varepsilon,t} = q_i \quad \text{on} \quad \Sigma_1 \tag{10}$$

And the radiating (non linear) condition

$$\frac{\partial}{\partial n}\, y_{\varepsilon,t} + (y_{\varepsilon,t})^4 = q_s \quad \text{on} \quad \Sigma_3^t \tag{11}$$

We suppose now that $y_{\varepsilon,t}|\Sigma_3^t$ has an upper bound (which is compatible with the fact that, from (11) and (7) $\frac{\partial}{\partial n}\, y_{\varepsilon,t} \leq 0$ on Σ_3^t).

It can be easily verified that $y_{\varepsilon,t}$ is continuous outside of $\overline{\Sigma_3^t}$ in $\bar{\Omega}_t$: for example by introducing, for any $\alpha > 0$, the function

$$g_\alpha(x_1,x_2,z) = y_{\varepsilon,t}\,(x_1,x_2,z)\ \rho_\alpha(z)$$

where $0 \leq \rho_\alpha \leq 1$ is a C^∞ function on $[0,L]$ such that $\rho(z) = 1$

for $0 \leq z \leq L-2\alpha$, and $\rho(z) = 0$ for $L-\alpha \leq z \leq L$. In particular $g_\alpha = y_{\varepsilon,t}$ in a neighbourhood of Σ_1 ; we have

$$g_\alpha = 0 \quad \text{on} \quad \{z = L-\alpha\} \cap \bar{\Omega}_t$$

$$\frac{\partial}{\partial n} g_\alpha = 0 \quad \text{on} \quad \Sigma_2^t \cap \{z \leq L-\alpha\}$$

$$\frac{\partial}{\partial n} g_\alpha = q_i \quad \text{on} \quad \Sigma_1$$

and

$$\Delta g_\alpha = \rho_\alpha''(z)\, y_{\varepsilon,t} + 2\, \rho_\alpha' \frac{\partial}{\partial z} y_{\varepsilon,t}$$

belongs to $L^2(\Omega_t)$

Then g_α is the solution of a linear well posed boundary problem on $\Omega_t \cap \{z < L-\alpha\}$ and we know that $g_\alpha \in C^0(\bar{\Omega}_t)$; then by the Maximum Principle (see Protter and Weinberger [1]) we know that the maximum for g_α on $\bar{\Omega}_t$ is achieved at a boundary point M at which $\frac{\partial}{\partial n} g\,(M) > 0$.

This point M can only be located on Σ_1. Then for each $\alpha > 0$, $y_{\varepsilon,t} \mid \Omega_t \cap \{z < L-2\,\alpha\}$ reaches its maximum on Σ_1. But since $y_{\varepsilon,t}$ is upper bounded on Σ_3^t we also have $y_{\varepsilon,t}$ reaching its maximum on Σ_1. Now it would be possible to obtain the continuity with respect to (t,ε) of $\max\{y_{\varepsilon,t}(x) : x \in \Sigma_1\} = \max\{g_\alpha : x \in \Sigma_1\}$. Thus this maximum is bounded for $(t,\varepsilon) \in [0,T] \times [0,\bar{\varepsilon}]$:

$$\left.\begin{array}{l} \exists M,\ \forall\, \varepsilon \in [0,\bar{\varepsilon}],\ \forall\, t \in [0,T],\ \forall\, x \in \bar{\Omega}_t \\ q_s^{1/4} \leq y_{\varepsilon,t}(x) \leq M \end{array}\right\} \tag{12}$$

Consider now

$$y_\varepsilon^t = y_{\varepsilon,t} \circ T_t\ . \tag{13}$$

It is the unique element of $B(\Omega) = B_o$ which minimizes on B_o the functional $E(t,\varphi) + \varepsilon\, F(t,\varphi)$
where

$$E(t,\varphi) = E_t(\varphi \circ T_t^{-1})$$
$$= \frac{1}{2} \int_\Omega <A(t).\nabla\varphi,\nabla\varphi> dx + \int_{\Sigma_3} (\frac{1}{5} |\varphi|^5 - q\,\varphi)\, J(t)\, d\Sigma \tag{14}$$
$$- \int_{\Sigma_1} q_i \varphi \, d\Sigma$$

and

$$F(t,\varphi) = F_t(\varphi \circ T^{-1}) = \int_\Omega [(\varphi - T_1)^+]^2 J(t)\, dx . \tag{15}$$

Obviously from (45) and (50), we have

$$M_e = q_1^{1/4} \leq y_\varepsilon^t(x) \leq M \ , \ \forall\, x \in \bar{\Omega},\ \forall\, t \in [0,T],\ \forall\, \varepsilon \in [0,\bar{\varepsilon}] \tag{16}$$

To obtain the coercivity of the second derivative $\varphi \mapsto d^2E(t,y_o^t\ ;\ 0,\varphi\ ;\ 0,\varphi)$ we need now to introduce the closed convex subset of $B(\Omega)$:

$$K = \{\varphi \in B(\Omega) : \frac{u_e}{2} \leq \varphi \leq M \quad \text{a.e. on} \quad \bar{\Omega}_t\} \tag{17}$$

From (16) we get $y_\varepsilon^t \in K$ for any ε and t.
We now turn to the verification of hypothesis H8, the continuity of $t \mapsto y_\varepsilon^t$ in $B(\Omega)$; $\varepsilon > 0$

<u>Lemma 1</u>. $\exists\, C > 0$, s.t. $\forall\, t \in [0,T]$, $\forall\, \varepsilon \in [0,\bar{\varepsilon}]$,

$$||y^t||_{B(\Omega)} \leq C \tag{18}$$

<u>Proof</u>. We have $||y^t||_B \leq ||y_{\varepsilon,t}||_{B_t} \quad ||T_t||_{W^{1,\infty}(\Omega)}$

But, $E_t(y_{\varepsilon,t}) \leq E_t(0) = 0$,

that is

$$\int_{\Omega_t} \frac{1}{2} |\nabla y_{\varepsilon,t}|^2 dx + \frac{1}{5} \int_{\Sigma_3^t} |y_{\varepsilon,t}|^5 d\Sigma$$
$$\leq \int_{\Sigma_1} \{|y_{\ ,t}|\, q_i + \varepsilon[(y_{\varepsilon,t} - T_1)^+]^2\}\, d\Sigma$$

By (16) we get $\leq [C + \varepsilon(C - T_1)^2]$ measure $(\Sigma_1) = a$

Then it is immediate that : $||y_\varepsilon^t||_{B_t} \leq \sqrt{a} + a^{1/5}$. ■

<u>Lemma 2</u>. $\forall\, \varepsilon \geq 0,\ ||y_\varepsilon^s - y_\varepsilon^t||_{B(\Omega)} \to 0$ as $s \to t$.

<u>Proof</u>. y_ε^s and y_ε^t are the two elements of B characterized by the variational equations :

$$\forall\, \varphi \in B,\quad dE(s,y_\varepsilon^s\ ;\ 0,\varphi) + \varepsilon\, dF(s,y_\varepsilon^s\ ;\ 0,\varphi) = 0$$

$$dE(y,t_\varepsilon^t\ ;\ 0,\varphi) + \varepsilon\, dF(t,y_\varepsilon^t\ ;\ 0,\ ) = 0$$

By substracting these equations, taking $z = y_\varepsilon^t - y_\varepsilon^s$ and $\varphi = z$ we get :

$$\begin{aligned} &\int_\Omega <A(t).\nabla z,\nabla z>dx + \int_{\Sigma_3} J(t)\,[(y_\varepsilon^t)^4 - (y_\varepsilon^s)^4]z\, d\Sigma \\ &+ \varepsilon \int_\Omega [(y_\varepsilon^t - M)^+ - (y_\varepsilon^s - M)^+]\, J(t)\, dx \\ &= - \int_\Omega <(A(t)-A(s)).\ \nabla y_\varepsilon^s,\ \nabla z>\, dx \\ &- \int_{\Sigma_3} (J(t)-J(s))\,(y_\varepsilon^s)^4 z\, d\Sigma - \varepsilon \int_\Omega (J(t)-J(s))(y_\varepsilon^s - M)^+ z\, dx \end{aligned} \qquad (19)$$

From (18) and (3), (4) it can easily be verified that the Right-Hand-Side of (19) goes to zero as s goes to t.

On the other side we have the monotony inequalities $(a^4-b^4)(a-b) \geq \frac{1}{8}\, |a-b|^5$ and

$$[(a-T_1)^+ - (b-T_1)^+](a-b) \geq 0\ ,$$

combining these two inequalities with the fact that $J(t) \geq 0$ on $\bar{\Omega}$ (for $J(t) \to 1$ in $C^0(\bar{\Omega})$ when $t \to 0$) we get in (19) :

$$\int_\Omega < A(t).\nabla z,\ \nabla z>\, dx \to 0\ ,\ s \to t$$

and

$$\int_{\Sigma_3} J(t)\ |z|^5\ d_{\Sigma_3} \to 0\ ;\ s \to t$$

Now going back to the moving domain Ω_t we get $||z \circ T_t^{-1}||_{B(\Omega_t)} \to 0$ but

$$||z \circ T_t^{-1}||_{B(\Omega_t)} \geq ||T_t||^{-1}_{W^{1,\infty}(\Omega)}\ ||z||_{B(\Omega)}$$

Then we get $||z||_{B(\Omega)} \to 0$ as $s \to T$. ■

5.1 Derivatives of E and F.

We recall (from J.P. ZOLESIO [1], [2]) that $t \mapsto A(t)$ and $t \mapsto J(t)$ are differentiable from $[0,T]$ in $L^{\infty}(\Omega)$ and that the derivatives are given by

$$A'(t) = \operatorname{div} V(t)\, I_d - (DV(t) + {}^{*}DV(t))$$

$$J'(t) = \operatorname{div} V(t).$$

Then for all φ in B we get the existence of

$$dE(t,\varphi\ ;\ 1,0) = \int_{\Omega} \frac{1}{2} <A'(t).\nabla\varphi,\nabla\varphi>\, dx + \int_{\Sigma_3} (\frac{1}{5}\, |\varphi|^5 - q_s\varphi)\, J'(t) d\Sigma$$

also we have, for $\varphi,\psi \in B(\Omega)$:

$$dE(t,\varphi\ ;\ 0,\psi) = \int_{\Omega} <A(t).\nabla\varphi,\nabla\psi>\, dx + \int_{\Sigma_3} (|\varphi|^3\varphi - q_s)\psi\, J(t)\, d\Sigma$$

and for $\varphi \in K,\ \xi,\psi \in B$

$$d^2E(t;\varphi;0,\psi;0,\xi) = \int_{\Omega} <A(t).\nabla\xi,\nabla\psi>\, dx + 4 \int_{\Sigma_3} |\varphi|^3\, \psi\xi\, J(t)\, d\Sigma.$$

Moreover :

$$d^2E(t,\varphi\ ;\ 0,\psi\ ;\ 0,\psi) \geq \int_{\Omega_t} |\nabla (\psi o T_t^{-1})|^2 dx + \frac{u_e^3}{2} \int_{\Sigma_3^t} (\psi o T_t^{-1})^2\, d\Sigma$$

$$\geq \operatorname{Min}\, (1, \frac{u_e^3}{2})\ ||\psi \text{ o } T_t^{-1}||_{H^1(\Omega_t)}$$

$$\geq \operatorname{Min}\, (1, \frac{u_e^3}{2})\ ||T_t||^{-1}_{W^{1,\infty}(\Omega)}\ ||\psi||^2_{H^1(\Omega)}$$

5.2 Characterization of the convex set S_t

The gradient of $E(t,.)$ at y_o^t is zero for y_o^t minimizes $E(t,0)$ on all the Banach space B, that is $dE(t,\, y_o^t\ ;\ 0,\varphi) = 0$, $\forall\, \varphi \in B$. Then :

$$\{\varphi \in V \text{ s.t. } dE(t,y_o^t\ ;\ 0,\varphi) = 0\} = H^1(\Omega)$$

Then to characterize S_t we just have to consider the tangent cone for this we have the

Lemma 3. $$T_{y_o^t}(K) = H^1(\Omega)\ .$$

Proof. We first obtain

$$\{\lambda(\varphi - y_o^t) \text{ s.t. } \lambda \geq 0, \varphi \in K\} = L^\infty(\Omega) \cap H^1(\Omega)$$

for $K \subset L^\infty(\Omega)$ and y_o^t an interior point (in $L^\infty(\Omega)$ topology to K). Then we conclude by density of $L^\infty(\Omega) \cap H^1(\Omega)$ in $H^1(\Omega)$. ■

We turn now to the verification of the hypothesis H5, H6 and H 16.

Let $p_n = p_\varepsilon^{t_n}$ converge weakly in $H^1(\Omega)$ to q (since p_ε^t is bounded in $H^1(\Omega)$, from Lemma 1, independently on $\varepsilon \geq 0$ and t).

Then this convergence is true in $H^s(\Omega)$, strongly for any $S < 1$ and the traces on Σ_3 converges in $H^{s-\frac{1}{2}}(\Sigma_3)$ then in $L^\alpha(\Sigma_3)$ for any $\alpha < 4$. In particular $(p_n)^2$ converges to q^2 strongly in $L^{3/2}(\Sigma_3)$. To verify H5, H6 and H16 it is now a direct application of the following.

Lemma 4. $\forall \varepsilon \geq 0$, for any sequence $t_n \to s$ there exists a subsequence t_m such that

$$y_\varepsilon^{t_m}|_{\Sigma_3} \to y_\varepsilon^t|_{\Sigma_3} \quad \text{in } L^p(\Sigma_3),\ m \to \infty,$$

for any p ; $1 \leq p < \infty$

(This subsequence converges in all the $L^p(\Sigma_3)$'s).

Proof : We have established that $y_\varepsilon^{t_n}$ converges to y_ε^o in $B(\Omega)$; then the traces on Σ_3 converges in $L^5(\Sigma_3)$. So there exists a subsequence which converges almost every where on Σ_3. But

$$|y_\varepsilon^{t_n}|_{\Sigma_3}| \leq M \quad ;$$

so this subsequence, written y^m for simplicity, verifies

$$|y^m|^p \to |y_\varepsilon^t|^p \quad \text{a.e. on } \Sigma_3$$

$$|y^m|^p \leq M^p \quad \text{a.e. on } \Sigma_3$$

By the Lebesgue convergence theorem we get the convergence of $|y^m|^p$ to $|y_\varepsilon^t|^p$ in $L^1(\Sigma_3)$ that is that y^m converges to y_ε^t in $L^p(\Sigma_3)$. ■

How Proposition 7 (in DELFOUR-PAYRE-ZOLESIO [1]) can be directly applied to the radiator problem and we get the

Theorem 4. The domain Ω being described in the first section, let $y(\Omega) \in B(\Omega)$ be the solution of

$$\operatorname*{Min}_{\varphi \in B(\Omega)} \int_\Omega \frac{1}{2} |\nabla\varphi|^2 dx + \int_{\Sigma_3} (\frac{1}{5} |\varphi|^5 - \varphi q_s) d\Sigma - \int_{\Sigma_1} q_i \varphi \, d\Sigma$$

For any admissible velocity field V (such that (39), (40)) let $y(\Omega_t)$ be the associated solution on Ω_t and

$$J(\Omega_t) = \int_{\Omega_t} [(y(\Omega_t) - T_1)^+]^2 \, d\Sigma$$

with $T_1 > q_s^{\frac{1}{4}}$.

Then the Eulerian derivative of J at Ω in the direction $V \in C^0([0,T], C^1(\mathbb{R}^3 ; \mathbb{R}^3))$ exists and is given by

$$dJ(\Omega ; V) \overset{\text{def}}{=} \lim_{t \searrow 0} (J(\Omega_t) - J(\Omega))/t$$

$$= \int_\Omega (y-T_1)^+ p \, dx + \int_\Omega \langle A'(0).\nabla y, \nabla p\rangle \, dx + \int_{\Sigma_3} J'(0)(y^4 - q_s) p \, d\Sigma$$

$$+ \frac{1}{2} \int_\Omega J'(0) \, [(y-T_1)^+]^2 dx$$

where $y = y(\Omega)$ and $p = p(\Omega)$ are respectively the element of $B(\Omega)$ and $H^1(\Omega)$ characterized by the problems

$$\int_\Omega \nabla y \, \nabla\varphi \, dx + \int_{\Omega_3} (|y|^3 y - q_s)\varphi \, d = \int_{\Sigma_1} q_i \, \varphi \, d \; , \; \forall \varphi \in B$$

$$\int_\Omega \nabla p \, \nabla\psi \, dx + 4 \int_{\Sigma_3} y^3 \, p\psi \, d\Sigma = \int_\Omega (y-T_1)^+ \psi \, d\nabla \; , \; \forall \psi \in H^1(\Omega)$$

and

$$A'(0) = \operatorname{div} V(0) \, I_d - (DV(0) + {}^*DV(0))$$

$$J'(0) = \partial_{x_1} V_{x_1}(0, x_1, x_2, L) + \partial_{x_2} V_{x_2}(0, x_1, x_2, L) \text{ on } \Sigma_3 .$$

R E F E R E N C E S

M.C. DELFOUR and J.P. ZOLESIO

[1], Dérivation d'un Min Max et application à la dérivation par rapport au contrôle d'une observation non différentiable de l'état, C.R. Acad. Sc. Paris 302 (1986), 571-574.

[2], Différentiability of a Min Max and Application to Optimal Control and Design Problems, Parts I and II, in Proc. IFIP WG 7.2 on Control Problems for Systems described by Partial Differential Equations, I. Lasiecka, ed. Gainesville, Flo. Fev. 1986, Optimization Software Inc., New-York, to appear.

[3], Shape sensitivity analysis via Min Max differentiability, October 1986.

M.C. DELFOUR, G. PAYRE and J.P. ZOLESIO

[1], Approximation of non-linear problems associated with radiating bodies in space, SIAM J. on Numerical Analysis, to appear.

[2], Shape optimal design of a radiating fin, in "Systems Modelling and Optimization", P. Thoft-Christensen, ed., pp.810-818, Springer-Verlag, Berlin, New-York 1984.

J. EKELAND and R. TEMAN

[1], Analyse convexe et problèmes variationnels, Dunod, Gauthier-Villars, Paris, Bruxelles, Montréal, 1974.

F. MIGNOT

[1], Contrôle dans les inéquations variationnelles elliptiques, J. Funct. Anal. 22 (1976), 130-185.

F. MIGNOT and J.P. PUEL

[1], Contrôle optimal d'un système gouverné par une inéquation variationnelle parabolique, C.R. Acad. Sc. Paris 298 (1984), Série I, 277-280.

[2], Optimal control in some variational inequalities, SIAM J. Control and Optim. 22 (1984), 466-476.

M.H. PROTTER and H.F. WEINBERGER

[1], Maximum principle in differential equations, Englewood Cleffs, N.J., Prentice-Hall, 1967.

SHI SHUZ HONG

[1], Optimal control of strongly monotone variational inequalities, Report CRM - 1346, Centre de Recherches Mathématiques, Université de Montréal, Montréal, Canada, February 1986.

J. SOKOLOWSKI

[1], Conical differentiability of projection on convex sets - an application sensitivity analysis of Signorini variational inequality, Technical Report, Institute of Mathematics of the University of Genova, 1981.

[2], Sensititivy analysis of contact problems with adhesive friction, Technical Report, University of Florida, Gainesville, Florida, 1986.

J. SOKOLOWSKI and J.P. ZOLESIO

[1], Dérivée par rapport au domaine de la solution d'un problème unilatéral, C.R. Acad. Sc. Paris, Sér. I 301 (1985), 103-106.

[2], Shape sensitivity analysis of unilatéral problems, SIAM J. on Math. Anal. 18 (1987).

J. ZOLESIO

[1], Identification de domaine. Thèse de doctorat d'état, Nice France, 1979.

[2], The material derivative, in "Optimization of Distributed Parameter Structures", E.J. Hang and J. Céa, eds. pp. 1457-1473, Sijchoff and Noordhoff, Alphen aan den Rijn, The Netherlands, 1980.

[3], Differentiability of repeated eigenvalue, in the same proceedings then [2].

On the design of the optimal covering of an obstacle

JAROSLAV HASLINGER AND PEKKA NEITTAANMÄKI

Abstract. We consider the problem of controlling the shape of the coincidence set in an obstacle problem. This so called packaging problem was introduced in [2] by Benedict, Sokolowski and Zolesio. In this work we prove the existence of at least one solution for the problem. Moreover, the validity of the penalty method for imposing the state constraints is shown. Sensitivity analysis shows that the discretized problem is only directionally differentiable. Numerical examples are presented.

1. INTRODUCTION

We shall consider the shape optimization problem introduced by Benedict, Sokolowski and Zolesio in [2]. Consider a membrane in possible contact with a rigid obstacle (see Fig. 1.1).

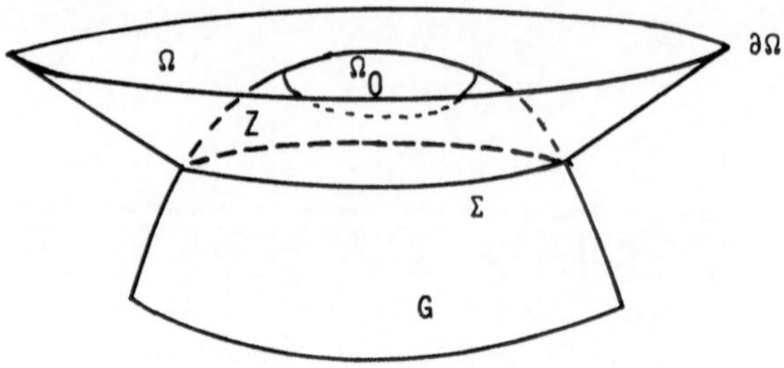

Fig. 1.1

Let φ describe the shape of the obstacle G. The system is covered by a partial differential inequality system:

$$-\Delta u(x) \geq f(x)\,, \qquad x \in \Omega \tag{1.1}$$

$$u(x) \geq \varphi(x)\,, \qquad x \in \Omega \tag{1.2}$$

$$(-\Delta u(x) - f(x))(u(x) - \varphi(x)) = 0\,, \qquad x \in \Omega \tag{1.3}$$

$$u(x) = 0\,, \qquad x \in \partial\Omega\,. \tag{1.4}$$

Above, u is the vertical displacement of the membrane and f is a given vertical force. Inequality (1.2) is the nonpenetrating condition. For any solution of the

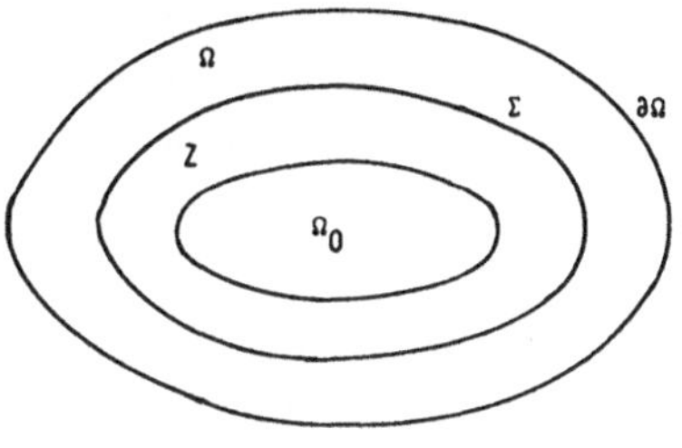

Fig. 1.2

problem (1.1)–(1.4) we may define the contact region Z as the subdomain of Ω, where $u = \varphi$ i.e. where the contact takes place. Since Σ, the boundary of this subdomain Z, is unknown before the contact problem is solved, it is called a free boundary. We suppose as in [2] that $\partial\Omega \cap \Sigma = \emptyset$ (see Fig. 1.2).

By the packaging problem we mean the design problem of minimizing the area of Ω on the condition that the contact range Z contains some specific subdomain Ω_0. So, our design problem reads

$$\begin{aligned} &\text{minimize meas}(\Omega) \\ &\text{subject to } Z \supset \Omega_0 \ . \end{aligned} \tag{1.5}$$

The main difficulty in the numerical realization of (1.5) is the presence of the state constraint $u = \varphi$ on Ω_0. One possibility to overcome this difficulty is to use the penalty approach. One can handle the state constraint also by using an appropriate indicator function (variational inequality approach [6], [7]). We shall utilize here the penalty approach. For convenience, we shall assume a special family of Ω (for example a sector of original Ω, say $\Omega(\alpha)$) which can be parametrized by a control parameter α from the set of admissible controls U_{ad}.

The paper is organized as follows. In chapter 2 we reformulate the problem (1.5) as a problem of finding the optimal parameter α in the sense of (1.5). We shall prove in chapter 2 (Th. 2.1) that if the set of admissible controls U_{ad} is nonempty then the packaging problem has at least one solution. In chapter 3 a penalty approach is applied. The penalized problem has at least one solution for all penalty parameters $\varepsilon > 0$ (Th. 3.1). In Th. 3.2 we prove a certain relation between the penalized problem and the original problem when ε tends to zero. In chapter 4 an approximation of the penalized problem by finite elements is discussed and a relation between the continuous and discrete problem is presented in Th. 4.1. Chapter 5 contains the sensitivity analysis of the problem by means of discrete material derivatives. We shall derive a formula for the directional derivative of the penalized cost function. It turns out that – opposite to the continuous case – in the discrete case the cost functional is not differentiable in general. Finally, chapter 6 contains some numerical examples.

This paper deals with a model case. The approach can be extended to many models simulated by variational inequalities (contact problems, dam problems, semiconductors etc.). In these fields the interpretation of the "packaging problem" is quite different. The goal of this work was just to give the methodological tools for handling these problems of great practical inportance.

2. Setting of the problem

Let $\Omega(\alpha) \subset \mathbf{R}^2$ be a bounded domain, given by

$$\Omega(\alpha) = \{(x_1, x_2) \in \mathbf{R}^2 \mid x_2 \in]0,1[\ ,\ 0 < x_1 < \alpha(x_2)\}\ ,$$

where $\alpha \in U_{ad}$ is a function describing the moving part $\Gamma(\alpha)$ of the boundary $\partial\Omega(\alpha)$:

$$\Gamma(\alpha) = \{(x_1, x_2) \mid x_1 = \alpha(x_2),\ x_2 \in]0,1[\}$$
$$\alpha \in U_{ad} = \{\alpha \in C^{0,1}([0,1]) \mid \hat{\alpha}_0 \le \alpha \le \hat{\beta}_0\ ,\ |\alpha'(x_2)| \le c_0\}\ ,$$

where $C^{0,1}$ denotes the set of Lipschitz functions and $\hat{\alpha}_0, \hat{\beta}_0, c_0$ are positive constants chosen in such a way that $U_{ad} \neq \emptyset$.

Notations $H^k(\Omega)$ ($k \ge 0$, integer) are used for classical Sobolev spaces of functions, the generalized derivatives of which up to the order k are square integrable in Ω; especially $L^2(\Omega) = H^0(\Omega)$. The norm (scalar product) of $H^k(\Omega)$ will be denoted by $\|\cdot\|_{k,\Omega}$ $((\cdot,\cdot)_{k,\Omega})$. The subspace $H^1_0(\Omega)$ of $H^1(\Omega)$ characterizes the homogeneous Dirichlet boundary condition $u = 0$ on $\partial\Omega$.

Let $\alpha \in U_{ad}$ be fixed. On any $\Omega(\alpha)$ we shall consider the following free boundary value problem:

$$\begin{cases} \text{find } u(\alpha) \in K(\Omega(\alpha)) \text{ such that} \\ (\nabla u(\alpha), \nabla(v - u(\alpha)))_{0,\Omega(\alpha)} \ge (f, v - u(\alpha))_{0,\Omega(\alpha)} \quad \forall v \in K(\Omega(\alpha))\ , \end{cases} \tag{P(α)}$$

where

$$K(\Omega(\alpha)) = \{v \in H^1_0(\Omega(\alpha)) \mid v \ge \varphi \text{ a.e. in } \Omega(\alpha)\}\ ,$$

$f \in L^2(\hat{\Omega})$, $\hat{\Omega} =]0, \hat{\beta}_0[\times]0,1[$.Concerning φ we suppose that $\varphi \in H^1(\hat{\Omega})$ is a given function such that $\varphi \le 0$ on $\partial\hat{\Omega}$ and in $]\hat{\alpha}_0, \hat{\beta}_0[\times]0,1[$.

Using Green's formula to (P(α)), we formally obtain the following relations for $u(\alpha)$:

$$\left.\begin{aligned} &-\Delta u(\alpha) \ge f \\ &u(\alpha) \ge \varphi,\ (-\Delta u(\alpha) - f)(u - \varphi) = 0 \end{aligned}\right\} \text{ a.e. in } \Omega(\alpha)$$
$$u(\alpha) = 0 \quad \text{on } \partial\Omega(\alpha)\ .$$

The set

$$Z(u(\alpha)) = \{x \in \Omega(\alpha) \mid u(\alpha)(x) = \varphi(x) \text{ a.e. }\}$$

will be called the contact region.

REMARK 2.1: By $(P(\alpha))$ we model under a vertical load f the deflection of a membrane that may come in contact with a rigid obstacle, described by a function φ.

In what follows we shall analyse the existence and the approximation of the solution to the so called packaging problem – the design problem of minimizing the area of $\Omega(\alpha)$, $\alpha \in U_{ad}$, such that the contact region $Z(u(\alpha))$ of the corresponding solution $u(\alpha)$ contains the specific region Ω_0. The mathematical formulation of this problem reads as follows:

$$\begin{cases} \text{find } \alpha^* \in \tilde{U}_{ad} \text{ such that} \\ J(\alpha^*) \leq J(\alpha) \quad \forall \alpha \in \tilde{U}_{ad} , \end{cases} \tag{P}$$

where

$$\tilde{U}_{ad} = \{\alpha \in U_{ad} \mid Z(u(\alpha)) \supseteq \Omega_0\}$$
$$J(\alpha) = \text{ meas } \Omega(\alpha) .$$

$\Omega(\alpha)$ is a domain over which $(P(\alpha))$ is solved. The main result of this section is

THEOREM 2.1. *Let $\tilde{U}_{ad} \neq \emptyset$. Then there exists at least one solution of* (P).

Before we prove this Theorem, we shall need some auxiliary results. In the sequel, the symbol $\tilde{v}$ will denote the extension of $v \in H_0^1(\Omega(\alpha))$ by zero from $\Omega(\alpha)$ on $\hat{\Omega}$.

LEMMA 2.1. *Let $\alpha_n \rightrightarrows \alpha$ (uniformly) in $[0,1]$ and let $v \in K(\Omega(\alpha))$ be given. Then there exist functions $V_j \in H^1(\hat{\Omega})$ and a subsequence $\{\alpha_{n(j)}\} \subset \{\alpha_n\}$ such that*

(i) $V_j \to \tilde{v}$, $\quad j \to \infty$ *in* $H^1(\hat{\Omega})$;
(ii) $V_j|_{\Omega_{n(j)}} \in K(\Omega_{n(j)}) \qquad (\Omega_{n(j)} \equiv \Omega(\alpha_{n(j)}))$.

PROOF: Let $v \in K(\Omega(\alpha))$ be given. As $v \in H_0^1(\Omega(\alpha))$, one can find $w_j \in \mathcal{D}(\Omega(\alpha))$ [1] such that

$$w_j \to v , \quad j \to \infty \qquad \text{in } H^1(\Omega(\alpha))$$

and

$$\tilde{w}_j \to \tilde{v} , \quad j \to \infty \qquad \text{in } H^1(\hat{\Omega}) . \tag{2.1}$$

Let

$$V_j = \sup\{\tilde{w}_j, \varphi\} .$$

[1] Let $C^\infty(\Omega)$ denote the set of all infinite times continuously differentiable functions and $\mathcal{D}(\Omega)$ the set of C^∞-functions which vanish in some neighbourhood of $\partial\Omega$.

Then $V_j \in H_0^1(\hat{\Omega})$, $V_j \geq \varphi$ a.e. in $\hat{\Omega}$. As the application $\sup\{\cdot,\cdot\}$ is continuous with respect to the $H^1(\hat{\Omega})$ norm, it follows from (2.1) and the properties of φ that

$$V_j \to \sup\{\tilde{v}, \varphi\} = \tilde{v} \ . \tag{2.2}$$

Thus (i) is proved. Let j_0 be fixed and denote $G_{j_0} = \text{supp}\, w_{j_0}$. As $\alpha_n \overrightarrow{\rightarrow} \alpha$ in $[0,1]$, there exists $n(j_0)$ such that $\Omega_{n(j_0)} \supset G_{j_0}$ and $V_{j_0}|_{\partial\Omega_{n(j_0)}} = 0$. Hence $V_{j_0}|_{\Omega_{n(j_0)}} \in K(\Omega_{n(j_0)})$ and (ii) is proved as well. □

REMARK 2.2: Sequence $\{V_j\}$, satisfying (i), (ii) can be chosen to be more regular. More precisely: there exists $\{W_j\}$, $W_j \in C^\infty(\overline{\hat{\Omega}})$, $W_j|_{\Omega_{n(j)}} \in \mathcal{D}(\Omega_{n(j)})$ such that (i), (ii) is satisfied.

Proof follows immediately from the fact that $\mathcal{D}(\Omega(\alpha_n)) \cap K(\Omega(\alpha_n))$ is dense in $K(\Omega(\alpha_n))$ with respect to $H^1(\Omega(\alpha_n))$ norm (see [3]). □

THEOREM 2.2. *Let $\alpha_n \overrightarrow{\rightarrow} \alpha$ in $[0,1]$ and let $u_n = u(\alpha_n)$ be solutions of* $(\mathrm{P}(\alpha_n))$. *Then there exist: a subsequence $\{u_{n''}\} \subset \{u_n\}$ and $U \in H_0^1(\hat{\Omega})$ such that*

$$\tilde{u}_{n''} \to U \ , \ n'' \to \infty \qquad \text{in } H^1(\hat{\Omega})\text{-norm}$$

and $u(\alpha) \equiv U|_{\Omega(\alpha)}$ solves $(\mathrm{P}(\alpha))$.

PROOF: It is easy to see that there exists a constant $c > 0$ such that

$$\|u_n\|_{1,\Omega_n} \leq c \qquad \forall n \ . \tag{2.3}$$

As

$$\|\tilde{u}_n\|_{1,\hat{\Omega}} = \|u_n\|_{1,\Omega_n} \leq c \ ,$$

one can extract a subsequence $\{\tilde{u}_{n'}\} \subset \{u_n\}$ such that

$$\tilde{u}_{n'} \rightharpoonup U \qquad \text{in } H_0^1(\hat{\Omega}) \ . \tag{2.4}$$

We show that $U|_{\Omega(\alpha)}$ solves $(\mathrm{P}(\alpha))$.

First of all we prove that $U|_{\Omega(\alpha)} \in K(\Omega(\alpha))$, i.e. $U|_{\Omega(\alpha)} \in H_0^1(\Omega(\alpha))$ and $U|_{\Omega(\alpha)} \geq \varphi|_{\Omega(\alpha)}$ a.e. in $\Omega(\alpha)$. Using (2.4), the fact that $\tilde{u}_{n'}|_{\hat{\Omega}\setminus\Omega_{n'}} \equiv 0$ and $\alpha_{n'} \overrightarrow{\rightarrow} \alpha$ in $[0,1]$ we easily obtain that $U|_{\hat{\Omega}\setminus\Omega(\alpha)} \equiv 0$ which implies $U|_{\Omega(\alpha)} \in H_0^1(\Omega(\alpha))$. On the other hand, $\tilde{u}_{n'} \geq \varphi$ a.e. in $\hat{\Omega}$. This and (2.4) give $U \geq \varphi$ a.e. in $\hat{\Omega}$.

Let us show that $U|_{\Omega(\alpha)}$ solves $(\mathrm{P}(\alpha))$. Let $v \in K(\Omega(\alpha))$. According to Lemma 2.1 there exist: a sequence $V_j \in H^1(\hat{\Omega})$ and a subsequence $\{\alpha_{n(j)}\} \subset \{\alpha_{n'}\}$ such that

$$V_j \to \tilde{v} \qquad \text{in } H^1(\hat{\Omega}) \ ; \tag{2.5}$$

$$V_j|_{\Omega_{n(j)}} \in K(\Omega_{n(j)}) \ . \tag{2.6}$$

Following the definition of $(\mathrm{P}(\alpha_{n(j)}))$:

$$(\nabla u_{n_j}, \nabla(v - u_{n_j}))_{0,\Omega_{n(j)}} \geq (f, v - u_{n_j})_{0,\Omega_{n(j)}} \qquad \forall v \in K(\Omega_{n(j)}) . \tag{2.7}$$

As $V_j|_{\Omega_{n(j)}} \in K(\Omega_{n(j)})$, one can substitute such a function into (2.7) and write

$$(\nabla u_{n_j}, \nabla(V_j - u_{n_j}))_{0,\Omega_{n(j)}} \geq (f, V_j - u_{n_j})_{0,\Omega_{n(j)}} \tag{2.8}$$

and also

$$(\nabla \tilde{u}_{n_j}, \nabla(V_j - \tilde{u}_{n_j}))_{0,\hat{\Omega}} \geq (f, V_j - \tilde{u}_{n_j})_{0,\hat{\Omega}} . \tag{2.9}$$

As $\{u_{n_j}\} \subset \{u_{n'}\}$, then using (2.4), (2.5) and passing to the limit in (2.9) with $j \to \infty$ we obtain

$$(\nabla U, \nabla(\tilde{v} - U))_{0,\hat{\Omega}} \geq (f, \tilde{v} - U)_{0,\hat{\Omega}} ,$$

or equivalently

$$(\nabla u, \nabla(v - u))_{0,\Omega(\alpha)} \geq (f, v - u)_{0,\Omega(\alpha)} .$$

As $v \in K(\Omega(\alpha))$ is arbitrary, $u(\alpha) \equiv U|_{\Omega(\alpha)}$ solves $(\mathrm{P}(\alpha))$.

Let us now show that there exists $\{u_{n''}\} \subset \{u_{n'}\}$ such that

$$\tilde{u}_{n''} \to U \qquad \text{in } H^1(\hat{\Omega}) .$$

According to Lemma 2.1 there exist a sequence $\chi_j \in H^1(\hat{\Omega})$ and a subsequence $\{\alpha_{n'(j)}\} \subset \{\alpha_{n'}\}$ such that

$$\chi_j \to U \qquad \text{in } H^1(\hat{\Omega}) ; \tag{2.10}$$

$$\chi_j|_{\Omega_{n'(j)}} \in K(\Omega_{n'(j)}) . \tag{2.11}$$

Then it holds (we set $n'' = n'(j)$):

$$\begin{aligned}
&\alpha \|U - \tilde{u}_{n''}\|^2_{1,\hat{\Omega}} \leq (\nabla(U - \tilde{u}_{n''}), \nabla(U - \tilde{u}_{n''}))_{0,\hat{\Omega}} \\
&= (\nabla U, \nabla(U - \tilde{u}_{n''}))_{0,\hat{\Omega}} - (\nabla \tilde{u}_{n''}, \nabla(U - \chi_j))_{0,\hat{\Omega}} - (\nabla \tilde{u}_{n''}, \nabla(\chi_j - \tilde{u}_{n''}))_{0,\hat{\Omega}} \\
&\leq (\nabla U, \nabla(U - \tilde{u}_{n''}))_{0,\hat{\Omega}} - (\nabla \tilde{u}_{n''}, \nabla(U - \chi_j))_{0,\hat{\Omega}} - (f, \chi_j - \tilde{u}_{n''})_{0,\hat{\Omega}} \\
&\to 0 \quad \text{if } n'' \to \infty
\end{aligned}$$

making use of (2.4) and (2.10). □

PROOF OF TH. 2.1: Let $q = \inf_{\alpha \in \tilde{U}_{ad}} J(\alpha)$ and denote by $\{\alpha_n\}$, $\alpha_n \in \tilde{U}_{ad}$ a minimizing sequence of the problem, i.e.

$$q = \lim_{n \to \infty} J(\alpha_n) .$$

As U_{ad} is compact, there exist a subsequence of $\{\alpha_n\}$ (still denoted by $\{\alpha_n\}$) and an element $\alpha^* \in U_{ad}$ such that

$$\alpha_n \rightrightarrows \alpha^* \qquad \text{in } [0,1] \, . \tag{2.12}$$

We prove that α^* is a solution of (**P**).

Let $u_n = u(\alpha_n) \in K(\Omega_n)$ solve $(\mathrm{P}(\alpha_n))$. According to Theorem 2.2 there exist: a subsequence $\{u_{n'}\} \subset \{u_n\}$ and an element $U \in H_0^1(\hat{\Omega})$, such that $U|_{\Omega(\alpha^*)}$ solves $(\mathrm{P}(\alpha^*))$ and

$$\tilde{u}_{n'} \to U \qquad \text{in } H^1(\hat{\Omega}) \, . \tag{2.13}$$

As $u_{n'} = \varphi$ a.e. in Ω_0, the same holds for $u(\alpha^*) \equiv U|_{\Omega(\alpha^*)}$ by virtue of (2.13), i.e. $\alpha^* \in \tilde{U}_{ad}$. Clearly,

$$\lim_{n' \to \infty} J(\alpha_{n'}) = J(\alpha^*) = q \, . \quad \square$$

3. Penalty method for the approximation of (P)

The main difficulty in the numerical realization of (**P**) is the presence of the state constraint $u(\alpha) = \varphi$ in Ω_0. To overcome this difficulty, the penalty approach for solving (**P**) is proposed (see [2]).

Let

$$J_\varepsilon(\alpha) = \text{ meas } \Omega(\alpha) + \frac{1}{\varepsilon} \int_{\Omega_0} (u(\alpha) - \varphi) \, dx \, , \quad \varepsilon > 0 \, ,$$

be a modified cost functional with a penalty term $\frac{1}{\varepsilon} \int_{\Omega_0} (u(\alpha) - \varphi) \, dx$. By $u(\alpha)$ we denote the solution of $(\mathrm{P}(\alpha))$.

The penalty form of (**P**) now reads as follows:

$$\begin{cases} \text{find } \alpha_\varepsilon^* \in U_{ad} \text{ such that} \\ J_\varepsilon(\alpha_\varepsilon^*) \le J_\varepsilon(\alpha) \quad \forall \alpha \in U_{ad} \, . \end{cases} \tag{$\mathbf{P}_\varepsilon$}$$

In the sequel we shall study the mutual relation between solutions of $(\mathbf{P}_\varepsilon)$ and (**P**) if $\varepsilon \to 0+$.

First of all, using the same approach as in the previous section, one can prove

THEOREM 3.1. *There exists at least one solution α_ε^* of $(\mathbf{P}_\varepsilon)$ for any $\varepsilon > 0$.*

Now, let $\varepsilon_j \to 0+$, $j \to \infty$ be a sequence of penalty parameters. The main result of this section is

THEOREM 3.2. *Let $\{\alpha_j^*\}$ be solutions of $(\mathbf{P}_{\varepsilon_j})$ and $u_j(\alpha_j^*)$ corresponding solutions of $(\mathrm{P}(\alpha_j^*))$. Then there exist: subsequences $\{\alpha_{j_k}^*\} \subset \{\alpha_j^*\}$, $\{u_{j_k}(\alpha_{j_k}^*)\} \subset \{u_j(\alpha_j^*)\}$ and elements $\alpha^* \in \tilde{U}_{ad}$, $U \in H^1(\hat{\Omega})$ such that*

$$\alpha_{j_k}^* \rightrightarrows \alpha^* \qquad \text{in } [0,1]\ ;$$
$$\tilde{u}_{j_k}(\alpha_{j_k}^*) \to U \qquad \text{in } H^1(\hat{\Omega})\ .$$

Moreover, α^* is a solution of $(\mathbf{P})$ and $U|_{\Omega(\alpha^*)}$ solves $(\mathrm{P}(\alpha^*))$.

PROOF: As U_{ad} is compact, there exist: a subsequence of $\{\alpha_j^*\}$ (still denoted by the same symbol) and an element $\alpha^* \in U_{ad}$ such that

$$\alpha_j^* \rightrightarrows \alpha^*\ ,\ j \to \infty \qquad \text{in } [0,1]\ . \tag{3.1}$$

Applying Th. 2.2 we see that there exist: a subsequence $\{\alpha_{j_k}^*\} \subset \{\alpha_j^*\}$ and an element $U \in H^1(\hat{\Omega})$ such that

$$\begin{cases} \alpha_{j_k}^* \rightrightarrows \alpha^* & \text{in } [0,1]\ ; \\ \tilde{u}_{j_k}(\alpha_{j_k}^*) \to U & \text{in } H^1(\hat{\Omega}),\ k \to \infty \end{cases} \tag{3.2}$$

and $U|_{\Omega(\alpha^*)}$ solves $(\mathrm{P}(\alpha^*))$. Let us prove that α^* solves $(\mathbf{P})$. First of all we show that $\alpha^* \in \tilde{U}_{ad}$ or equivalently $U|_{\Omega_0} = \varphi$ a.e. in Ω_0. From the definition of $(\mathbf{P}_{\varepsilon_{j_k}})$ it follows that

$$\begin{aligned} &\text{meas } \Omega(\alpha_{j_k}^*) + \frac{1}{\varepsilon_{j_k}} \int_{\Omega_0} (u(\alpha_{j_k}^*) - \varphi)\, dx \\ &\le \text{ meas } \Omega(\alpha) + \frac{1}{\varepsilon_{j_k}} \int_{\Omega_0} (u(\alpha) - \varphi)\, dx \end{aligned} \tag{3.3}$$

holds for any $\alpha \in U_{ad}$. Substituting an element $\alpha \in \tilde{U}_{ad}$ $(\tilde{U}_{ad} \neq \emptyset)$ into the right hand side of (3.3) we are led to

$$0 \le \int_{\Omega_0} (u(\alpha_{j_k}^*) - \varphi)\, dx \le \text{ const } \cdot \varepsilon_{j_k}\ .$$

This and $(3.2)_2$ imply

$$\int_{\Omega_0} (u(\alpha_{j_k}^*) - \varphi)\, dx \to \int_{\Omega_0} (U - \varphi)\, dx = 0\ .$$

As at the same time $U \ge \varphi$ in Ω_0, we conclude that $U = \varphi$ a.e. in Ω_0. It remains to verify that α^* solves $(\mathbf{P})$. As $u(\alpha_{j_k}^*) \ge \varphi$ in Ω_0 it follows from (3.3) that

$$\begin{aligned} \text{meas } \Omega(\alpha_{j_k}^*) \le &\text{ meas } \Omega(\alpha_{j_k}^*) + \frac{1}{\varepsilon_{j_k}} \int_{\Omega_0} (u(\alpha_{j_k}^*) - \varphi)\, dx \\ \le &\text{ meas } \Omega(\alpha) \qquad \text{for any } \alpha \in \tilde{U}_{ad}\ . \end{aligned}$$

Letting $k \to \infty$, we obtain the assertion of Theorem 3.2. □

4. Approximation of $(\mathbf{P}_\varepsilon)$

In this Section we shall discuss the finite element approximation of the penalized form $(\mathbf{P}_\varepsilon)$, assuming $\varepsilon > 0$ fixed.

Let $D_h : 0 = x_2^{(0)} < x_2^{(1)} < \cdots < x_2^{(N)} \equiv 1$ be a partition of $[0,1]$. The discretization of U_{ad} is defined as follows:

$$U_{ad}^h = \{\alpha_h \in C([0,1]) \mid \hat{\alpha}_0 \le \alpha_h \le \hat{\beta}_0,\ |\alpha_h'| \le c_0,\ \alpha_h|_{[x_2^{(i-1)}, x_2^{(i)}]} \in P_1\}\,,$$

i.e. U_{ad}^h contains all functions from U_{ad}, which are piecewise linear over D_h.

Let $\alpha_h \in U_{ad}^h$ be given. By $T_h(\alpha_h)$ we denote a triangulation of the polygonal domain $\Omega(\alpha_h)$, satisfying the usual requirements concerning the mutual position of its triangles. Nevertheless, some additional assumptions will be necessary:

(i) the triangulation $T_h(\alpha_h)$ as a function of $\alpha_h \in U_{ad}^h$ depends continuously on α_h for any $h > 0$ fixed

(ii) the family $\{T_h(\alpha_h)\}$ is uniformly regular with respect to $h \in]0,1[$, $\alpha_h \in U_{ad}^h$, i.e. a constant $\rho > 0$ exists such that

$$\vartheta_0(h, \alpha_h) \ge \rho \qquad \forall h \in]0,1[,\ \alpha_h \in U_{ad}^h\,,$$

where $\vartheta_0(h, \alpha_h)$ is the minimal interior angle of triangles belonging to $T_h(\alpha_h)$.

Moreover, in the sequel we assume that Ω_0 is a polygonal domain, covered by a finite number of $T_i \in T_h(\alpha_h)$. The domain $\Omega(\alpha_h)$ with a given triangulation $T_h(\alpha_h)$ will be denoted by $\Omega_h(\alpha_h)$ (see Fig. 4.1 and Fig. 5.1).

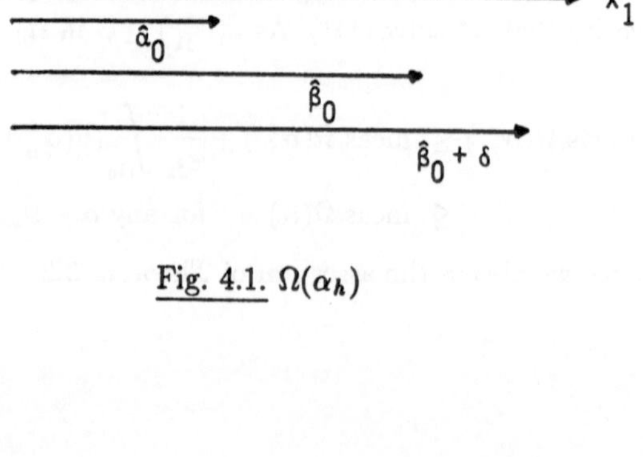

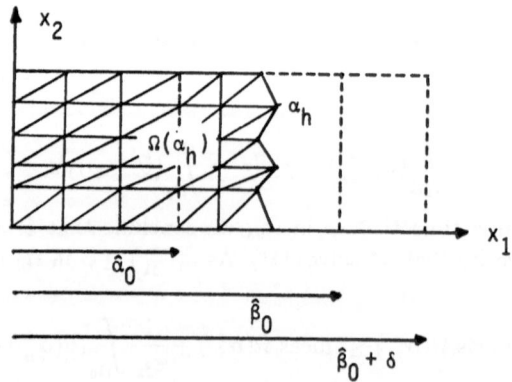

Fig. 4.1. $\Omega(\alpha_h)$

With any $T_h(\alpha_h)$, the closed convex set $K_h(\Omega_h(\alpha_h))$ will be associated:

$$K_h(\Omega_h(\alpha_h)) = \{v_h \in C(\overline{\Omega_h(\alpha_h)}) \mid v_h|_{T_i} \in P_1(T_i) \ \forall T_i \in T_h(\alpha_h),$$
$$v_h(a_i) \geq \varphi(a_i) \ \forall a_i \in N_h, \ v_h = 0 \text{ on } \partial\Omega_h(\alpha_h)\} .$$

$K_h(\Omega_h(\alpha_h))$ contains all continuous, piecewise linear functions over $T_h(\alpha_h)$, satisfying the inequality constraints at all interior nodes (the family of which is denoted by N_h) of $T_h(\alpha_h)$, only. Moreover, we shall suppose that $\varphi \in H^{1+\tilde{\varepsilon}}(\hat{\hat{\Omega}})$ for some $\tilde{\varepsilon} > 0$, where $\hat{\hat{\Omega}} =]0, \hat{\beta}_0 + \delta[\times]0,1[$ with $\delta > 0$, $\varphi \leq 0$ in $]\hat{\alpha}_0, \hat{\beta}_0 + \delta[\times]0,1[$ (see Fig. 4.1). By $\hat{T}_h(\alpha_h)$ we denote "the continuation" of $T_h(\alpha_h)$ on $\hat{\hat{\Omega}} \setminus \Omega_h(\alpha_h)$, which is also uniformly regular with respect to $h \in]0,1[$, $\alpha_h \in U^h_{ad}$. If v is a function defined on $\Omega(\alpha)$, the symbol $\tilde{\tilde{v}}$ denotes its extension by zero from $\Omega(\alpha)$ on $\hat{\hat{\Omega}}$.

The state problem is now approximated by means of the classical Ritz-Galerkin method:

$$\left\{\begin{array}{l} \text{find } u_h(\alpha_h) \in K_h(\Omega_h(\alpha_h)) \text{ such that} \\ (\nabla u_h(\alpha_h), \nabla(v_h - u_h(\alpha_h)))_{0,\Omega_h(\alpha_h)} \geq (f, v_h - u_h(\alpha_h))_{0,\Omega_h(\alpha_h)} \\ \qquad \forall v_h \in K_h(\Omega_h(\alpha_h)) . \end{array}\right. \qquad (\mathrm{P}(\alpha_h)_h)$$

Let $\varepsilon > 0$ be fixed. The approximation of $(\mathbf{P}_\varepsilon)$ now reads as follows:

$$\left\{\begin{array}{l} \text{find } \alpha^*_{\varepsilon h} \in U^h_{ad} \text{ such that} \\ J_\varepsilon(\alpha^*_{\varepsilon h}) \leq J_\varepsilon(\alpha_h) \quad \forall \alpha_h \in U^h_{ad} , \end{array}\right. \qquad (\mathbf{P}_{\varepsilon h})$$

where

$$J_\varepsilon(\alpha_h) = \text{ meas } \Omega(\alpha_h) + \frac{1}{\varepsilon} \int_{\Omega_0} (u_h(\alpha_h) - r_h\varphi)\, dx ,$$

where $u_h(\alpha_h)$ solves $(\mathrm{P}(\alpha_h)_h)$ and $r_h\varphi$ denotes the piecewise linear approximation of φ over triangulation of $\overline{\Omega}_0$.

It is not difficult to prove that $(\mathbf{P}_{\varepsilon h})$ has at least one solution $\alpha^*_{\varepsilon h}$.

Next, we shall study the relation between solutions of $(\mathbf{P}_\varepsilon)$ and $(\mathbf{P}_{\varepsilon h})$, $h \to 0+$. We first prove

Lemma 4.1. *Let $\alpha_h \rightrightarrows \alpha$, $h \to 0+$ in $[0,1]$, $\alpha_h \in U^h_{ad}$, $\alpha \in U_{ad}$. Let $u_h(\alpha_h)$ be solutions of $(P(\alpha_h)_h)$. Then there exist:* a subsequence $\{u_{h_j}(\alpha_{h_j})\} \subset \{u_h(\alpha_h)\}$ *and an element $U \in H^1(\hat{\Omega})$ such that*

(k) $\tilde{u}_{h_j}(\alpha_{h_j}) \to U$ *in* $H^1(\hat{\Omega})$;
(kk) $U|_{\Omega(\alpha)}$ *is the solution of* $(\mathrm{P}(\alpha))$.

Proof: From the definition of $(\mathrm{P}(\alpha_h)_h)$ it follows that there exists a constant $c > 0$ such that

$$\|u_h(\alpha_h)\|_{1,\Omega(\alpha_h)} \leq c , \qquad (4.1)$$

and also

$$\|\tilde{u}_h(\alpha_h)\|_{1,\hat{\Omega}} \le c\,. \tag{4.2}$$

Due to (4.2) there exist a subsequence $\{\tilde{u}_{h_j}(\alpha_{h_j})\} \subset \{\tilde{u}_h(\alpha_h)\}$ and an element $U \in H^1(\hat{\Omega})$ such that

$$\tilde{u}_{h_j}(\alpha_{h_j}) \rightharpoonup U \qquad \text{in } H^1(\hat{\Omega})\,. \tag{4.3}$$

First of all we show that $U|_{\Omega(\alpha)} \ge \varphi$ a.e. in $\Omega(\alpha)$ or equivalently

$$\int_{\Omega(\alpha)} (U - \varphi)\chi\, dx \ge 0$$

for any $\chi \ge 0$, $\chi \in L^2(\Omega(\alpha))$. Let a function χ with the previous property be given. It is easy to prove the existence of functions $\chi_h \in L^2(\Omega_h(\alpha_h))$ that are piecewise constant over $T_h(\alpha_h)$ and

$$\begin{cases} \chi_h \ge 0 & \text{in } \Omega_h(\alpha_h) \\ \tilde{\chi}_h \to \tilde{\chi} & \text{in } L^2(\hat{\Omega})\,. \end{cases} \tag{4.4}$$

Then

$$\begin{aligned}(u_{h_j}(\alpha_{h_j}) - r_{h_j}\varphi, \chi_{h_j})_{0,\Omega_{h_j}(\alpha_{h_j})} &= \left(\tilde{u}_{h_j}(\alpha_{h_j}) - r_{h_j}\varphi, \tilde{\chi}_{h_j}\right)_{0,\hat{\Omega}} \\ &= \sum_{\mathrm{T}_i \in \hat{T}_h(\alpha_h)} \left(\tilde{u}_{h_j}(\alpha_{h_j}) - r_{h_j}\varphi, \tilde{\chi}_{h_j}\right)_{0,\mathrm{T}_i}\,,\end{aligned} \tag{4.5}$$

where $r_{h_j}\varphi$ denotes the piecewise linear interpolation of φ over $\hat{T}_h(\alpha_h)$. As $\hat{T}_h(\alpha_h)$ is uniformly regular with respect to $h > 0$, $\alpha_h \in U^h_{ad}$, one has

$$\|r_{h_j}\varphi - \varphi\|_{1,\hat{\Omega}} \le ch^{\bar{\varepsilon}}\, \|\varphi\|_{1+\bar{\varepsilon},\hat{\Omega}}\,. \tag{4.6}$$

Any integral appearing on the right hand side of (4.5) can be exactly evaluated by means of the quadrature formulae, using values of functions at vertices of T_i. From the definition of $K_{h_j}(\Omega_{h_j}(\alpha_{h_j}))$, $r_{h_j}\varphi$ and $(4.4)_1$ we conclude

$$\left(\tilde{u}_{h_j}(\alpha_{h_j}) - r_{h_j}\varphi, \tilde{\chi}_{h_j}\right)_{0,\hat{\Omega}} \ge 0 \qquad \text{for any } h_j > 0\,. \tag{4.7}$$

Letting $h_j \to 0+$ and using (4.3), (4.4) and (4.6) we see that

$$(U - \varphi, \chi)_{0,\Omega(\alpha)} = \left(\tilde{U} - \varphi, \tilde{\chi}\right)_{0,\hat{\Omega}} \ge 0\,,$$

i.e. $U|_{\Omega(\alpha)} \in K(\Omega(\alpha))$. Let us prove that $U|_{\Omega(\alpha)}$ solves $(\mathrm{P}(\alpha))$. From the definition of $(\mathrm{P}(\alpha_{h_j})_{h_j})$ it follows that

$$(\nabla u_{h_j}, \nabla(v_{h_j} - u_{h_j}))_{0,\Omega_{h_j}(\alpha_{h_j})} \geq (f, v_{h_j} - u_{h_j})_{0,\Omega_{h_j}(\alpha_{h_j})} \tag{4.8}$$

holds for any $v_{h_j} \in K_{h_j}(\Omega_{h_j}(\alpha_{h_j}))$. Let $v \in K(\Omega(\alpha))$ be given and let $\{W_i\}$ be a sequence of functions with properties given by Remark 2.2. For i fixed, the function $W_i|_{\Omega_{h_j}} \in K(\Omega(\alpha_{h_j})) \cap C^\infty(\overline{\Omega}(\alpha_{h_j}))$ for any h_j sufficiently small. By w_{ih_j} we denote the piecewise linear Lagrange interpolate of W_i. Clearly, $w_{ih_j} \in K_{h_j}(\Omega_{h_j}(\alpha_{h_j}))$ for h_j sufficiently small, i.e. such a function can be substituted into (4.8):

$$(\nabla u_{h_j}, \nabla(w_{ih_j} - u_{h_j}))_{0,\Omega_{h_j}(\alpha_{h_j})} \geq (f, w_{ih_j} - u_{h_j})_{0,\Omega_{h_j}(\alpha_{h_j})} \tag{4.9}$$

or

$$(\nabla \tilde{u}_{h_j}, \nabla(\tilde{w}_{ih_j} - \tilde{u}_{h_j}))_{0,\hat{\Omega}} \geq (f, \tilde{w}_{ih_j} - \tilde{u}_{h_j})_{0,\hat{\Omega}} \, . \tag{4.10}$$

Passing to the limit with $h_j \to 0+$ and then $i \to \infty$, we finally obtain:

$$(\nabla U, \nabla(\tilde{v} - U))_{0,\hat{\Omega}} \geq (f, \tilde{v} - U)_{0,\hat{\Omega}} \tag{4.11}$$

or equivalently

$$(\nabla U, \nabla(v - U))_{0,\Omega(\alpha)} \geq (f, v - U)_{0,\Omega(\alpha)} \, .$$

Thus $U|_{\Omega(\alpha)}$ solves $(\mathrm{P}(\alpha))$. The proof that there exists a subsequence of $\{u_{h_j}(\alpha_{h_j})\}$ (which is still denoted by the same symbol) such that

$$\tilde{u}_{h_j} \to U \qquad \text{in } H^1(\hat{\Omega})$$

proceeds in the same way as in the proof of Th. 2.2. □

Now we are able to prove <u>the main result of this section</u>

THEOREM 4.1. *Let $\{\alpha^*_{\varepsilon h}\}$ be solutions of $(\mathbf{P}_{\varepsilon h})$ and $u_h(\alpha^*_{\varepsilon h})$ solutions of $(\mathrm{P}(\alpha^*_{\varepsilon h})_h)$. Then there exist subsequences $\{\alpha^*_{\varepsilon h_j}\} \subset \{\alpha^*_{\varepsilon h}\}$, $\{u_{h_j}(\alpha^*_{\varepsilon h_j})\} \subset \{u_h(\alpha^*_h)\}$ and elements $\alpha^*_\varepsilon \in U_{ad}$, $U_\varepsilon \in H^1(\hat{\Omega})$ such that*

$$\alpha^*_{\varepsilon h_j} \rightrightarrows \alpha^*_\varepsilon \, , \; j \to \infty \qquad \text{in } [0,1] \, ; \tag{4.12}$$

$$\tilde{u}_{h_j}(\alpha^*_{\varepsilon h_j}) \to U_\varepsilon \qquad \text{in } H^1(\hat{\Omega}) \, . \tag{4.13}$$

Moreover, α^*_ε is a solution of $(\mathbf{P}_\varepsilon)$ and $U|_{\Omega(\alpha^*_\varepsilon)}$ solves $(\mathrm{P}(\alpha^*_\varepsilon))$.

PROOF: As $U^h_{ad} \subset U_{ad}$ and U_{ad} is compact, there exists a subsequence of $\{\alpha^*_{\varepsilon h}\}$ (still denoted as $\{\alpha^*_{\varepsilon h}\}$) and an element $\alpha^*_\varepsilon \in U_{ad}$ such that

$$\alpha^*_{\varepsilon h} \rightrightarrows \alpha^*_\varepsilon \, , \; h \to 0+ \qquad \text{in } [0,1] \, . \tag{4.14}$$

Applying Lemma 4.1 we have that there exist subsequences $\{\alpha^*_{\varepsilon h_j}\} \subset \{\alpha^*_{\varepsilon h}\}$, $\{u_{h_j}(\alpha^*_{\varepsilon h_j})\} \subset \{u_h(\alpha^*_{\varepsilon h})\}$ and an element $U_\varepsilon \in H^1(\hat{\Omega})$ such that $U_\varepsilon|_{\Omega(\alpha^*_\varepsilon)}$ solves $(\mathrm{P}(\alpha^*_\varepsilon))$. Moreover

$$\tilde{u}_{h_j}(\alpha^*_{\varepsilon h_j}) \to U_\varepsilon \qquad \text{in } H^1(\hat{\Omega}) \,. \tag{4.15}$$

We prove that α^*_ε is a solution of $(\mathbf{P}_\varepsilon)$.

Let $\alpha \in U_{ad}$ be arbitrarily given. Then there exists a sequence $\{\alpha_{h_j}\}$, $\alpha_{h_j} \in U^{h_j}_{ad}$ (h_j are indices for which (4.15) holds) such that

$$\alpha_{h_j} \rightrightarrows \alpha \,,\; j \to \infty \qquad \text{in } [0,1] \,. \tag{4.16}$$

At the same time we may assume (applying Lemma 4.1 once again) that

$$\tilde{u}_{h_j}(\alpha_{h_j}) \to U \qquad \text{in } H^1(\hat{\Omega}) \,, \tag{4.17}$$

where $U|_{\Omega(\alpha)}$ solves $(\mathrm{P}(\alpha))$. The rest of the proof follows immediately from the definition of $(\mathbf{P}_{\varepsilon h_j})$, (4.12), (4.13), (4.16) and (4.17). □

5. Sensitivity Analysis

Let $h > 0$ be fixed. The state problem $(\mathrm{P}(\alpha_h)_h)$ expressed in the matrix form reads as follows:

$$\begin{cases} \text{find } x(\alpha) \in K(\alpha) \text{ such that} \\ \mathcal{L}(x(\alpha)) \le \mathcal{L}(x) \quad \forall x \in K(\alpha) \,, \end{cases} \tag{5.1}$$

where

$$\mathcal{L}(x) = \frac{1}{2}(x, A(\alpha)x)_{\mathbf{R}^n} - (F(\alpha), x)_{\mathbf{R}^n}$$

is a quadratic function, given by a symmetric, positive definite matrix $A(\alpha)$ (stiffness matrix of our problem) and $F(\alpha)$ is a linear term, arising from the discretization of the right hand side f. Both A and F depend on the discrete design variable $\alpha \in \mathbf{R}^{N+1}$. In our special geometry, the vector α is given by x_1 coordinates of the so called principle moving points $A_i = (\alpha_h(x_2^{(i)}), x_2^{(i)})$, $\alpha_h \in U^h_{ad}$, $\alpha = (\alpha_0, \alpha_1, \dots, \alpha_N)$, $\alpha_i = \alpha_h(x_2^{(i)})$. These points will move in x_1 direction only. The position of other nodes of $T_h(\alpha_h)$ will be fixed (indicated by □) or uniquely determined by A_i (the so called associated moving points, indicated by ○), see Fig. 5.1.

The closed convex set $K(\alpha)$ is given by

$$K(\alpha) = \{x \in \mathbf{R}^n \mid x_i \ge \varphi_i(\alpha) \quad \forall i = 1, \dots, n\} \,.$$

Here $\varphi_i(\alpha) = \varphi(z_i(\alpha))$, where $z_i(\alpha)$ is the cartesian coordinate of the i-th node of $T_h(\alpha_h)$, depending in general on the design variable α (see [5]).

It is well known that the mapping $\alpha \mapsto x(\alpha)$ is only directionally differentiable, i.e. the finite limite

$$\lim_{t \to 0+} \frac{x(\alpha + t\tilde{\alpha}) - x(\alpha)}{t}$$

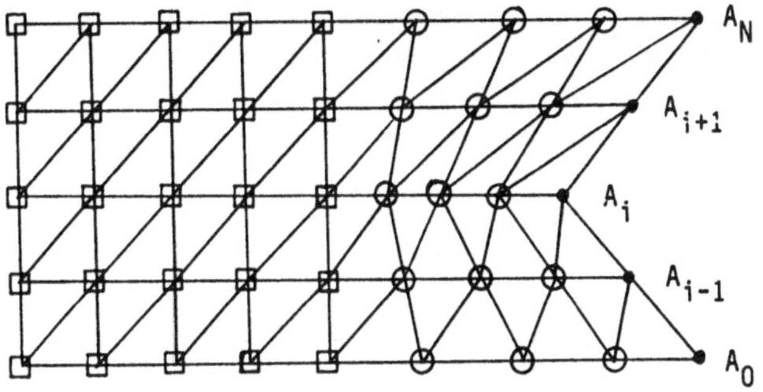

Fig. 5.1. $\Omega_h(\alpha_h)$

exists for any $\tilde{\alpha} \in \mathbf{R}^{N+1}$ (see [1], [4]).

Moreover $x'(\alpha, \tilde{\alpha})$ can be characterized as the unique solution of the following minimization problem

$$\begin{cases} \text{find } x'(\alpha,\tilde{\alpha}) \in K(\alpha,\tilde{\alpha}) \text{ such that} \\ L(x'(\alpha,\tilde{\alpha})) \le L(z) \quad \forall z \in K(\alpha,\tilde{\alpha}) , \end{cases} \tag{5.2}$$

where

$$L(z) = \frac{1}{2}(z, A(\alpha)z) - (F'(\alpha) - A'(\alpha)x(\alpha), z)$$

and

$$K(\alpha,\tilde{\alpha}) = \{z \in \mathbf{R}^n \mid z_i = \varphi_i'(\alpha) \quad \forall i \in \mathrm{I}^+ \\ \qquad z_i \ge \varphi_i'(\alpha) \quad \forall i \in \mathrm{I}^0\} .$$

Here

$$A'(\alpha) = (a_{ij}'(\alpha)), \quad a_{ij}'(\alpha) = \nabla_\alpha a_{ij}(\alpha) \cdot \tilde{\alpha} ,$$
$$F'(\alpha) = (F_i'(\alpha)), \quad F_i'(\alpha) = \nabla_\alpha F_i(\alpha) \cdot \tilde{\alpha} ,$$
$$\varphi_i'(\alpha) = \nabla_\alpha \varphi_i(\alpha) \cdot \tilde{\alpha}$$

are directional derivatives of A, F and φ_i respectively and

$$\mathrm{I}^+ = \{i \mid x_i(\alpha) = \varphi_i(\alpha),\ a_{ij}(\alpha)x_j(\alpha) - J_i(\alpha) > 0\}$$
$$\mathrm{I}^0 = \{i \mid x_i(\alpha) = \varphi_i(\alpha),\ a_{ij}(\alpha)x_j(\alpha) - J_i(\alpha) = 0\} .$$

If I^0 is empty, then the mapping $\alpha \mapsto x(\alpha)$ is once continuously differentiable.

Let Γ be the set of indices associated with the nodes of $T_h(\alpha_h)$ lying in the polygonal domain $\overline{\Omega}_0$, where the contact with the obstacle is required. Assuming the equidistant partition of $[0,1]$, i.e. $h = x_2^{(i+1)} - x_2^{(i)}$ $\forall i = 0,\dots,N-1$, the cost functional J in problem (**P**) is equal to

$$E(\alpha) = \frac{h}{2}\sum_{i=0}^{N} \alpha_i \, .$$

The matrix form of $(\mathbf{P}_{\varepsilon h})$ reads now as follows:

$$\begin{cases} \text{find } \alpha^* \in U \text{ such that} \\ E_\varepsilon(\alpha^*) \le E_\varepsilon(\alpha) \quad \forall \alpha \in U \, , \end{cases} \tag{5.3}$$

where

$$E_\varepsilon(\alpha) = \frac{h}{2}\sum_{i=0}^{N} \alpha_i + \frac{1}{2\varepsilon}\left(\tilde{x}(\alpha) - \tilde{\varphi}(\alpha), \tilde{x}(\alpha) - \tilde{\varphi}(\alpha)\right)_{\mathbf{R}^{\mathrm{card}(\Gamma)}}$$

with $\tilde{x}(\alpha)$ $(\tilde{\varphi}(\alpha))$ being the restriction of the solution $x(\alpha)$ of (5.1) ($\varphi(\alpha)$ resp.) on grid lying in $\overline{\Omega}_0$, i.e.

$$\begin{aligned} \tilde{x}_i(\alpha) &= x_i(\alpha) \qquad \forall i \in \Gamma \\ \tilde{x}_i(\alpha) &= 0 \qquad \forall i \notin \Gamma \end{aligned}$$

and

$$U = \{\alpha \in \mathbf{R}^{N+1} \mid \hat{\alpha}_0 \le \alpha_i \le \hat{\beta}_0 \quad \forall i = 0,\dots,N \, ;$$
$$- c_0 \le \frac{\alpha_{i+1} - \alpha_i}{h} \le c_0 \quad i = 0,\dots,N-1\} \, .$$

The directional derivative of E_ε at the point α and the direction $\tilde{\alpha} \in \mathbf{R}^{N+1}$ is given by

$$E'_\varepsilon(\alpha,\tilde{\alpha}) = \frac{h}{2}\sum_{i=0}^{N} \tilde{\alpha}_i + \frac{1}{\varepsilon}\left(\tilde{x}(\alpha) - \tilde{\varphi}(\alpha), \tilde{x}'(\alpha) - \tilde{\varphi}'(\alpha)\right)_{\mathbf{R}^{\mathrm{card}(\Gamma)}} \, . \tag{5.4}$$

From (5.2) we get that $E'(\alpha,\tilde{\alpha})$ is linear in $\tilde{\alpha}$ if I^0 is empty. (If $I^0 \ne \emptyset$ this is not true, see Example 5.1.) In this case the standard adjoint state technique can be applied: Let us first arrange the second term on the right hand side of (5.4). By introducing Lagrange multipliers the problem (5.1) can be equivalently characterized through the existence of non-negative λ_i, $i = 1,\dots,n$ such that

$$a_{ij}(\alpha)x_j(\alpha) = F_i(\alpha) + \lambda_i(\alpha) \, , \qquad i = 1,\dots,n \tag{5.5}$$
$$(\lambda(\alpha), x(\alpha) - \varphi(\alpha))_{\mathbf{R}^n} = 0 \, . \tag{5.6}$$

Let Γ_+ be a subset of Γ such that

$$i \in \Gamma_+ \Leftrightarrow x_i(\alpha) > \varphi_i(\alpha) .$$

Then

$$\begin{aligned}&(\tilde{x}(\alpha) - \tilde{\varphi}(\alpha), \tilde{x}'(\alpha) - \tilde{\varphi}'(\alpha))_{\mathbf{R}^{\mathrm{card}(\Gamma)}} = \\ &\sum_{i\in\Gamma}(x_i(\alpha) - \varphi_i(\alpha))(x_i'(\alpha) - \varphi_i'(\alpha)) = \\ &\sum_{i\in\Gamma_+}(x_i(\alpha) - \varphi_i(\alpha))(x_i'(\alpha) - \varphi_i'(\alpha)) .\end{aligned} \tag{5.7}$$

Let I$\subseteq \{1, 2, \ldots, n\}$ be such that

$$i \in \mathrm{I} \Leftrightarrow x_i(\alpha) > \varphi_i(\alpha) .$$

Note that I$\supseteq \Gamma_+$. As a consequence of (5.6) and the fact that $\lambda_i(\alpha) \geq 0\ \forall i = 1, \ldots, n$, the corresponding Lagrange multipliers $\lambda_i(\alpha)$ are equal to zero for any $i \in$ I. From (5.5) we see that

$$a_{ij}(\alpha)x_j(\alpha) = F_i(\alpha) \qquad \forall i \in \mathrm{I} . \tag{5.8}$$

As $x_j(\alpha)$ are known for any $j \notin$ I (namely $x_j(\alpha) = \varphi_j(\alpha)$), then the elimination of all $x_j(\alpha)$, $j \notin$ I leads to a new linear system:

$$\tilde{\tilde{A}}(\alpha)\tilde{\tilde{x}}(\alpha) = \tilde{\tilde{F}}(\alpha) , \tag{5.9}$$

where $\tilde{\tilde{x}}(\alpha)$ is a vector, containing all non-active components of $x(\alpha)$, $\tilde{\tilde{x}}(\alpha) \in \mathbf{R}^{\mathrm{card(I)}}$. Derivatives of $\tilde{\tilde{x}}(\alpha)$ can be found by a classical procedure. Moreover, the mapping $\alpha \mapsto \tilde{\tilde{x}}(\alpha)$ is once continuously differentiable and the same holds for a function $\alpha \mapsto E_\varepsilon(\alpha)$ as follows from (5.4) and (5.7).

We shall close this chapter with a counterexample due to Tiihonen [8]:

EXAMPLE 5.1: Let us consider the problem:

$$\begin{cases} -u^t(x)'' = -1 , & x \in (0,t) \\ u^t(x) \geq 0 , & x \in (0,t) \\ u^t(0) = 0 , u^t(t) = 1 \end{cases} \tag{5.10}$$

The solution of (5.10) has the form

$$u^t(x) = (\frac{1}{t} - \frac{t}{2})x + \frac{1}{2}x^2 \qquad \text{for } t \leq \sqrt{2}$$

$$u^t(x) = \begin{cases} \frac{1}{2}(x - t + \sqrt{2})^2 & \text{if } x \geq t - \sqrt{2} \\ 0 & \text{if } x \leq t - \sqrt{2} \end{cases} \qquad \text{for } t \geq \sqrt{2} .$$

The derivative with respect to the design parameter t is

$$\frac{\partial}{\partial t}u^t(x) = (-\frac{1}{t^2} - \frac{1}{2})x \qquad \text{for } t \le \sqrt{2}$$

$$\frac{\partial}{\partial t}u^t(x) = \begin{cases} t - x - \sqrt{2} & \text{if } x \ge t - \sqrt{2} \\ 0 & \text{if } x \le t - \sqrt{2} \end{cases} \qquad \text{for } t \ge \sqrt{2}\,.$$

For $t = \sqrt{2}$ both expressions reduce to $-x$. Thus u^t is continuously differentiable in t.

Let $\{0, \frac{t}{3}, \frac{2t}{3}, t\}$ be the partition of $[0, t]$. The discrete state inequality reads:

$$\begin{bmatrix} -2 & 1 \\ 1 & -2 \end{bmatrix} \begin{bmatrix} x_1 \\ x_2 \end{bmatrix} \ge \frac{t^2}{18} \begin{bmatrix} 1 \\ 1 \end{bmatrix} - \begin{bmatrix} 0 \\ 1 \end{bmatrix}, \qquad x_1, x_2 \ge 0.$$

For $t \le \sqrt{6}$ we have that

$$x_1 = \frac{1}{3}(1 - \frac{t^2}{6})\,,$$
$$x_2 = \frac{1}{3}(2 - \frac{t^2}{6})\,,$$

i.e. the contact condition $x_i \ge 0$ is not active.

For $t \ge \sqrt{6}$ we have

$$x_1 = 0\,,$$
$$x_2 = \frac{1}{2}(1 - \frac{t^2}{18})\,.$$

Thus it can be seen that neither $x_1(t)$ nor $x_2(t)$ are differentiable in t at $t = \sqrt{6}$. Consequently, we can note that also $\sum_i x_i^2(t)$ is nondifferentiable.

We can say in general that if $x(\alpha)$ is the FE-solution of some obstacle problem with the obstacle $\varphi(\alpha)$, then $\sum_{i \in I}(x_i(\alpha) - \varphi_i(\alpha))^2$ is not differentiable in α.

We recall that in the continuous setting of the problem (**P**) the cost functional is differentiable for every α [2].

6. Numerical examples

EXAMPLE 6.1: Let $\Omega_0 =]0.25, 0.5[\times]0.25, 0.75[$, $f(x, y) = -1.0$ and $\varphi(x, y) = -0.05\,x$.

The constraint constants of the control set U_{ad} are $\hat{\alpha}_0 = 0.6$, $\hat{\beta}_0 = 0.8$ and $c_0 = 2.0$. The penalty parameter $\varepsilon = 10^{-4}$.

We shall use FE-grid shown in Fig. 4.1 with $h = 1/16$. Thence we have 17 design nodes A_i. In minimization of $E_\varepsilon(\alpha)$ we have applied the E04VCE routine of NAG-subroutine library and the nonsmooth optimization algorithm (bundle method). Both give the same results.

Let the initial guess be $A_i = (0.8, ih)$, $i = 0, \ldots, 16$. For the initial design $E_\varepsilon(\alpha^0) = 0.800$. After 14 SQP-iterations of E04VCE we get the results indicated in Fig. 6.1. Figure 6.1(a) shows the triangulation of final $\Omega(\alpha_h)$ (the

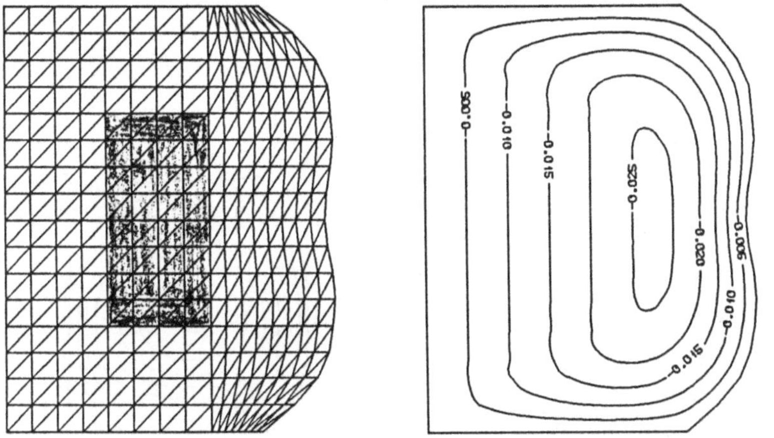

a) b)

Fig. 6.1.

darkened area in Fig. 6.1(a) indicates Ω_0). Figure 6.1(b) shows the contour plots of the solution.

The value of $E_\varepsilon(\alpha^{14})$ is 0.784. The state constraint $u_h = \varphi$ on Ω_0 is satisfied. The constraints are not active.

EXAMPLE 6.2: Let $\Omega_0 =]0.25, 0.5[\times]0.625, 0.75[$, $f(x, y) = 2\sin(2\pi y)$ and $\varphi(x, y) = -0.015$. The constraint parameters are $\hat{\alpha}_0 = 0.6$, $\hat{\beta}_0 = 1.0$ and $c_0 = 2.0$. Let $\varepsilon = 10^{-4}$.

We shall use similar triangulation as above. Moreover, in minimization we shall apply NAG-subroutine E04VCE.

For the initial guess we choose $A_i = (0.6, ih)$, $i = 0, \ldots, 7$, $A_8 = (0.65, 8h)$, $A_9 = (0.7, 9h)$, $A_{10} = (0.75, 10h)$, $A_i = (0.8, ih)$, $i = 11, \ldots, 16$. For the initial design $E_\varepsilon(\alpha^0) = 0.853$. Figure 6.2 shows the results after 7 SQP-iterations.

The value of $E_\varepsilon(\alpha^7)$ is 0.792. Again the state constraint $u_h = \varphi$ on Ω_0 (darkened part of $\Omega(\alpha^7)$) is satisfied. Some of the Lipschitz constraints and box constraints are active.

In the fortcoming paper [6] more numerical examples will be given. Moreover, in [6] the comparison with the variational inequality approach will be given.

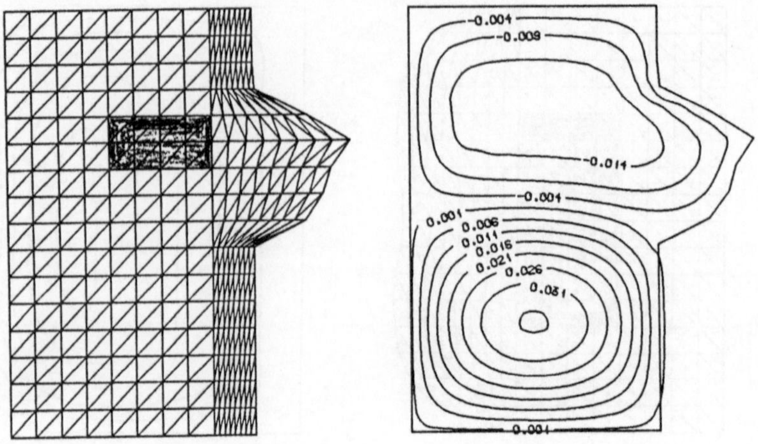

a) b)

Fig. 6.2.

ACKNOWLEDGEMENTS

The authors are indebted to R. Mäkinen for his assistance in numerical tests.

REFERENCES

1. M.P. Bendsøe, N. Olhoff and J. Sokolowski, *Sensitivity analysis of problems of elasticity with unilateral constraints*, Math.Report no. 1984–10, Matematisk Institut, Danmarks Tekniske Hojskole (1984).
2. B. Benedict, J. Sokolowski and J.P. Zolesio, *Shape optimization for contact problems*, in "Lecture Notes in Control and Inform. Sciences 59," Proc. of 11th IFIP Conference on System Modelling and Optimization (P. Topft-Christensen ed.), Springer-Verlag, New York, Berlin, Heidelberg, 1984, pp. 789–799.
3. R. Glowinski, "Numerical methods for nonlinear variational problems," Springer-Verlag, New York, Berlin, Heidelberg, Tokyo, 1984.
4. J. Haslinger, V. Horak and P. Neittaanmäki, *Shape optimization in contact problem with friction*, Numer. Funkt. Anal. and Optimiz. (1986), 557–587.
5. J. Haslinger and P. Neittaanmäki, *On optimal shape design of systems governed by mixed Dirichlet-Signorini boundary value problems*, Math. Meth. Appl. Sci. **8** (1986), 157–181.
6. J. Haslinger, R. Mäkinen, P. Neittaanmäki and D. Tiba, in preparation.

7. J. Haslinger, P. Neittaanmäki and D. Tiba, *On state constrained optimal shape design problems*, Preprint 54, Dept. Math. Univ. Jyväskylä, 1986. To appear in Proc. of Oberwolfach Conference Optimal Control of PDE's, Birkhauser 1986.
8. T. Tiihonen, private communication.

J. Haslinger, Faculty of Mathematics and Physics, Charles University, KAM MFF UK, Malostranské 2/25, CS-11800, Prague, Czechoslovakia
P. Neittaanmäki, University of Jyväskylä, Department of Mathematics, Seminaarinkatu 15, SF-40100 Jyväskylä, Finland

Exponential local stability of first order strictly hyperbolic systems with nonlinear perturbations on the boundary

Irena Lasiecka

Department of Mathematics
University of Florida, Gainesville

Introduction.

Let Ω be an open bounded domain in R^n with sufficiently smooth boundary Γ. Consider

$$(1.1)\quad \begin{cases} y_t(t,x) = A(x,\partial)y(t,x) + F_1\ y(t,\cdot)(x) & x\in\Omega;\ t>0 \\ y(0,x) = y_o(x)\in[L_2(\Omega)]^k;\ x\in\Omega \\ My|_\Gamma = F_2 y(t,\cdot)(x) + G(y(t,\cdot))(x) & x\in\Gamma;\ t>0. \end{cases}$$

Here $A(x,\partial)$ is strictly hyperbolic operator of the form $A(x,\partial) = \sum_{j=1}^{n} A_j(x)\partial_j + B(x)$ with A_j, B smooth $k\times k$ matrix valued functions defined on Ω, $F_1 \in \mathscr{L}([L_2(\Omega)]^k \to [L_2(\Omega)]^k)$, $F_2 \in \mathscr{L}([L_2(\Omega)]^k \to [L_2(\Gamma)]^r)$, and the boundary operator M is given by $My = M(x)y$ where $M(x)$ is smooth $e\times k$ matrix valued function such that rank $M(x) = r$; $x \quad \Gamma$. Finally, G is a nonlinear continuous operator from $[L_2(\Omega)]^k$ into $[L_2(\Gamma)]^r$.

The main goal of the present paper is to study the problem of local uniform stability of the hyperbolic system (1.1) with nonlinear perturbation on the boundary represented by the operator G. More precisely, our aim is to address the following question: Does the presence of bounded nonlinear perturbation on the boundary affect the overall stability of the system? Of course, in this context we think of F_1 and F_2 as stabilizing feedbacks for the linear part of (1.1) while G should be considered as an arbitrary (uncontrolled) "small" at the origin(*) nonlinear perturbation. This sort of problem is well known in the case of ordinary differential equations in R^n; in order to achieve local stability for a certain class of nonlinear perturbations, it is enough to stabilize the linear part of the system. The same conclusions can be obtained as a straight forward generalization to the case of ODE in Banach spaces with nonlinear perturbation represented by the action of the bounded operator. Our results are qualitatively similar, the major novelty of a problem being that by considering nonlinear term on the boundary we deal automatically with the situation where the nonlinear operator G gives rise to unbounded perturbation (if (1.1) considered as an abstract ODE in some appropriate Banach space). To our knowledge the literature has considered so far unbounded nonlinear perturbations only in case of systems described by analytic semigroups ([K-1], [D-L]). Instead, our main focus here is on hyperbolic dynamics which excludes the analyticity of the underlined semigroup.

In order to study the stability properties of (1.1), we shall first recast (1.1) as an equation in Banach space with unbounded perturbation (Sect. 2). Next we shall develop an "abstract theory" adequate to treat some nonlinear, unbounded perturbations of an arbitrary C_0-semigroup (Thm 1 and 2 in sect. 3). The abstract assumptions of Thm. 2 which describe the degree of unboundedness of the perturbations are motivated by canonical example (1.1). However, they are more general than the boundary perturbation case considered in (1.1). An application of the general theory to hyperbolic problems (1.1) will yield the desired stability results (Thm. 3 in sect. 4.) Another application of the theory to a simple first order nonlinear scalar hyperbolic equation defined on nonreflexive space L_1 and describing age-dependent population dynamics model will be given in sect. 5.

(*) "smallness" is usually expressed by requiring that $||G(y)|| \to 0$ when $||y|| \to 0$.

perturbations, it is enough to stabilize the linear part of the system. The same conclusions can be obtained as a straight forward generalization to the case of ODE in Banach spaces with nonlinear perturbation represented by the action of the bounded operator. Our results are qualitatively similar, the major novelty of a problem being that by considering nonlinear term on the boundary we deal automatically with the situation where the nonlinear operator G gives rise to unbounded perturbation (if (1.1) considered as an abstract ODE in some appropriate Banach space). To our knowledge the literature has considered so far unbounded nonlinear perturbations only in case of systems described by analytic semigroups ([K-1], [D-L]). Instead, our main focus here is on hyperbolic dynamics which excludes the analyticity of the underlined semigroup.

In order to study the stability properties of (1.1), we shall first recast (1.1) as an equation in Banach space with unbounded perturbation (Sect. 2). Next we shall develop an "abstract theory" adequate to treat some nonlinear, unbounded perturbations of an arbitrary C_o-semigroup (Thm 1 and 2 in sect. 3). The abstract assumptions of Thm. 2 which describe the degree of unboundedness of the perturbations are motivated by canonical example (1.1). However, they are more general than the boundary perturbation case considered in (1.1). An application of the general theory to hyperbolic problems (1.1) will yield the desired stability results (Thm. 3 in sect. 4.) Another application of the theory to a simple first order nonlinear scalar hyperbolic equation defined on nonreflexive space L_1 and describing age-dependent population dynamics model will be given in sect. 5.

2. Abstract formulation of (1.1).

In this section we shall rewrite equation (1.1) as an "abstract" ODE in Banach space. To this end let us introduce operator $A: [L_2(\Omega)]^k \to [L_2(\Omega)]^k$ given by

$$(2.1) \quad \begin{cases} Au = A(x,\partial)u \quad \text{for } u \in D(A) \\ D(A) = \{u \in [L_2(\Omega)]^k;\ Au \in [L_2(\Omega)]^k;\ Mu|\Gamma = 0\}. \end{cases}$$

It is well known [R-1] [O-M] that A generates strongly continuous semigroup e^{At} on $[L_2(\Omega)]^k$. We introduce next the "Dirichlet" map (natural extension from the boundary Γ into the interior Ω) defined by $Dg \equiv v$, where:

$$(2.2) \quad \begin{cases} A(x,\partial)v = \lambda_o v \quad \text{in } \Omega \\ Mv|\Gamma = g \quad \text{on } \Gamma \end{cases}$$

for some positive $\lambda_o > 0$.

The following regularity result proved in [Ch-L] will play a key role in the sequel: there exists $\lambda_o > 0$ large enough so that (2.2) admits unique solution $v = Dg$ and

$$(2.3) \quad \begin{cases} D \in \mathscr{L}([L_2(\Gamma)]^r, [L_2(\Omega)]^k) \\ Dg|\Gamma \in \mathscr{L}([L_2(\Gamma)]^r; [L_2(\Gamma)]^k). \end{cases}$$

With the help of operators A and D, the solution to the open loop problem

$$(2.4) \quad \begin{cases} y_t = A(x,\partial)y \\ y(0) = 0 \\ My|\Gamma = g \end{cases}$$

can be written in a semigroup form as

(2.5) $$y(t) = A\int_0^t e^{A(t-z)}Dg(z)dz - \lambda_o \int_0^t e^{A(t-z)}Dg(z)dz \equiv (Lg)(t)$$

where λ_o is fixed (large enough).

It was shown in [Ch-L] that:

(2.6) $$L \in \mathscr{L}(L_2(0T;(L_2(\Gamma)^r), C(0T;[L_2(\Omega)]^k)).$$

One can also rewrite (2.5) as an abstract ODE

(2.7) $$\begin{cases} \dot{y}(t) = Ay + (A-\lambda_o)Dg \\ y(0) = 0 \end{cases}$$

where the equation is understood in the sense of the dual space to $D(A^*)$.

With the operator $B:(L_2(\Gamma)^r \to D(A^*)'$ defined by

(2.8) $$Bg \equiv (A-\lambda_o)Dg \quad \text{on } D(A^*)';$$

(2.7) hence (2.4) can be written as:

(2.7') $$\begin{cases} \dot{y}(t) = Ay + Bg \quad \text{on } D(A^*)' \\ y(0) = 0. \end{cases}$$

(2.7') is an abstract version of (2.4). The specific feature of (2.7') is that the operator B is not bounded from $U \to H$ where $U \equiv (L_2(\Gamma))^r$ and $H \equiv (L_2(\Omega))^k$. Even more, $D(B) = 0$ when B considered as acting between U and H. However, by the virtue of (2.3) we have

(2.9) $$(A-\lambda_o)^{-1}B \in \mathscr{L}(U,H).$$

Having established the semigroup representation for the open loop problem (2.4) we are now in a position to express (1.1) as the following abstract ODE:

$$(2.10)\qquad \begin{cases} \dot{y}(t) = Ay(t) + F_1 y(t) + BF_2 y(t) + BG(y(t)) \text{ on } D(A^*)' \\ y(0) = y_o \quad H \end{cases}$$

where $F_1 \in \mathcal{L}(H,H)$, $F_2 \in \mathcal{L}(H,U)$, G is continuous from H to U and A and B are given by (2.1), (2.8).

The next section will be devoted to the study of stability properties of the abstract ODE in the form (2.10).

3. Stability properties of abstract ODE with unbounded perturbations.

Motivated by the example 1.1 in section 1 we are led to consider the following abstract model:

$$(3.1)\quad \begin{cases} y_t = Ay(t) + F_1y(t) + BF_2y(t) + BG(y(t)) \\ y(0) = y_o \in X. \end{cases}$$

Here A is the generator of C_o-semigroup on a Banach space X, $F_1: X \to X$ and $F_2: X \to U$ are stabilizing feedbacks with U another Banach space. $B: U \to X'$ is assumed to satisfy $R(\lambda,A)B \in \mathscr{L}(U,X)$ Finally $G: X \to U$ is a nonlinear continuous operator from X into U such that $G(0) = 0$ and

$$(3.2)\quad ||G'(y)||_{X\to U} \to 0 \text{ when } ||y||_X \to 0$$

Assuming that the stabilizing feedbacks F_1 and F_2 are selected in such a way as to guarantee the exponential stability of the linear system (i.e. with $G\equiv 0$), the main goal of the present section is to study the stability properties of the overall system (3.1) for <u>all</u> perturbations G subject to (3.2). As we have mentioned already in the introduction, the major novelty of the problem under study is that the perturbation G is acting on the system through the <u>unbounded</u> operator B and that the semigroup e^{At} is an arbitrary C_o-semigroup (versus analytic case).

Our main abstract result is formulated in the Theroem below.

<u>Theorem 1:</u>

Assume that

(H-1) $A_F \equiv A + F_1 + BF_2$ generates an exponentially stable semigroup i.e.

$$||e^{A_Ft}||_{X\to X} < Ce^{-\alpha_o t}; \; \alpha_o > 0; \; t > 0,$$

(H-2) $A_F^{-1}B \in \mathscr{L}(U,X)$

(H-3) $\quad \int_0^T |B^* e^{A_F^* t} x|_U dt \leq C_T ||x||_{X^*}$ for some $T > 0$.

Then there exists $R > 0$ such that for all $||y_o||_X \leq R$ the solution $y(t)$ corresponding to (3.1) exists globally in $C(0\infty; X)$, it is unique and it satisfies

$$||y(t)||_X \leq Ce^{-\alpha t}||y_o||_X \quad \alpha < \alpha_o; \; t > 0.$$

Remarks.

1. Condition limiting "the unboundedness" of the operator B is the hypothesis (H-3). (H-3) is expressed in terms of B^* and A_F^*. Thus the degree of unboundedness of B depends on the structure of stabilizing feedback operator $F \equiv F_1 + BF_2$. This is in contrast with the standard theory of bounded perturbation where the exponential stability of the nonlinear system (with G subject to (3.2)) holds with any feedback stabilizing the linear part of the system.

2. Notice also that the hypothesis (H-3) holds automatically if A_F is a generator of analytic semigroup and B is A_F-bounded. In this sense our theorem generalizes perturbation stability results formulated for analytic case in [K-1].

Proof of Theorem 1

We shall start with the following Lemma:

Lemma 3.1

Under the hypothesis of Theorem 1 we have

$$\int_0^\infty |B^* e^{A_F^* t} x|_U e^{\alpha t} dt \leq \frac{C_T e^{\alpha T}}{1 - e^{-(\alpha_o - \alpha)T}} ||x||_{X^*} \leq C||x||_{X^*}.$$

Proof:

From (H-3) we obtain

$$(3.3)\qquad \int_0^T |B^* e^{A_F^* t} x|_U e^{\alpha t}\, dt \le C_T e^{\alpha T} ||x||_{X^*} = \hat{C}_T ||x||_{X^*}$$

where we take $\alpha < \alpha_o$.

Next we compute

$$\int_T^{2T} |B^* e^{A_F^* t} x|_U e^{\alpha t}\, dt = \int_T^{2T} |B^* e^{(A_F^*+\alpha)(t-T)} e^{(A_F^*+\alpha)T} x|_U\, dt =$$

$$= \int_0^T B^* e^{(A_F^*+\alpha)t}\, e^{(A_F^*+\alpha)T} x|_U\, dt \le \hat{C}_T ||e^{(A_F^*+\alpha)T} x||_{X^*}$$

where in the last inequality we have used (3.3) applied to the adjoint semigroup.

Generally we have

$$\int_{(n-1)T}^{nT} |B^* e^{A_F^* t} x|_U e^{\alpha t}\, dt \le \hat{C}_T ||e^{(A_F^*+\alpha)(n-1)T} x||_{X^*}.$$

Thus

$$\int_0^{nT} |B^* e^{A_F \alpha t} x|_U e^{\alpha t}\, dt \le \hat{C}_T[1 + ||e^{(A_F^*+\alpha)T} x||_{X^*} t + \dots ||e^{(A_F^*+\alpha)(n-1)T} x||_{X^*}]$$

by (H-1)

$$\le \hat{C}_T[1 + e^{-(\alpha_o-\alpha)T} + \dots e^{-(\alpha_o-\alpha)(n-1)T}] ||x||_{X^*} \le \hat{C}_T \sum_{n=0}^{\infty} (e^{-(\alpha_o-\alpha)T})^n ||x||_{X^*}$$

$$= \hat{C}_T ||x||_{X^*} \frac{1}{1-e^{-(\alpha_o-\alpha)T}}$$

which completes the proof of the Lemma.

Next define

L_α: $C(0\infty;U) \to C(0\infty;X)$ by

$$(L_\alpha u)(t) \equiv A_F \int_o^t e^{(A_F+\alpha)(t-z)} A_F^{-1} B\, u(z)dz.$$

Since by the virtue of (H-2), $A_F^{-1}L_\alpha \in \mathscr{L}(C(0\infty;U), \to C(0\infty;X))$, L_α is closed. Moreover, L_α is weak* densely defined on $C(0\infty;\, U)$ as $C^1(0\alpha;U) \subset \mathscr{D}(L_\alpha)$. This can be easily verified by integrating by parts term $\int_o^t \frac{d}{dz} e^{(A_F+\alpha)(t-z)} A_F^{-1} Bu(z)dz$.

Moreover, we shall prove

<u>Lemma</u> 3.2.

$$L_\alpha \in \mathscr{L}(C(0,\infty,U);\; C(0\infty;X))$$

<u>Proof:</u>

Assume first that $u \in C^1(0\infty;U)$.

Then

$$(L_\alpha u)(t) = A_F^{-1}Bu(t) - e^{(A_F+\alpha)t} A_F^{-1}Bu(0) - \int_o^t e^{(A_F+\alpha)(t-z)} A_F^{-1}B\dot{u}(z)dz$$

and $L_\alpha u \in C[0,\infty;H]$.

Notice also that

(3.4) $\qquad |(L_\alpha u)(t)|_X = ((L_\alpha u)(t),\, x_t^*)_{X,X^*}$ for some $x_t^* \in X^*$ such that $||x_t^*||_{X^*} = 1$. With $x_{nt}^* \in D(A_F^*)$ and $u \in C^1(0\infty;U)$ we compute

$$((L_\alpha u)(t),\, x_{nt}^*)_{X,X^*} = (A_F \int_o^t e^{(A_F+\alpha)(t-z)} A_F^{-1}Bu(z)\, dz,\; x_{n,t}^*)_{X,X^*}$$

$$= \int_o^t (e^{(A_F+\alpha)(t-z)} A_F^{-1}Bu(z),\; A_F^* x_{n,t}^*)_{X,X^*}\, dz$$

$$= \int_o^t \langle u(z),\; B^* e^{(A_F^*+\alpha)(t-z)} x_{n,t}^*\rangle_U dz.$$

Hence

$$|((L_\alpha u)(t),\, x_{nt}^*)_{X,X^*}| \le \sup_{t>0} |u(t)|_U \int_o^\infty |B^* e^{(A_F^*+\alpha)(t-z)} x_{n,t}^*|_U dz,$$

and by Lemma 3.1

(3.5) $\leq C \sup_{t>0} |u(t)|_U \, ||x_{n,t}^*||_{X^*}$

where C does not depend on t.

Now for every x_t^* such that $||x_t^*||_{X^*} = 1$ and (3.4) holds we select a sequence $x_{nt}^* \in D(A^*)$ such that $x_{nt}^* \to x_t^*$ in the w*-topology of X* as $n \to \infty$.

Thus with $u \in C^1[0\infty;U]$ we obtain

$\langle L_\alpha u(t), x_{nt}^*\rangle_{X,X^*} \to \langle (L_\alpha)(t), x_t^*\rangle_{X,X^*}$, and from (3.4) and (3.5)

$|(L_\alpha u)(t)|_X = \lim_{n\to\infty} \langle L_\alpha u(t), x_{nt}^*\rangle_{X,X^*} \leq$

$C \sup_{t>0} |u(t)_U \, ||x_t^*||_{X^*} = C|u|_{C[0,\infty;U]}$.

A standard closedness and density argument extends this result to all $u \in C[0\infty;U]$. The proof of the Lemma is thus completed. ■

To continue with the proof of Theorem 1, we shall construct the solution of (3.1) by constructing a fixed point of

(3.6) $$y(t) = e^{A_F t} y_o + \int_o^t e^{A_F(t-z)} B\, Gy(z)\, dz$$

setting $v(t) \equiv e^{\alpha t} y(t)$ we rewrite (3.6) as

(3.7) $$v(t) = e^{(A_F+\alpha)t} y_o + L_\alpha(e^{\alpha \cdot} G(e^{-\alpha \cdot}(\cdot)))(t).$$

Denoting the RHS of (3.7) by (Fv)(t), we see that solving (3.7) is equivalent to finding v such that

(3.8) $Fv = v$.

Thus in order to prove Theorem 1 it is enough to assert that F has the unique fixed point on the space Z defined by

$$Z \equiv \{z \in C[0\infty;X]: \sup_{t<0} |z(t)|_X < R_o\} \text{ for some } R_o > 0.$$

By using Lemma 1, hypothesis (H-1) and following the same arguments as in [L-1] one can show for any $y_o \in X$ such that

$||y_o||_H < \frac{2R_0}{M}$ where R_o is sufficiently small we have

(3.9) $\quad FZ \subset Z$ and

$$(3.10) \qquad |F_{v_1} - F_{v_2}|_{C[0\infty;X]} < \zeta |v_1 - v_2|_{C[0,\infty]} \qquad \zeta < 1.$$

Thus the conclusion in Theorem 1 follows by the virtue of the standard Fixed Point Theorem.

■

Below we shall give sufficient conditions for the hypothesis H-3 to hold, expressed in terms of A and B, in the special case when the stabilizing feedback operators F_1 and F_2 are bounded.

Theorem 2.

Assume that $F_1 \in \mathscr{L}(X,X)$, $F_2 \in \mathscr{L}(X,U)$. Moreover assume that

$$(\text{H-3}') \qquad \int_0^T |B^* e^{A^* t} x|_U dt < C_T ||x||_{X^*}$$

where $C_T \to 0$ when $T \to 0$.
Then the hypothesis (H-3) holds true.

Remark.
The main obstacle in applying Theorem 1 (or 2) to a concrete pde problem is the verification of condition (H-3). Later we shall see that in the case of hyperbolic problems with nonlinearity in the boundary conditions, condition (H-3) amounts to the fact that the "traces" on the boundary corresponding to the solutions $e^{A_F^* t}$ are bounded in $L_1(0T;U)$. The main difficulty here is that the above regularity of the traces does not follow from the interior regularity of the solutions generated by A_F^*, and it is to be established as an independent regularity result. Thus, the advantage of hypothesis (H-3') over (H-3) is that in order to establish the above mentioned

regularity for the feedback semigroup, it is enough to have these properties for the generic semigroup e^{A^*t}. This is usually a simpler task to accomplish.

Proof of Theorem 2.

Define:

$$w(t) = e^{A_F^*t} x. \quad \text{Then}$$

$$(3.11) \qquad w(t) = e^{A^*t} x + \int_0^t e^{A^*(t-z)} F_1^* w(z)dz + \int_0^t e^{A^*(t-z)} F_2^* B^* w(z)dz.$$

Let $\Lambda : L_1(OT;D(B^*)) \to$ itself be defined by the right hard side of equation (3.11). We shall prove that Λ has a fixed point. Indeed

$$|\Lambda w(t)|_{D(B^*)} < |B^* e^{A^*t} x|_U + \int_0^t |B^* e^{A^*(t-z)} [F_1^* + F_2^* B^*] w(z) dz|_U .$$

Hence

$$\int_0^T |\Lambda w(t)|_{D(B^*)} dtH < \int_0^T |B^* e^{A^*t} x \quad |_U dt +$$

$$+ \int_0^T \int_z^T |B^* e^{A^*(t-z)} (F_1^* + F_2^* B^*) \cdot w(z)|_U dtdz$$

by (H-3')

$$< C_T ||x||_{X^*} + C_T \int_0^T |(F_1^* + F_2^* B^*) w(z)|_U dz < C_T ||x||_{X^*} +$$

$$+ C_T ||w||_{L_1[OT:D(B^*)]} .$$

Now we shall prove the Λ is a contraction on $L_1(Ot_o, D(B^*)$ for some $t_o > 0$. In fact,

$$\int_0^{t_o} |(\Lambda w_1 - \Lambda w_2)(t)|_{D(B^*)} dt < \int_0^{t_o} \int_0^t |B^* e^{A^*(t-z)} F_1^* + F_2^* B^*)(w_1 - w_2)(z)|_U dzdt$$

$$< \int_0^{t_o} \int_z^{t_o} |B^* e^{A^*(t-z)} (F_1^* + F_2^* B^*)(w_1 - w_2)(z)| dtdz$$

by (H-3')

$$\leq C_{t_o}\int_o^{t_o} |(F_1^* + F_2^*B^*)(w_1-w_2)(z)|_U dz \leq C_{t_o}C_T||w_1-w_2||_{L_1[Ot_o;D(B^*)]}$$

Taking t_o small enough gives contraction on $L_1(Ot_o,D(B^*))$. Thus by fixed point theorem we obtain the existence and uniqueness of w(t) on $L_1(Ot_o,D(B^*))$. Repeating the same argument finitley many times gives global existence of w(t) and

(3.12) $\quad B^*w \in L_1[OT;U]$.

Thus by (3.11) and closed graph theorem

$$\int_o^T |B^*e^{A_F^*t}x|_U dt \leq C_T||x||_{X^*}$$

which is the desired conclusion. ■

<u>Remark</u>: Notice that in the process of proving Theorem 2, we showed that (H-3') and the boundedness of F_1,F_2 are sufficient for A_F to generate C_o semigroup on X.

4. Stability of first order hyperbolic systems (1.1).

In this section we shall return to the nonlinear hyperbolic problem (1.1), introduced in section 1. We have shown in section 2 that (2.11) is an abstract model for (1.1) with $X \equiv [L_2(\Omega)]^k$; $U \equiv (L_2(\Gamma))^r$. Thus we are in a position to apply our general results formulated in Theorems 1 and 2. To this end we need to verify the hypothesis (H-2) and (H-3')(since F_1 and F_2 in (1.1) are assumed to be bounded!). Let us begin with the hypothesis (H-3').

The adjoint operator A^* corresponding to A given by (2.1) has the form:

(4.1a) $$A^*y = A^*(x,\partial)y \quad y \in D(A^*)$$

where $A^*(x,\partial)$ is the formal adjoint to $A(x,\partial)$ and $D(A^*)$ is given by

(4.1b) $$D(A^*) = \{y \in [L_2(\Omega)]^k;\ (A_N^+)^{-1}S^T A_N^- y^- + y^+ = 0\}.$$

Here A_N^+ and A_N^- correspond to the partition of the matrix

$A_N \equiv \sum_{j=1}^{n} A_j(x)n_j$ (with $\vec{n} = (n,\dots n_n)$ outward normal to Γ)

according to the sign of the eigenvalue of $A_N(x)$. In fact, by the virtue of strict hyperbolicity we can assume without loss of

generality that $A_N = \begin{bmatrix} A_N^- & 0 \\ 0 & A_N^+ \end{bmatrix}$ where A_N^-(resp A_N^+) are

$r\times r$ (resp. $k-r\times k-r$) matrices with negative (resp. positive) eigenvalues. Accordingly, the boundary operator M can be written as $M = [I,S]$ where I is $(r\times r)$ identity matrix and $S(x)$ is an $r \times k-r$ smooth matrix valued function.

In order to verify the validity of hypothesis (H-3') we need to characterize the operator B^*. By the virtue of (2.8) we have

(4.2) $\qquad B^*u \equiv D^*(A^*-\lambda_o)u \quad \text{for } u \in D(A^*).$

On the other hand it can be shown (see for example [Ch-L] or [D-S-L]) that:

$$(4.3) \qquad D^*(A^*-\lambda)u = A_N^- u^-|_\Gamma \qquad u \in D(A^*).$$

Remark: As we have already mentioned in section 3, the operator B*, in the special case of boundary hyperbolic problems, is a trace operator--intrinsically unbounded and uncloseable on $L_2(\Omega)$).

If we set $w(t) \equiv e^{A^*t}x$, then

$$(4.4) \qquad B^*e^{A^*t}x = D^*(A^*-\lambda)w(t) = A_N^- w^-(t)|_\Gamma$$

where

$$(4.5) \qquad \begin{cases} \frac{d}{dt} w(t) = A^*(x,\partial)w(t) \\ w(0) = x \\ w^+ + (A_n^+)^{-1}S^T A_N^{-1} w^- = 0 \text{ on } \Gamma. \end{cases}$$

The regularity results for the problem (4.5) given in [R-1] in particular yield:

$$(4.6) \qquad ||w||_{L_2[0T;(L_2(\Gamma)^k]} \leq C\,||x||_{[L_2(\Omega)]^k}\,.$$

Thus

$$(4.7) \qquad \int_0^T ||B^*e^{A^*t}\, x\,||_U^2\, dt =$$

$$\int_0^T ||A_N^-\, w^-(t)||^2_{(L_2(\Gamma))^r}\, dt \leq C||x||^2_{[\,L_2(\Omega)]^k}$$

which inequality a posteriori implies (H-3').

Next we shall establish the validity of (H-2). Notice first that by the same arguments as those used for the proof of Theorem 2 one can show that (4.7) implies

(4.8) $$\int_0^T \|B^* e^{A_F^* t} x\|_U^2 \, dt \leq C\|x\|^2_{[L_2(\Omega)]^k} = C\|x\|_X^2 .$$

where $A_F = A + F_1 + (A-\lambda_o)DF_2$.

Introduce the operator $L_F : L_2(OT;U) \to L_2(OT;X)$ given by

$$(L_F g)(t) = \int_o^t e^{A_F(t-z)} Bg(z)dz.$$

By the virtue of (4.8) and arguing along the same lines as in [L-T-1] one can prove that

(4.9) $$L_F \in \mathscr{L}(L_2(Ot;U) \to ([Ot;X]).$$

On the other hand (4.9) is equivalent to the wellposedness of the following problem:

(4.10) $$\begin{cases} y_t = A_F y + Bg \\ y(0) = 0 \end{cases}$$

where the map $g \to y$ is continuous from $L_2(OT;U)$ into $C[OT;X]$. (4.10) is equivalent in our case to

(4.10') $$\begin{cases} y_t = A(x,\partial)y + F_1 y & \text{in } \Omega \times (OT) \\ M_y|_\Gamma - F_2 y|_\Gamma = g & \text{on } \Gamma \times (0,T) \\ y(0) = 0 \end{cases}$$

By using (with minor modifications) the arguments leading to the result given in Theorem 3.2 [D-L-S], one can show that the stationary problem corresponding to (4.10') is wellposed i.e.: for any $\lambda \in \rho(A_F)$ equation

(4.11) $$\begin{cases} A(x,\partial)y + F_1 y = \lambda y & \text{in } \Omega \\ My|\Gamma - F_2 y|\Gamma = g & \text{on } \Gamma \end{cases}$$

admits unique solution y such that the map $g \to y$ is continuous from $U \to X$. On the other hand (4.11) is equivalent to

(4.12) $$A_F y + Bg = \lambda y.$$

Since $0 \in \rho(A_F)(e^{A_F t}$ is exponentially stable), we can take in (4.12) $\lambda = 0$. Thus $y = -A_F^{-1}Bg$ and $||y||_X \leq C||g||_U$ which completes the proof of (H-2).

The Theorem below summarizes the discussion of this section.

Theorem 3.

Assume that (i) $F_1 \in \mathscr{L}[(L_2(\Omega))^k \to (L_2(\Omega))^k]$, $F_2 \in \mathscr{L}([L_2(\Omega)]^k, [L_2(\Gamma)]^r)$;

(ii) the nonlinear operator

$G : [L_2(\Omega)]^k \to (L_2(\Gamma)]^r$ is continuous and such that: $G(0) = 0$ and

$||G'(y)||_{[L_2(\Omega)]^k \to [L_2(\Gamma)]^r} \to 0$ when $||y||_{[L_2(\Omega)]^k} \to 0$.

Then for all $y_o \in [L_2(\Omega)]^k$, the solution y to (1.1) exists, it is unique in $C[0t; [L_2(\Omega)]^k]$.

If in addition we assume that the operators F_1 and F_2 are such that the linear semigroup generated by (1.1) with $G \equiv 0$ is exponentially stable, with the margin of stability equal to α_o, then there exists $R_o > 0$ such that for all $||y_o||_{L_2(\Omega)} \leq R_o$, the solution $y(t)$ to (1.1) (with $G \neq 0$) satisfies:

$$||y(t)||_{[L_2(\Omega)]^k} \leq Ce^{-\alpha t}||y_o||_{[L_2(\Omega)]^k} \qquad \begin{array}{l}\alpha<\alpha_o \\ t > 0.\end{array} \blacksquare$$

5. Applications to the model of age-dependent population dynamics.

As an expample of the class of problems discussed in sect. 3, 4, we shall consider a simple model describing population dynamics. Consider

$$(5.1)\quad \begin{cases} y_t(x,t) + y_x(x,t) + \mu(x)y(x,t) = 0 & x \in (0,\infty);\ t > 0 \\ y(x,0) = y_o(x) & \\ y(0,t) = \int_o^\infty \beta(x)y(x,t)\,dx + G(y(\cdot,t)). & \end{cases}$$

Here $y(x,t)$ is the density of the population with respect to age x at time t, y_o $C(0,\infty)$, $\mu < \mu(x) < \bar{\mu}$ is the age mortality modulus, β $C(0,\infty)$, $\beta(x) < \bar{\beta}$ is the age fertility modulus and y_o is the known initial age distribution. The operator $G : L_1(0,\infty) \to R^1$ such that $G(0) = 0$ and $|G'(y)| \to 0$ when $|y|_{L_1(0,\infty)} \to 0$ represents a nonlinear perturbation on the boundary . Let us introduce the expression

$$\Pi(a,b) \equiv \exp\left[- \int_a^b \mu(x)dx\right],$$

which describes the probability of survival from age b to a. It is well-known by Sharpe-Lotka Theorem (see [W-1)) that if $\lambda = \lambda_1$ the real solution of

$$(5.2)\qquad 1 = \int_o^\infty e^{-\lambda a}\beta(x)\Pi(x,0)dx$$

is negative, then the solution $y(x,t)$ corresponding to the linear part of (5.1) (i.e. with $G = o$) decays exponentially in $L_1(0,\infty)$ norm. The main goal of this section is to establish that the similar stability result holds in presence of nonlinear perturbation. In order to formulate our results, we need to select an appropriate state space. The natural state space for population dynamics models is $L_1(0,\infty)$. Notice that, if one would to treat the problem within the framework of $L_2(0,\infty)$ space, then the relevant stability properties will follow directly from the results of section 3 (Thm. 3), as (5.1) can be viewed as a simple strictly hyperbolic first order equation. Instead, our aim is to provide the stability results formulated in L_1 topology. To accomplish this we shall use the results of Theorems 1,2, after rewriting eq (1.1) as an abstract ODE.

Let $Ay \equiv -y_x - \mu(x)y \quad y \in D(A)$

(5.3) $\qquad D(A) = \{y \in L_1(0\infty);\ y \in L_1(0\infty),\ y(0) = 0\}$

It is well known that A generates continuous semigroup e^{At} on $L_1(0\infty)$. Let F_2: $L_2(0\infty) \to R^1$ be given by

(5.4) $\qquad F_2 y \equiv \int_0^\infty \beta(x)y(x)dx.$

Clearly $F_2 \in \mathscr{L}(L_1(0\infty);\ R^1)$.

Next we define, as in Sect. 3, the operator $D : R^1 \to L_1(0,\infty)$ as $Dg = v$ iff

$$\begin{cases} v_x + \mu(x)v = 0 \\ v(0) = g \end{cases}$$

It is straight forward to verify that

(5.5) $\qquad v(x) = (Dg)x = g \exp\left(- \int_0^x \mu(x)dx\right)$ and

(5.6) $\qquad D \in \mathscr{L}(R^1,\ L_1(0\infty))$

With the above notation our abstract model for (5.1) is

$$(5.7) \qquad \begin{cases} y_t = Ay - ADF_2 y - ADG(y) \\ y(0) = y_o \quad L_1(0\infty) \equiv X \end{cases}$$

where A,D,F_2 are given by (5.3), (5.4), (5.5). Thus we are in the situation described in section 3 as (5.7) is a special case of (3.1) with

$-B \equiv AD$ (formally); $X = L_1(0\infty)$; $U = R^1$.

By applying Theorem 1 to our case we shall obtain the following result:

Theorem 4.

Assume that λ_1 the real solution of (5.2) is negative. Then there exists $R_o > 0$ such that for all $||y_o||_{L_1(0\infty)} < R_o$ the solution of (5.1) $y(t)$ satisfies

$$||y(t)||_{L_1(0\infty)} < Me^{-\alpha t}||y_o||_{L_1(0,\infty)} \qquad \alpha > 0.\ t > 0.$$

Proof.

The assertion of Thm 4 will follow from Thm 1 as soon as we verify hypothesis H1-H3. Notice first that by Sharpe-Lotka Theorem [W-1] the semigroup generated by $A_F \equiv A - (ADF_2) = A(I-DF_2)$ is exponentially stable on $L_1(0,\infty)$. Thus the hypothesis (H-1) is fullfilled.

As for (H-2) we write

$$A_f^{-1}B = [A(I-DF_2)]^{-1}AD = (I-DF_2)^{-1}D.$$

Since 1 is not an eigenvalue of DF_2 (as 0 is not an eigenvalue of A_F), and DF_2 being bounded and of finite rank, it is compact, then $(I-DF_2)^{-1}$ is bounded on $L_1(0\infty)$. This fact together with (5.6) concludes the proof of (H-2).

To assert the validity of (H-3) we shall use Theorem 2. In fact, since $F_2 \in \mathscr{L}(L_1(0\infty); R^1)$ is is enough to prove that

(5.8) $\quad \int_0^T |D^*A^* e^{A^*t}x|_{R^1} dt < C_T|x|_{L_\infty(0,\infty)} \quad x \ D(A^*)$ where $C_T \to 0$ with $T \to 0$.

It is straight forward to verify that

(5.9) $A^*y = y_x - \mu(x)y \quad y \in D(A^*).$

$D(A^*) = W_1^\infty(0,\infty)$

where A^* generates weakly * continuous semigroup e^{A^*t} on $L_\infty(0,\infty)$. As for D^*A^* (which is weak * densly defined on $L_\infty(0,\infty)$) we compute with $y \in D(A^*)$

$$(D^*A^*y,g)_{R^1} = (A^*y,Dg)_{L_1L_\infty} = \int_0^\infty (y_x-\mu(x)y)(Dg)(x)dx$$

$$= y(\infty)(Dg)(\infty) - y(0)(Dg)(0) - \int_0^\infty(\frac{\partial}{\partial x}(Dg)(x) + \mu(x)Dg(x))y(x)dx = -y(0)g$$

Thus

(5.10) $D^*A^*y = -y(0)$ for $y \in D(A^*)$.

To prove (5.8) we set

$w(t) = e^{A^*t}w_o$. Then

$$(5.11) \quad \begin{cases} w_t = w_x - \mu(x)w & x \in (0,\infty) \\ w(0) = w_o \end{cases}$$

It is known the solution to (5.11) when restricted to $w_o \in C(0,\infty)$ is strongly continuous in x, hence $|w(t,x)| < C|w_o|_{L_\infty(0,\infty)}$ for all $x \in (0,\infty)$. and $w_o \in C(0,\infty)$.

Thus

$$|D^*A^*e^{A^*t}y| = |(e^{A^*t}y)(0)| = |w(t,0)| < C|y|_{L_\infty(0,\infty)}.$$

for $y \in C(),\infty)$ hence in particular for $y \in D(A^*) = W_1^\infty(0,\infty)$ This completes the proof of (5.8) hence of (H-3). The proof of the Theorem is thus completed.

References

[Ch-L] S. Chang, I. Lasiecka Riccati equations for nonsymmetric and nondissipative hyperbolic systems with L_2-boundary controls. J. Math. Anal. Appl. vol. 116, No. 2, (1986) pp. 378-414.

[D-L-S] W. Desch, I. Lasiecka, W. Schappacher. Feedback boundary problems for linear semigroups. Israel J. of Mathematics. Vol. 51, No. 3, (1985) pp. 77-207.

[K-1] H. Kielhofer. Stability and semilinear evolution equations in Hilbert space. Arch. Rational Math. Anal 57(1974) pp. 150-165.

[L-1] I. Lasiecka. Stabilization of hyperbolic and parabolic equations with nonlinearly perturbed boundary conditions. In preparation.

[L-T] I. Lasiecka, R. Triggiani. Regularity of hyperbolic equations under $L_2(OT; L_2(\Gamma)$ -Dirichlet boundary terms. Appl. Math. and Optimiz. Vol. 10, (1983) pp. 275-286.

[M-O] A. Majda, S. Osher. Initial boundary value problems for hyperbolic equations with uniformly characteristic boundary. Comm. Pure Appl. Math. 28 (1975) pp. 607-676.

[R-1] J. Rauch. L^2 is a continuable initial condition for Kreiss' mixed problems. Comm. Pure Appl. Math. 25 (1972) pp. 265-285.

[W-1] G. F. Webb. A semigroup proof of the Sharpe-Lotka Theorem. Lecture Notes in Mathematics. 1076 Infinite Dimensional Systems Springer-Verlag 1984.

FREE BOUNDARIES AND NON-SMOOTH SOLUTIONS TO SOME FIELD EQUATIONS : VARIATIONAL CHARACTERIZATION THROUGH THE TRANSPORT METHOD

J.J. MOREAU

Laboratoire de Mécanique Générale des Milieux Continus
Université des Sciences et Techniques du Languedoc
34060 MONTPELLIER-Cédex, France

1. INTRODUCTION

Some methods primarily devised for the optimization of domains are currently used also in the numerical treatment of problems, arising from Physics or Engineering, which involve unknown boundaries. This requires, of course, that the location of the said boundaries could be characterized variationally.

To the author's knowledge, it is in the dynamics of inviscid fluids that variational statements of such a sort have been first proposed. In that field of applications, the unknown surfaces may represent the free boundary through which a liquid confines an atmosphere with negligible inertia and given pressure ; they may also describe a jet boundary, in a possibly compressible flow, i.e. a discontinuity locus of the hydrodynamic field, separating two parts of the fluid with preserved material identity (shock waves do not fall into the scope of this lecture). Results in that line have been known since the fifties [1] [2] . In recent papers [3] [4] , the author has shown that, when considered from the viewpoint of the dynamics of the whole material in presence, the determination of such surfaces is a problem whose nonlinearity has the same intimate structure as the nonlinearity of the conventional equations of fluid dynamics holding in the regions of smooth flow. This is made clear by expressing dynamics in terms of Schwartz's distributions ;

in doing it, one puts forward the vector distribution *divergence* of a second order *tensor measure* associated with the time-space distribution of mass and velocity. For instance, the free boundary of a liquid appears as a surface, interior to the investigated region of time-space, across which the material density abruptly drops to the zero value corresponding to the assumedly mass-less atmosphere. Expressing the dynamics of the whole in terms of distributions encompasses the usual equations, verified in the liquid domain, as well as the conditions to be satisfied on the free surface.

Some details on this aspect of dynamics are given in Sect. 2 below ; for brevity only steady flows are considered.

The next step consists in giving a variational significance to the divergence operator acting on second order tensor measures. When the traditional calculus of variation is applied, with a view to characterize the possible solutions to some field equations, an alteration of the investigated field is performed by adding to it a term, arbitrary in a certain class of functions. This additive variation has to be smooth enough for the familiar trick of integration by part to work ; such a procedure leaves invariant the location of possible singularities and so is unable to characterize it. For this reason, we have instead proposed the *transport method* (called in [3] and [4] the method of *horizontal variations* ; it seems preferable to abandon this denomination which could generate confusion with some other uses of the word "horizontal", in Differential Geometry).

In this method every alteration of the investigated object is effected by transporting it along an arbitrarily chosen smooth vector field, say φ, with compact support in the considered region of $\mathbb{R}^n$. This vector field may be viewed as the velocity field of some imagined continuous medium Λ, called a *carrier*. When the technique is used in a problem of continuum mechanics, one should keep in mind that the carrier has nothing to do with the material in presence ; in particular, the real variable indexing the evolution of Λ in $\mathbb{R}^n$ is denoted by τ, not to be confused with the time t of Dynamics, when the latter figures among the problem variables. For every position of the carrier, a certain real functional, involving the transported object, is calculated. The result is a real function of τ ; expressing that its derivative vanishes at $\tau = 0$, whatever is the test vector field φ , yields the expecced characterization. Such a derivation, applied to a geometric object transported along a vector field is known in Differential Geometry as a *Lie derivation*. For the applications we have in view, it seems more efficient to describe the transport process in terms of the classical kinematics of continua, using but a little of the formalism of Differential Geometry.

Section 3 provides the necessary information about the concepts of vector and tensor distributions on a differential manifold, with emphasis on the special case where distributions are actually measures.

The transport of such objects by what we have called a carrier is described in Section 4.

This yields in Section 5 the very simple formula by which the divergence operation, acting on a doubly contravariant tensor measure, is interpreted variationally in terms of transport. The real functional to be extremized is *the integral of the Euclidean trace* of the investigated tensor measure. The reader specially interested in Mechanics could additionally refer to [3] , where Hamilton's principle of the least action is connected with this formula.

Section 6 gives some examples of applications of the preceding to hydrodynamical situations.

Then comes in Section 7 the study of the *second derivative* of the considered real functional, in the course of any twice differentiable transport, if the investigated tensor measure makes the first derivative vanish. This provides a necessary condition for the functional to achieve a local minimum. The calculation results in some positivity property, concerning the investigated tensor measure, which tends to explain the preeminence of measures, in that context, over distributions of higher order.

The final Section 8 summarizes the logical pattern of the transport method and sketches its application to more general situations.

2. STEADY FLOW OF AN INVISCID FLUID

In some region of a tridimensional reference frame, with orthonormal Cartesian coordinates x_1, x_2, x_3, the steady motion of an inviscid fluid is considered. Let u_1, u_2, u_3 denote the components of the velocity field ; let p and ρ be the pressure and density scalar fields. The components of the gravity field are given, equal to the partial derivatives $U_{,i}$ of some real function U. Then, under the usual smoothness assumption for the investigated fields, the Euler time-independent equations of fluid dynamics write down as

$$\rho\, u_j\, u_{i,j} = -\, p_{,i} + \rho\, U_{,i} \ , \tag{2.1}$$

to be joined with the equation of mass conservation

$$(\rho u_j)_{,j} = 0 \ . \tag{2.2}$$

By combination, this yields (δ_{ij} is the Kronecker symbol)

$$(\rho u_i u_j + p\, \delta_{ij})_{,j} = \rho U_{,i} \ . \tag{2.3}$$

The left-hand side may be seen as the i-component of the vector field *divergence of the symmetric tensor field* with components in brackets.

If on the contrary the involved functions are not smooth enough for the partial derivatives to exist in the elementary sense, it is generally admitted that such partial differential equations as above have to be understood with reference to the partial derivatives of *distributions*. This is an abuse of language ; in Schwartz's theory of distributions, a real function f of the x-variables cannot constitute a distribution. But, as soon as f is locally integrable relative to the Lebesgue measure ℓ, there is defined the real measure $f\ell$ i.e. the measure admitting f as *density* relative to ℓ. A real measure on some region of x-space is a special sort of Schwartz distribution ; then (2.3) will precisely be replaced by

$$(\rho\, u_i u_j \ell + p\, \delta_{ij} \ell)_{,j} = \rho U_{,i} \ell \ . \tag{2.4}$$

In order that the *tensor measure* with components in brackets make sense, one supposes $\rho \in L^{\infty}_{loc}$, $u_i \in L^2_{loc}$, $p \in L^1_{loc}$ (the components $U_{,i}$ of the gravity field are essentially smooth functions).

Similarly (2.2) expresses that the *vector measure*, admitting as Cartesian components the three real measures $\rho u_i \ell$, has zero divergence in the sense of distributions.

Before showing how this formulation of hydrodynamics may encompass some free boundary situations, let us recall an elementary calculation rule for distributions.

Assume that the considered domain Ω of the x-space is divided by a surface S determining two subdomains Ω^+ and Ω^- . At every point of S , denote by n_i the components of the normal unit, directed toward Ω^+ and assumed continuous on S . If a function f is C^1 in Ω^+ and Ω^- and possesses respective one-sided limits f^+ and f^- at every point of S , the measure $f\ell$ is a distribution on Ω whose partial derivative relative to x_i is easily found equal to the sum of the two following measures :

(a) $f_{,i}\ell$, diffuse in Ω .

(b) $(f^+ - f^-)n_i s$ concentrated on S .

Here s denotes the area measure of S , a nonnegative real measure on Ω , with S as support.

Coming back to hydrodynamics, suppose that Ω^- contains an inviscid fluid, while Ω^+ corresponds to an atmosphere with $\rho = 0$. Mass conservation is expressed by $(\rho u_j \ell)_{,j} = 0$; through the above calculation rule, this is equivalent to (2.2) being satisfied in Ω^+ and Ω^- and to the vanishing of the measure concentrated on S whose density relative to the area measures equals $[(\rho u_j)^+ - (\rho u_j)^-]n_j$. Since $\rho = 0$ in Ω^+ , the latter simply yields $\rho^- u_j^- n_j = 0$, as expected for a steady flow. After that, the same calculation rule is applied to (2.4) . Considering, on one hand, the diffuse part, one obtains that (2.3) holds in Ω^- and Ω^+ ; in particular this yields that p equals a constant throughout Ω^+ . As for the part of its left-hand member concentrated on S , equation (2.4) yields

$$\bar{\rho}\, u_i u_j n_j + \bar{p}\, n_i = \rho^+ u_i^+ u_j^+ n_j + p^+ n_i .$$

Since $\rho^+ = 0$ and $\bar{\rho}\, u_j n_j = 0$, this is just the pressure condition $p^+ = p^-$.

3. VECTOR AND TENSOR MEASURES ON A MANIFOLD

A possible way of constructing measure theory consists in taking the Riesz representation theorem as a definition. From that standpoint, by a real (signed) measure on a locally compact topological space X (one may precise : a Radon measure) is meant a real linear functional, meeting some continuity requirements we shall recall later, defined on the space $\mathcal{D}^o(X)$ of the continuous real functions with compact support in X . The treatise of Bourbaki [5] is developed in that line, up to include the more sophisticated matter of measures with values in topological linear spaces.

In particular, the considered locally compact space may be an n-dimensional C^k-differential manifold M , with $k \geq 1$. Then the above amounts to define measures as *distributions* of a special sort. Let ℓ be an integer, $0 \leq \ell \leq k$; denote by $\mathcal{D}^\ell(M)$ the linear space of the C^ℓ real functions on M with compact support. For every compact subset K of M , denote by $\mathcal{D}^\ell_K(M)$ the subspace of $\mathcal{D}^\ell(M)$ consisting of the functions f with support contained in K . A Banach norm $\|f\|_{K,\ell}$ is defined on $\mathcal{D}^\ell_K(M)$ as the sup of the absolute values of f and of its partial derivatives up to order ℓ at all points of M ; of course, this is conditioned by the choice of coordinates in the C^k manifold M , but in view of $\ell \leq k$, any admissible change of coordinates (more precisely the change of a covering of M with local charts for another one) replaces the said norm by an equivalent one. By definition, a real distribution of order ℓ (strictly speaking, one should say "of order $\leq \ell$") is a linear functional on $\mathcal{D}^\ell(M)$ whose restriction to every $\mathcal{D}^\ell_K$ is continuous.

Real distributions of order ℓ on M make a topological linear space denoted by $\mathcal{D}'^\ell(M)$; real measures on M are the elements of $\mathcal{D}'^o(M)$.

In the author's view , the above duality construction serves the purpose of Mechanics very well. Generally, dual linear spaces have been a basic ingredient of Classical Mechanics much before the concept was mathematically formalized. This duality is in fact the essence of the method of "virtual power" or "virtual work" which has played a central role in Mechanics since the 17 th century at least (some authors trace it back to Aristotle).

Another definite advantage of this approach of measures is that it readily adapts to the introduction of *vector measures* on the manifold M. With every point x of M is associated the dual pair of n-dimensional linear spaces M'_x, the *tangent* space, and M'^{*}_x the *cotangent* space. By definition a *field of covectors* on M (or section of the cotangent fiber bundle) is an assignement associating with every $x \in M$ an element, say $v(x)$ of M'^{*}_x. Basically, the choice of a (local) coordinate system $(x^1, x^2, \ldots, x^n)$ in M induces respective bases in the linear spaces M'_x and M'^{*}_x. Then the covector field v may be described by n functions $v_i(x^1, x^2, .., x^n)$ expressing the components of $v(x)$ in M'^{*}_x. Since, by axiom, any change of (local) coordinate system in M is C^k, $k \geq 1$, the concept of the continuity of v is coordinate-free, as well as that of the support of such a field. We shall denote by $\mathcal{D}^{o}(M,M'^{*})$ the linear space of the continuous covector fields on M with compact support and by $\mathcal{D}^{o}_K(M,M'^{*})$ the subspace consisting of those fields whose support is contained in some compact subset K of M. The sup of the absolute values of the components of $v \in \mathcal{D}^{o}_K(M,M'^{*})$ constitutes a Banach norm $\|v\|_K$ on this space ; changing the coordinate system (more precisely changing the covering of M by local charts) replaces this norm by an equivalent one.

By definition, a *vector measure* on M is a real linear functional on $\mathcal{D}^{o}(M,M'^{*})$ whose restriction to each $\mathcal{D}^{o}_K(M,M'^{*})$ is continuous. More generally, such a duality procedure has been used by G. de Rham when constructing his theory of *currents* on C^∞ manifolds [9] .

Vector measures on M make a topological linear space denoted by $\mathcal{D}'^{o}(M,M')$. Observe that, in contrast with the special case where M is an open subset of $\mathbb{R}^n$, a vector measure on an arbitrary C^k manifold M can by no means be seen as an additive vector function of sets since it does not make sense to add vectors localized at different points of the manifold.

When applied to a covector field $v \in \mathcal{D}^o(M,M'^*)$, a vector measure $m \in \mathcal{D}'^o(M,M')$ yields, by definition, a real number denoted by $\ll v,m \gg$ or, more expressively by $\int \langle v,dm \rangle$. A soon as m is fixed the meaning of this symbol may be extended to a larger class of covector fields than $\mathcal{D}^o(M,M'^*)$, said integrable relative to m. From the Radon-Nikodym theorem one easily deduces that every $m \in \mathcal{D}'^o(M,M')$ can be (non uniquely) represented under the form $m = m'_\mu \mu$ where μ is a nonnegative real measure on M and m'_μ is a locally μ-integrable vector field. This means that for every $v \in \mathcal{D}^o(M,M'^*)$, one has

$$\int \langle v,dm \rangle = \int \langle v(x),m'_\mu(x) \rangle \, d\mu \ ;$$

here $\langle .,. \rangle$ denotes, at every point x of M , the real-valued duality pairing of the cotangent and tangent spaces.

Incidentally, a vector measure $m \in \mathcal{D}'^o(M,M')$ is said *divergence-free* if, for every $\varphi \in \mathcal{D}^1(M)$, one has $\int \langle \mathrm{grad}\, \varphi \ , dm \rangle = 0$; we denote by $\mathrm{grad}\, \varphi$ the gradient field of φ (or differential of φ) , naturally an element of $\mathcal{D}^o(M,M^*)$. In [6] this concept has been applied to classical hydrodynamics, yielding a generalization of the Kelvin-Helmholtz theorem on vorticity which encompasses more recent results as the conservation of the *helicity* of a flow. More generally, the divergence of a vector measure on M may be defined as a scalar distribution of order 1 . The point to be stressed is that this operation makes sense in the simple framework of the differential manifold M , without reference to any metric or connection ; this contrasts with the divergence of vector fields.

A similar duality device may be generally used in defining *tensor distributions*, in particular *tensor measures*, on the C^k manifold M . For instance, there exist four sorts of *second order tensor fields* on M : they are assignements associating respectively with every $x \in M$ an element of $M'_x \otimes M'_x$, $M'_x \otimes M'^*_x$, $M'^*_x \otimes M'_x$, or $M'^*_x \otimes M'^*_x$. This allows for the definition of spaces $\mathcal{D}^\ell$, $\mathcal{D}^\ell_K$ consisting of such fields and, through duality, the definition of spaces of tensor distributions of order ℓ with specified tensorial type. In particular, a *doubly contravariant tensor measure* on M is an element of $\mathcal{D}'^{o}(M,M'\otimes M')$, the dual space of $\mathcal{D}^{o}(M,M'^{*}\otimes M'^{*})$; the latter consists of continuous doubly covariant tensor fields on M, with compact supports. As in the case of vector measures, the Radon-Nikodym theorem may be used in order to prove that, for every $T \in \mathcal{D}'^{o}(M,M'\otimes M')$, there exist (non uniquely) a nonnegative real measure μ and a doubly contravariant locally μ-integrable tensor field T'_{μ} such that $T = T'_{\mu}\,\mu$. Through the use of local coordinates $(x^1,\dots,x^n)$ in X, one defines the tensor field $x \to T'_{\mu}(x) \in M'_x \otimes M'_x$ by its components T'^{ij}_{μ}, which are elements of $L^1_{loc}(X,\mu;\mathbb{R})$; one may even choose μ in order that they belong to $L^{\infty}_{loc}(X,\mu;\mathbb{R})$.

4. CARRIERS AND LIE DERIVATIVES

For all the sequel, X denotes a fixed n-dimensional C^k manifold, $k \geq 2$; in usual applications, X simply reduces to an open subset of $\mathbb{R}^n$.

Let $\varphi \in \mathcal{D}^1(X,X')$, i.e. φ is a C^1 vector field of X, with compact support. We are to look at it as *the Eulerian velocity field of some continuum* Λ *in motion over* X. This precisely means that every element, or *particle*, of Λ is a moving point in X, say $\tau \to \xi(\tau)$ verifying the differential equation

$$\frac{d\xi}{d\tau} = \varphi(\xi(\tau)) . \qquad (4.1)$$

Through the use of (local) coordinates in X, the study of this differential equation in the manifold may be reduced to the similar problem in $\mathbb{R}^n$, for which standard theory is available. The assumptions made about φ secure that, for every $\xi_o \in X$, there exists a unique solution $\tau \to \xi(\tau)$ to (4.1), defined for τ ranging over the whole real line, such that $\xi(0) = \xi_o$. Furthermore, standard results concerning the dependence on initial conditions of the solutions to differential equations, entail that, for every fixed τ, the mapping $\xi_o \to \xi(\tau)$, commonly denoted by $\exp \tau\varphi$, is C^1 of X to itself. Since this mapping admits $\exp(-\tau)\varphi$ as inverse, it constitutes a C^1 diffeomorphism of X, leaving invariant every point of the subset $X \setminus \text{support } \varphi$.

Let us express as $\tau \to p(\tau,\lambda)$ the motion of a particle λ of Λ relatively to X ; under the usual wording of the kinematics of continua, for every τ , the mapping $p_\tau : \lambda \to p(\tau,\lambda)$ is called *the placement of* Λ *into* X *at time* τ. The above statements equivalently mean that, for every τ , the placement $p_\tau : \Lambda \to X$ is one-to-one and that the "transplacement" $p_\tau \circ p_0^{-1}$ is a C^1 diffeomorphism of X . We shall in turn reformulate this by saying that the set Λ may be equipped with the structure of a C^1 manifold, in such a way that, for every τ , the placement p_τ is a C^1 diffeomorphism of Λ onto X .

Commonly in Mechanics, time derivatives are denoted by a dot. In accordance, $\dot{p}(\tau,\lambda)$ will refer here to the derivative of the mapping $\tau \to p(\tau,\lambda)$ of $\mathbb{R}$ into X ; for every τ and λ , this derivative is an element of the tangent space X'_x , $x = p(\tau,\lambda)$. The differential equation (4.1) , with which all began, manifests itself as the identity

$$\forall \tau \in \mathbb{R} \ , \ \forall \lambda \in \Lambda \quad : \quad \dot{p}(\tau,\lambda) = \varphi(p(\tau,\lambda)) \ . \tag{4.2}$$

We shall call a *carrier* such moving differential manifold as Λ , elaborated from a given C^1 vector field φ in X . For simplicity, we started with φ independent of τ , i.e. the motion of Λ over X is a steady flow. It is sometimes useful to consider more generally a vector field $x \to \varphi(\tau,x)$ in X , depending on τ, at least for τ ranging over some open real interval I containing zero ; then φ will be supposed C^1 in τ and x jointly (equivalently $(\tau,x) \to (1,\varphi(\tau,x))$ is a C^1 vector field of the product manifold $I \times X$) ; in addition the support of $x \to \varphi(\tau,x)$, for every $\tau \in I$, will be assumed contained in a τ-constant compact subset of X . Again this allows one to equip Λ with the structure of a C^1 manifold, in such a way that every placement $\lambda \to p(\tau,\lambda)$, $\tau \in I$, is a C^1 diffeomorphism of Λ onto X .

With a C^1 manifold, as are Λ or X above, one may associate various linear spaces, respectively consisting of real functions, vector or tensor fields, distributions of order ≤ 1 , etc. For every element of any of these spaces, there is a natural definition of its *image* under any C^1 diffeomorphism of the considered manifold onto another one ; this image is an object of the same nature attached to the target manifold. The spaces $\mathcal{D}^o$, $\mathcal{D}^1$, $\mathcal{D}'^o$, $\mathcal{D}'^1$ introduced in the preceding section generate examples of this. As another instance, consider the linear space $C^1(\Lambda,\mathbb{R})$ of the continuously differentiable real functions on the C^1 manifold Λ and some C^1 diffeomorphism $p : \Lambda \to X$. For every real function $\kappa : \Lambda \to \mathbb{R}$, the image under p is naturally defined as the function $k = \kappa \circ p^{-1} : X \to \mathbb{R}$, which belongs to $C^1(X,\mathbb{R})$ iff $\kappa \in C^1(\Lambda,\mathbb{R})$. Similarly may be considered a *vector field* on Λ , say $\lambda \to \alpha(\lambda) \in \Lambda'_\lambda$. Its image under p is defined as $x \to p'_\lambda(\alpha(\lambda))$, with $\lambda = p^{-1}(x)$ and $p'_\lambda : \Lambda'_\lambda \to X'_x$ the tangent linear mapping to p at point λ . This image belongs to $C^o(X,X')$ iff $\alpha \in C^o(\Lambda,\Lambda')$. For a *covector field* on Λ , the image is analogously defined, only using instead of p'_λ its inverse transpose.

Let us come back to the kinematical setting ; then p_τ , $\tau \in I$ is the τ-depending C^1 diffeomorphism of Λ onto X , generated by the given velocity field φ . Let α be any of the objects we associated above with the C^1 differential structure of Λ . Assume α independent of τ ; then its image under p_τ , say a^τ , is a τ-dependent object of the same nature related to X . Generally a τ-varying object of X derived in that way from a τ-constant object of Λ is said *convected by the moving continuum*, or carrier, Λ . This belongs to the vocabulary of usual physics : a function α which assigns to every particle of the moving continuum a time-independent real value is commonly called a *convected quantity*. When observed from the "reference manifold" X , such a function α : $\Lambda \to \mathbb{R}$ is reflected as $a^\tau : X \to \mathbb{R}$; in the familiar case of C^1 real functions, one elementarily characterizes convection by the transport equation $\partial a^\tau/\partial\tau + \langle\varphi,\mathrm{grad}\, a^\tau\rangle = 0$.

Symmetrically, let us consider now a τ-constant object, say a , related to X . Its image under p_τ^{-1} is a τ-dependent object of the same nature, related to Λ, say a^τ . Recall that the various classes of objects we agreed to consider constitutes each a linear spaces ; there is usually no difficulty in endowing those respective linear spaces with topologies and, due to the assumptions made at the start about the vector field φ, to show that the derivative $d\alpha^\tau/d\tau$, at $\tau = 0$, exists, an element of the same linear space. In turn, this derivative admits an image under the placement p_o; this is an object, related to the C^k-structure of X, of the same nature as a. Traditionnally this object is denoted by $L_\varphi a$ and called the *Lie derivative of* a along φ.

We finish this section by recalling an example of Lie derivative which plays an essential role in the classical kinematics of continua. As a τ-constant object of the geometry of X, let us take a second order *doubly covariant symmetric tensor field* g assumed to be C^1 at least. Practically, this will be the tensor field involved in the definition of a *Riemannian metric* on X (or a pseudo-Riemannian metric, in Relativity theory) through the writing $ds^2 = g_{ij}dx^i dx^j$. For every placement p_τ of the moving continuum Λ, the image of g under p_τ^{-1} is a doubly covariant symmetric tensor field on Λ that we shall denote by γ^τ. Through the writing $d\sigma^2 = \gamma^\tau_{ij} d\lambda^i d\lambda^j$, there is defined a metric on Λ, actually the *metric induced on* Λ *by its placement at time* τ in the Riemann space X. Saying that $d\sigma^2$ depends on τ means that the continuum Λ is not expected to move "rigidly". For every particle $\lambda \in \Lambda$, $\gamma^\tau(\lambda)$ is a τ-dependent element of the symmetrized tensor product $\Lambda'^*_\lambda \otimes_s \Lambda'^*_\lambda$. Under the smoothness assumptions previously made, one easily finds that the τ-derivative $\dot\gamma^\tau$ exists at $\tau = 0$; thereby is defined $\dot\gamma^o \in C^o(\Lambda, \Lambda'^* \otimes_s \Lambda'^*)$. The image of $\dot\gamma^o$ under the placement $p_o : \Lambda \to X$ constitutes the Lie derivative $L_\varphi g$, a C^o doubly covariant symmetric tensor field on X.

The calculation of $L_\varphi g$ is performed, under diverse notations, in any textbook on the Mechanics of Continua ; we shall come back to this in Section 7. Most books are restricted to the usual case where X equals an open subset of a Euclidean space ; then some orthonormal Cartesian coordinates are used as x^i. If the (possibly τ-dependent) velocity field φ is described by its components $\varphi_i(x^1,\dots,x^n)$ relative to this Cartesian frame, the components of the tensor $L_\varphi g$ are found equal to $\partial\varphi_i/\partial x^j + \partial\varphi_j/\partial x^i$. In this setting of orthonormal Cartesian coordinates, it is usual to define the differential operator def, acting on differentiable vector fields, by writing $e = \mathrm{def}\,\varphi$ for $e_{ij} = (\varphi_{i,j} + \varphi_{j,i})/2$. Then

$$L_{\varphi} g = 2 \,\mathrm{def}\, \varphi \,. \tag{4.3}$$

Due to coordinates being orthonormal, in our Euclidean space, no distinction has to be made here between covariance and contravariance. In the elementary kinematics of continua, e is usually called the *spatial strain rate* tensor associated with the Eulerian velocity field φ . In fact, knowing $2e = L_{\varphi} g$ enables one to calculate what we have denoted above by $\dot{\gamma}^{\tau}$; thereby may be computed the time-rate of change of the length of any "infinitesimal material curve element", as soon as is known the image of this element in X under the placement p_{τ} .

Actually an expression of $L_{\varphi} g$ formally as simple as (4.3) may be written in the general case where g defines a regular pseudo-Riemannian metric on X ; one finds (cf. [7])

$$(L_{\varphi} g)_{ij} = \varphi_{i|j} + \varphi_{j|i}$$

where $\varphi_i(x^1,\dots,x^n)$ are the covariant components of the possibly τ-dependent velocity field φ and where $|j$ refers to the *covariant derivative* in the j direction, relative to the pseudo-Riemannian connection.

5. THE METRIC TRACE INTEGRAL OF A TENSOR MEASURE

As before, X denotes a C^k manifold, $k \geq 2$. A doubly covariant symmetric C^1 tensor field g is supposed given in X ; it is intended to define a Riemannian metric on X but, at the present stage, we do not need it to satisfy any positivity condition.

One considers a doubly contravariant symmetric tensor measure T in X such that the integral

$$J = \langle\!\langle g, T \rangle\!\rangle = \int g_{ij} \, dT^{ij} \tag{5.1}$$

makes sense, i.e. the real measure $g_{ij} dT^{ij}$ is bounded. The real number J will be called the *metric trace integral* of T , relative to g . This name is suggested by the special case where X

equals an open subset of some Euclidean space and g the tensor associated with the Euclidean metric ; then, if $x^1,\dots,x^n$ are *orthonormal* Cartesian coordinates, one has $g_{ij} = 1$ for $i = j$ and zero otherwise, therefore

$$J = \int g_{ij}\, dT^{ij} = \int dT^{ii} = \int \text{trace } dT .$$

In this special situation, no distinction has to be made between upper and lower indices ; T^{ii} is a scalar measure, easily proved invariant under any orthogonal change of coordinates. The integral J is meaningful provided this measure is bounded.

Let us define a carrier Λ by its velocity field $\varphi \in \mathcal{D}^1(X,X')$, possibly depending on $\tau \in I$, an open real interval containing zero, in a C^1 way.

Incidentally, the possible lack of boundedness of the real measure $g_{ij}T^{ij}$ comes from g not having a compact support. The essentials of the calculations we shall perform below may be adapted to such a situation by restricting the integral to a compact subset of X containing, for every $\tau \in I$, the support of φ .

Denote by T^τ the τ-dependent doubly contravariant symmetric tensor measure on X , equal to T for $\tau = 0$ and *convected* by the carrier. According to the system of definitions developed in Section 4 , this means that a doubly contravariant τ-constant tensor measure Θ on the manifold Λ is introduced as the image of T under p_o^{-1} ; by definition, T^τ equals for every τ the image of Θ under p_τ . *We are to calculate the* τ*-derivative, at* $\tau = 0$ *, of the function* $\tau \to J(\tau)$ *, the metric trace integral of* T^τ .

As in Section 4 , γ^τ denotes the image of g under p_τ^{-1} . The definitions of images of fields and measures or distributions are precisely devised in order to preserve the various duality pairings ; in particular, for every $\tau \in I$, one has $\langle\langle g,T^\tau\rangle\rangle = \langle\langle\gamma^\tau,\Theta\rangle\rangle$. We have seen that, under the smoothness assumptions made, concerning the carrier motion, the τ-derivative $\dot{\gamma}^\tau$ exists, a continuous doubly covariant symmetric tensor field in Λ with compact support. Then a standard argument of derivation under the integral symbol (see e.g. [4]) yields

$$\frac{d}{d\tau} \langle\!\langle g, T^\tau \rangle\!\rangle = 2\langle\!\langle \mathrm{def}\ \varphi, T^\tau \rangle\!\rangle \quad , \tag{5.2}$$

a formula to be applied at time $\tau = 0$.

In the case where X equals some open subset of an Euclidean space, with orthonormal Cartesian coordinates, no distinction is made between upper and lower indices and, in view of the symmetry of T , one has

$$2\langle\!\langle \mathrm{def}\ \varphi, T^\tau \rangle\!\rangle = \langle\!\langle \varphi_{i,j} + \varphi_{j,i} \, , \, T^{ij} \rangle\!\rangle$$

$$= 2\langle\!\langle \varphi_{i,j} \, , \, T^{ij} \rangle\!\rangle .$$

Now the definition of partial derivatives in the theory of distributions, yields

$$\langle\!\langle \varphi_{i,j} \, , \, T^{ij} \rangle\!\rangle = -\langle\!\langle \varphi_i \, , \, T^{ij}{}_{,j} \rangle\!\rangle . \tag{5.3}$$

Therefore the linear functional $\varphi \to -\langle\!\langle \varphi_{i,j} \, , \, T^{ij} \rangle\!\rangle$ is a *vector distribution of order one,* whose Cartesian component of rank i equals $T^{ij}{}_{,j}$. This vector distribution will naturally be called *the divergence of the tensor measure* T , by analogy with the divergence of a tensor *field*, commonly used in the Mechanics of Continua. In the latter case, s^{ij} denote the components of some C^1 (symmetric) tensor field, relative to some orthonormal Cartesian frame ; partial derivation is understood in the traditional sense, so that the divergence $s^{ij}{}_{,j}$ is a C^0 vector field. Here is the connection between the two concepts : let ℓ denote the Lebesgue measure in our (locally) Euclidean space X ; one easily checks that $s^{ij}\ell$ are the components of a tensor measure whose divergence, in distribution sense, equals the vector measure with components $s^{ij}{}_{,j}\ell$.

As a result of the above calculations, we may formulate the following:

PROPOSITION 5.1 A bounded symmetric tensor measure T *in the locally Euclidean manifold* X *has zero divergence if and only if, for every carrier, as defined in Section 4, the real function* $\tau \to \int \mathrm{trace}\ dT^\tau$ *has zero derivative at* $\tau = 0$ *; here* T^τ *denotes* *the doubly contravariant symmetric tensor measure convected by the carrier, equal to* T *for* $\tau = 0$.

More generally, (5.2) with (5.3) gives a variational meaning to the divergence operator, acting on symmetric tensor measures. This may be applied in characterizing variationally the solutions to some field equations which assert that such a divergence equals some vector distribution of adequate form. An example of this sort will be presented in the next Section.

Of course, it is not necessary to restrict oneself to locally Euclidean manifolds ; if X is pseudo-Riemannian everythings works the same way, provided we define the operator

$$\text{div} \; : \; \mathcal{D}'^{o}(X, X' \otimes_s X') \to \mathcal{D}'^{1}(X, X'^{*})$$

as the negative transpose of

$$\text{def} \; : \; \mathcal{D}^{1}(X, X') \to \mathcal{D}^{o}(X, X'^{*} \otimes_s X'^{*}) \quad ;$$

recall that def denotes the operator $\varphi \to (L_\varphi g)/2$.

6. HYDRODYNAMICAL EXAMPLE

This Section is to show the ability of the transport method to provide a variational treatment of equations in which the divergence of a tensor measure appears jointly with other terms. The setting is that of Section 2 ; we write again equation (2.4) under the form

$$(\rho u_i u_j \ell)_{,j} + (p\ell)_{,i} - \rho U_{,i}\ell = 0 \ . \qquad (6.1)$$

The tensor measure with components $\rho u_i u_j \ell$ has a peculiar structure ; relatively to the nonnegative real measure $m = \rho\ell$, it admits as density the tensor field with components $u_i u_j$, namely the tensor product of the vector u by itself. In order that such a tensor measure be transported by the carrier Λ in the way described in the preceding Section, i.e. that it equal for every τ the image under p_τ of some τ-constant doubly contravariant tensor measure on Λ, it is enough : firstly to have the measure m convected by Λ, secondly to have the vector field u convected. The former means that a τ-dependent real measure m^τ on the space X of the x-coordinates, reducing to m for $\tau = 0$, equals the image under p_τ of some τ-constant real measure μ on Λ. This is the same as the familiar situation of the Mechanics of Continua : then μ denotes the mass measure, defined independently of time on a material continuum ; the image m^τ of this real measure under the placement mapping $p_\tau : \Lambda \to X$ admits a density, say ρ^τ, relatively to the Lebesgue measure ℓ. Elementarily, the law of transport for ρ^τ consists in the following : the value of ρ^τ corresponding to any determined particle λ of Λ verifies $\rho^\tau J^\tau = \text{const}$, where J^τ denotes the Jacobian determinant of p^τ at point λ (relative to an arbitrarily chosen coordinate system in Λ) .

The metric trace integral of the above tensor measure equals $I(\tau) = \int u_i^\tau u_i^\tau \rho^\tau d\ell$; that is, formally, twice the kinetic energy of the investigated portion of fluid. As a result of Section 5 , by calculating the τ-derivative at $\tau = 0$ of this expression one obtains

$$\frac{1}{2}\frac{dI}{d\tau} = - \langle\!\langle (\rho u_i u_j \ell)_{,j} , \varphi^i \rangle\!\rangle \ , \qquad (6.2)$$

where $\varphi \in \mathcal{D}^1(X,X')$ denotes the velocity field of the carrier Λ.

The transport method also gives a variational meaning to the last term in (6.1). Assuming as above that the measure $m^\tau = \rho^\tau \ell$ is convected by the carrier, one considers the integral $G(\tau) = \int U \rho^\tau d\ell$. In order to calculate its τ-derivative, one uses the fact that m^τ equals the image of a τ-constant measure on Λ. When U is evaluated by following up the motion of a particle $\lambda \in \Lambda$ in the reference manifold X, one finds

$$\frac{d}{d\tau} U(p_\tau(\lambda)) = U_{,i}\, \varphi^i \;;$$

therefore, at $\tau = 0$,

$$\frac{dG}{dt} = \int_\Lambda U \frac{d}{d\tau}(p_\tau(\lambda))\, d\mu = \int_X U_{,i}\, \varphi^i \rho\, d\ell \,. \tag{6.3}$$

There finally remains to treat the middle term in (6.1). Mechanically, the dynamical equation of inviscid flows have to be exploited in conjunctions with the knowledge of some *compressibility law* for the investigated fluid. We shall make here the traditional assumption of a *barotropic flow*, i.e. for every fluid particle, a certain relation between the density ρ and the pressure p is asserted, without explicit recourse to temperature. This holds in particular in the very usual situation where the fluid may be admitted to evolve isentropically. Let us write down this relation as

$$p = p(\kappa,\sigma) \tag{6.4}$$

with $\sigma = \rho^{-1}$ denoting the volume of unit mass. Here κ is a variable of arbitrary mathematical nature, assumed to be a constant for every fluid particle ; this may refer as suggested above to the entropy of the particle, but, since our main motivation lies in the treatment of sharp inhomogeneities, one has also to be prepared to make κ account for the chemical nature, possibly different in various parts of the flow. Traditionally, there is introduced a primitive of $\sigma \to p(\kappa,\sigma)$, say $P(\kappa,\sigma)$.

We shall apply the transport method to the integral $\int P(\kappa,\sigma)\rho\, d\ell$, supposing as before that the measure $m = \rho\ell$ is convected, i.e. $m^\tau = \rho^\tau \ell$ with $\rho^\tau J^\tau = \text{const.}$ for every $\lambda \in \Lambda$.

On the other hand we agree to effect the transport with κ kept independent of τ for every λ and, naturally, with σ related to ρ by $\sigma^\tau = (\rho^\tau)^{-1}$. Under these assumptions one calculates the τ-derivative of $\int P(\kappa,\sigma^\tau)\rho^\tau d\ell$. By applying to the carrier Λ the dilatation formula, classical in the Kinematics of Continua, one has, if σ^τ is evaluated for a τ-constant element of Λ ,

$$\frac{d\sigma^\tau}{d\tau} = \sigma^\tau \varphi^i{}_{,i} \ .$$

Then, keeping in mind that $\rho^\tau d\ell$ equals the image under p_τ of the τ-constant measure $d\mu$ on Λ , one finds, for $\tau = 0$,

$$\frac{d}{d\tau}\int_X P(\kappa^\tau,\sigma^\tau)\rho^\tau d\ell = \int_\Lambda \frac{\partial P}{\partial\sigma}\frac{d\sigma^\tau}{d\tau}\, d\mu$$

$$= \int_X p(\kappa,\sigma)\varphi^i{}_{,i}\sigma^\tau\rho^\tau d\ell = -\int_X p_{,i}\varphi^i d\ell$$

since φ has compact support.

Adding up the various terms, one obtains (cf. [3] , Proposition 10.3):

PROPOSITION 6.1 A vector field u *, and three functions* p,ρ,κ *(with* $\sigma = \rho^{-1}$*) make a solution of (6.1) and (6.4) if and only if, for every carrier with velocity field in* $\mathcal{D}^1(X,X')$ *, the* τ*-derivative of the following functional vanishes at* $\tau = 0$

$$B(\tau) = \int \left(\frac{1}{2}\, u_i^\tau u_i^\tau + P(\kappa^\tau,\sigma^\tau) + U\right)\rho^\tau d\ell \quad ;$$

here it is assumed that u^τ *,* κ^τ *,* σ^τ *,* ρ^τ *, reducing to the above for* $\tau = 0$ *, are transported by the carrier in the way defined in the preceding.*

Let us finish by stressing that the condition $(\rho u_j \ell)_{,j}$ of mass conservation is not involved in the above variational characterization. In that respect, one has to recall the observation made in Section 3 : the concept of a divergence-free vector measure belongs to the geometry of C^1 manifolds ; hence it is preserved under any C^1 diffeomorphism such as the placement p_τ of the carrier Λ into X . Consequently, for a vector measure convected by a carrier, the vanishing of divergence is a τ-invariant property. Now the law of transport defined in the preceding implies that the vector measure with components $\rho u_j \ell$ is convected. It will be pointed out in the sequel that the essence of the transport method consists in characterizing some investigated object as a critical point for some real functional in an infinite dimensional manifold. The elements of this manifold result from each other through the transport by carriers. In the present case all of them are divergence-free vector measures ; hence the vanishing of divergence does not constitute a constraint regarding the variational procedures.

An analogous remark applies to the fact that, since the investigated flow is supposed steady, the variable κ must assume a constant value along each streamline. This property is evidently preserved when u^τ and κ^τ are transported by any carrier in the imposed way.

7. THE SECOND VARIATION RATE

Let us come back to the setting of Proposition 5.1 . Some symmetric doubly contravariant tensor measure T on the locally Euclidean manifold X is supposed to have zero divergence ; equivalently, in view of the proposition, the trace integral $J(\tau)$ of T^{τ} has zero time-derivative at $\tau = 0$, whatever is the carrier Λ , with velocity field $\varphi \in \mathcal{D}^{1}(X,X')$.

Incidentally, since the distribution T^{ij} is by assumption a measure, the vanishing of the left-hand member of (5.3) for every $\varphi \in \mathcal{D}^{1}(X,X')$ is secured, in view of a density argument, as soon as the same holds for φ ranging over the subset $\mathcal{D}^{\infty}(X,X')$, the C^{∞} vector fields with compact support in X .

We are now to investigate *the second derivative of the function* $\tau \to J(\tau)$. This of course requires some additional assumption concerning the carrier motion ; we shall suppose that φ is twice differentiable in τ and x , i.e. the vector field $(1,\varphi)$ of the product manifold $I \times X$ is C^{2} .

As before, γ^{τ} denotes the image of the tensor field g under p_{τ}^{-1} . For every $\lambda \in \Lambda$, some element $\dot{\gamma}^{\tau}(\lambda)$ of $\Lambda_{\lambda}'^{*} \otimes_{s} \Lambda_{\lambda}'^{*}$ constitutes the derivative of $\tau \to \gamma^{\tau}(\lambda)$ if and only if, for every pair α,β of τ-constant elements of Λ_{λ}' , one has

$$\frac{d}{d\tau}(\gamma_{ij}^{\tau}(\lambda)\alpha^{i}\beta^{j}) = \dot{\gamma}_{ij}^{\tau}\,\alpha^{i}\beta^{j} . \tag{7.1}$$

The respective images $a(\tau)$, $b(\tau)$ of α and β under $p_{\tau}'(\lambda)$ (the tangent linear mapping to p_{τ} at point λ) are moving vectors in X , associated with the moving point $\tau \to p_{\tau}(\lambda)$. Since X is locally Euclidean, we shall make use in it of *Cartesian coordinates* x^{i} , nonnecessarily orthonormal. By the definition of γ^{τ} , one has, for every $\tau \in I$,

$$\gamma_{ij}^{\tau}(\lambda)\alpha^{i}\beta^{j} = g_{k\ell}a^{k}(\tau)b^{\ell}(\tau) . \tag{7.2}$$

In the terminology of Section 4 , the moving vectors $a(\tau)$ and $b(\tau)$ are convected by the carrier ; this is known to be expressed by

$$\frac{d}{d\tau}a^{k}(\tau) = \varphi^{k}_{\ ,i}(\tau,x)a^{i}(\tau) , \tag{7.3}$$

with $x = p_\tau(\lambda)$, and the similar law for $b(\tau)$ (such a law is nothing but the classical formula concerning the derivation, relative to initial data, for the solutions to the differential equation (4.1)). Since x^i are Cartesian coordinates in X , g is a constant, hence the right-hand side of (7.2) possesses the following τ-derivative

$$g_{k\ell}\,\varphi^k_{\ ,i}\,a^i b^\ell + g_{k\ell}\,a^k\varphi^\ell_{\ ,j}\,b^j$$
$$= \varphi_{\ell,i}\,a^i b^\ell + \varphi_{k,j}\,a^k b^j = (\varphi_{\ell,k} + \varphi_{k,\ell})a^k b^\ell , \tag{7.4}$$

where $\varphi_1,\ldots,\varphi_n$ denote the *covariant components* of the velocity field φ. This proves the existence of $\dot{\gamma}^\tau$, namely the image, under the tangent mapping to p^{-1} , of the element $2e$ of $X'^*_x \otimes_s X'^*_x$ with components $\varphi_{\ell,k} + \varphi_{k,\ell}$. What we have done here is only establishing that the tensor field 2e equals the Lie derivative $L_\varphi g$, the classical fact from which Proposition 5.1 was deduced. We shall now apply the same technique in calculating the τ-derivative of the expression (7.4) . For fixed $\lambda \in \Lambda$,

$$\frac{d}{d\tau}\,e_{k\ell}(\tau,p_\tau(\lambda)) = e_{k\ell,\tau} + e_{k\ell,m}\frac{d}{d\tau}\,p^m_\tau(\lambda)$$
$$= e_{k\ell,\tau} + e_{k\ell,m}\varphi^m .$$

Recall that the components of e , as well as the components of φ , essentially are functions of $\tau,x^1,\ldots,x^n$; the notation $,\tau$ refers to the partial derivatives of such functions with respect to τ , evaluated at point $x = p_\tau(\lambda)$. Then, using (7.3) , one calculates the derivative of $\tau \to e_{k\ell}(\tau,p_\tau(\lambda))a^k(\tau)b^\ell(\tau)$ as follows

$$(e_{k\ell,\tau} + e_{k\ell,m}\varphi^m)a^k b^\ell + e_{k\ell}(\varphi^k_{\ ,i}a^i b^\ell + a^k\varphi^\ell_{\ ,j}b^j)$$

i.e., after renaming some indices of summation,

$$\frac{d}{d\tau}\,(e_{k\ell}a^k b^\ell) = h_{k\ell}\,a^k b^\ell \tag{7.5}$$

with

$$h_{k\ell} = e_{k\ell,\tau} + e_{k\ell,m}\,\varphi^m + e_{m\ell}\varphi^m_{\ ,k} + e_{k\,m}\,\varphi^m_{\ ,\ell} .$$

Since (7.5) holds whatever is the couple α,β of elements of Λ'_λ one starts with, this proves the existence of the second derivative $\ddot{\gamma}^\tau$ of the function $\tau \to \gamma^\tau(\lambda)$; it equals the image of the tensor $2h\,(\tau, p_\tau(\lambda))$ under the tangent mapping to p_τ^{-1} .

We shall use this to establish :

PROPOSITION 7.1 *Let* T *, a bounded symmetric tensor measure in the locally Euclidean manifold* X*, have zero divergence. Let* T^τ *denote the doubly contravariant tensor measure convected by the carrier* Λ *, equal to* T *for* $\tau = 0$ *. Then the real function* $\tau \to J(\tau) = \int \text{trace}\; dT^\tau$ *, whose first derivative vanishes at* $\tau = 0$ *in view of Proposition 5.1 , admits, as soon as the carrier velocity is* C^2*, a second derivative at* $\tau = 0$ *equal to*

$$\ddot{J}(0) = 2\int g_{ij}\,\varphi^i{}_{,k}\varphi^j{}_{,\ell}\,dT^{k\ell} . \tag{7.6}$$

Proof. In view of the preceding, $\ddot{\gamma}^\tau$ is a continuous doubly covariant tensor field of Λ , with compact support ; T^τ equals the image in X of a τ-constant doubly contravariant tensor measure Θ in Λ. Therefore the standard procedure of derivation under the integral symbol yields

$$\ddot{J}(\tau) = \int \ddot{\gamma}^\tau_{ij}\; d\Theta^{ij} = 2\int h_{k\ell}\,dT^{k\ell} .$$

Due to $T^{k\ell} = T^{\ell k}$, the expression $h_{k\ell}$, symmetric in k and ℓ , may be replaced by

$$\varphi_{k,\ell\tau} + \varphi_{k,\ell m}\varphi^m + \varphi_{k,m}\varphi^m{}_{,\ell} + \varphi_{m,k}\varphi^m{}_{,\ell}$$
$$= (\varphi_{k,\tau} + \varphi_{k,m}\varphi^m)_{,\ell} + \varphi_{m,k}\varphi^m{}_{,\ell} .$$

Since T has zero divergence

$$\int (\varphi_{k,\tau} + \varphi_{k,m}\varphi^m)_{,\ell}\,dT^{k\ell} = 0 \quad ;$$

furthermore $\varphi_{m,k} = g_{mi}\varphi^i{}_{,k}$, so (7.6) is proved. □

One may contemplate as follows the use of Proposition 7.1 . Suppose that a doubly contravariant symmetric tensor measure, with bounded trace, is given in X . The totality of the tensor measures which can be obtained from this one through the transport by carriers with $\mathcal{D}^1$ velocity fields constitutes, roughly speaking, an infinite-dimensional manifold, say M. Proposition 5.1 states that, in this manifold, the elements with zero divergence are the critical points of the trace integral, a real function. Exploring the vicinity of such a critical point through the transport by carriers with $\mathcal{D}^2$ velocity fields is the object of Proposition 7.1 . In particular, the nonnegativity of $\ddot{J}$ is a necessary condition for the functional to achieve a *local minimum* at the considered point. The following Proposition allows one to discuss this condition.

PROPOSITION 7.2 *As before,* X *denotes a locally Euclidean manifold. The expression (7.6) is nonnegative for every* $\varphi \in \mathcal{D}^1(X,X')$ *(equivalently for every* $\varphi \in \mathcal{D}^\infty(X,X')$*) if and only if one these equivalent conditions is satisfied :*

(a) For every $\psi \in \mathcal{D}^0(X,X')$ *(equivalently for every* $\psi \in \mathcal{D}^\infty(X,X')$*) one has*

$$\int \psi_i \psi_j \, dT^{ij} \geq 0 \ .$$

(b) There exist (non uniquely) a nonnegative scalar measure μ *and functions* $T'^{ij}_{\mu} \in L^\infty_{loc}(X,\mu)$ *such that* $T^{ij} = T'^{ij}_{\mu}\mu$ *and that for* μ*-almost every* $x \in X$ *the quadratic form* $\xi \to \xi_i \xi_j T'^{ij}_{\mu}(x)$ *is nonnegative in* $\mathbb{R}^n$.

Proof. It is understood that ψ_i denote the covariant components of the vector field ψ , while the components T^{ij} of the tensor measure are contravariant. Thus, the properties (a) and (b) are invariant under any change of coordinates. Let us take profit of this by assuming that *orthonormal* Cartesian coordinates are used in X ; then the expression (7.6) reduces to $2 \int \varphi^i{}_{,k} \varphi^i{}_{,\ell} \, dT^{k\ell}$.

Suppose this expression is nonnegative for every $\varphi \in \mathcal{D}^\infty(X,X')$; in particular, one may fix $i_o \in \{1,\ldots,n\}$ and assume $\varphi^i = \theta \in \mathcal{D}^\infty(X,\mathbb{R})$ for $i = i_o$ and $\varphi^i = 0$ otherwise. Then the quadratic functional defined on $\mathcal{D}^\infty(X,X')$ as $\psi \to \int \psi_k \psi_\ell \, dT^{k\ell}$ is nonnegative for every vector field ψ which equals the gradient of some $\theta \in \mathcal{D}^\infty(X,\mathbb{R})$. Through the use of standard test functions (see e.g. [8]) this may be proved to imply that the quadratic functional is nonnegative for every $\psi \in \mathcal{D}^\infty(X,X')$ and therefore, by density, for every $\psi \in \mathcal{D}^0(X,X')$; this is property (a) . Conversely, the special form that expression (7.6) takes on in orthonormal coordinates makes this expression visibly nonnegative as soon as (a) holds.

Trivially (b) implies (a) . Conversely, we have observed in Section 3 that a representation of the form $T^{ij} = T'^{ij}_\mu \mu$, with $T'^{ij}_\mu \in L^\infty_{loc}(X,\mu)$ exists for every tensor measure. Suppose that (a) holds ; take $\psi_i = \xi_i \sqrt{\theta}$, with $\theta \in \mathcal{D}^0(X,\mathbb{R}^+)$ and $\xi = (\xi_1,\ldots,\xi_n) \in \mathbb{R}^n$; then

$$\int \theta \; \xi_i \xi_j \; T'^{ij}_\mu \, d\mu \geq 0 \; .$$

This shows that $\xi_i \xi_j \; T'^{ij}_\mu(x) \geq 0$ for every $x \in X$ with the possible exception of a μ-negligible subset $N(\xi)$. If ξ ranges over a countable dense subset Ξ of $\mathbb{R}^n$, the union of $N(\xi)$, $\xi \in \Xi$, is μ-negligible. Hence, for μ-almost every x in X , the quadratic form $\xi \to \xi_i \xi_j \; T'^{ij}_\mu(x)$ in nonnegative on Ξ ; by density it is nonnegative on the whole of $\mathbb{R}^n$.

□

REMARK 1. We have defined in the foregoing some infinite-dimensional manifold M as the totality of the doubly contravariant tensor measures in X which result from one of them through the transport by carriers with $\mathcal{D}^1$ velocity fields. Clearly properties (a) and (b) are possessed by every element of M as soon as this is true for one of them. Such is the case for the tensor measure considered in Section 6 , namely $T^{ij} = u^i u^j \mu$, with μ a nonnegative real measure.

REMARK 2. We restricted ourselves in the preceding to the transport of tensor measures. Actually, supposing again for simplicity that the manifold X is locally Euclidean, one may take as T^{ij} the components of a symmetric tensor distribution of order m . If the velocity field of a carrier Λ belongs to $\mathcal{D}^{m+1}(X,X')$, the transport of the tensor distribution T makes sense, generating a moving tensor distribution T^τ in X. Suppose, on the other hand, the real expression $J = \ll g_{ij}, T^{ij} \gg$ meaningful ; this is true in particular if T has compact support. Thereby a real function $\tau \to J(\tau)$ is defined for τ in a neighborhood of zero. The derivative $\dot{J}$ of this function may be calculated as in Section 5 ; again it turns out that $\dot{J}(0)$ vanishes iff $T^{ij}{}_{,j} = 0$. When this holds, one may come to the calculation of the second derivative ; like in Proposition 7.1 one finds

$$\ddot{J}(0) = 2 \ll g_{ij}\, \varphi^i{}_{,k}\, \varphi^j{}_{,\ell}, T^{k\ell} \gg ,$$

a meaningful expression since $\varphi \in \mathcal{D}^{m+1}(X,X')$. Similarly to Proposition 7.2, this expression is found nonnegative iff, for every $\psi \in \mathcal{D}^\infty(X,X')$, one has $\ll \psi_{,i}\psi_{,j}, T^{ij} \gg \;\geq 0$. *Now this property implies that the tensor distribution* T *is actually a measure*. The proof [8] is easily based on the classical fact that any nonnegative real distribution equals a measure.

In the applications of the transport method, the nonnegativity of $\ddot{J}(0)$ appears as a natural "stability" requirement. For this reason the presentation of the method may practically be restricted to the case of tensor measures.

8. CONCLUSION

The logical pattern of the foregoing may be summarized as follows.

In a *reference manifold* X (this equals in practical instances an open subset of Euclidean $\mathbb{R}^n$) some *investigated object* U is required to satisfy a certain system of partial differential equations. Nonsmooth solutions are expected, i.e. the unknown object U is an element of some space $\mathcal{J}$ of distributions with prescribed tensorial type and distributional order.

Here are the essentials of the transport method.

A *reference field* r is specified in X. The preceding sections have been restricted to the case where r equals the doubly covariant tensor field g defining in X some Riemannian metric (and for simplicity most calculations have been performed with a Euclidean metric). Generally r may be a field whose tensoriel type and order of differentiability match the tensorial type and distributional order of U ; in other words, for every U in $\mathcal{J}$, the real valued linear functional $U \to \langle\langle r,U\rangle\rangle$ makes sense.

A carrier Λ is defined in $\mathcal{X}$ by its velocity field, say $\varphi \in \mathcal{D}^k(\mathcal{X},\mathcal{X}')$, with k large enough for the transport of elements of $\mathcal{J}$ to be defined. For instance, $k = 1$ allows for the transport of tensor measures ; generally the transport of tensorial objects with distributional order $k-1$ is meaningful. As indicated in Section 4, φ may also depend on the formal time $\tau \in I$ (I is an open real interval containing zero) with continuous differentiability up to order k in $I \times \mathcal{X}$; in that case it is assumed that the support of $x \to \varphi(\tau,x)$ is contained in a τ-constant compact subset of $\mathcal{X}$.

Since φ belongs to $\mathcal{D}^k$, the carrier Λ may be endowed with the structure of a C^k-manifold, in such a way that, for every $\tau \in I$, the placement mapping $p_\tau : \Lambda \to \mathcal{X}$ is a C^k diffeomorphism. Assume that the tensor field r is C^{k-1} in $\mathcal{X}$; then its image under p^{-1} is a C^{k-1} tensor field in Λ , say ρ^τ . For every $\lambda \in \Lambda$, the value $\rho^\tau(\lambda)$ constitutes a τ-depending element of the tensorial product of some copies of the tangent and/or cotangent space to Λ at point λ . The smoothness assumptions made imply that the τ-derivative $\dot{\rho}^\tau(\lambda)$ exists and that $\lambda \to \dot{\rho}^\tau(\lambda)$ is a C^{k-1} field on Λ , of the same tensorial type as ρ^τ , with compact support. Taking, for $\tau = 0$, the image of $\dot{\rho}^\tau$ under p yields a C^{k-1} tensor field on $\mathcal{X}$, with compact support ; this is, by definition, the *Lie derivative* $L_\varphi r$.

The reference field r has been assumed to belong to $C^k(\mathcal{X},\mathcal{T})$, where $\mathcal{T}$ denotes a certain tensorial type, possibly astrained to some conditions of symmetry or skew-symmetry. Therefore, $\varphi \to L_\varphi r$ is a continuous linear mapping, say D , of $\mathcal{D}^k(\mathcal{X},\mathcal{X}')$ to $\mathcal{D}^{k-1}(\mathcal{X},\mathcal{T})$. In the preceding sections, restricted to the case $r = g$, D equalled twice the operator def ; the *divergence* operator, acting on symmetric tensor measures, emerged as the negative transpose of def. Generally, the transpose of D is a linear operator acting on distributions, say $D^* : \mathcal{D}'^{k-1}(X,T^*) \to \mathcal{D}'^{k}(X,X'^*)$; here T^* refers to the tensorial type dual of T . The transport method provides a variational characterization of the solutions U to $D^*U = 0$, or more generally to $D^*U = f$, provided the right-hand member f has an adequate form.

For more comments on the prospect of this method, let us again restrict ourselves to the special case $r = g$, with X equal to an open subset of a Euclidean space. Using orthonormal Cartesian coordinates in this space, one looks for a symmetric tensor measure T^{ij} with zero divergence, i.e. $T^{ij}_{\ ,j} = 0$ in X . We have interpreted Proposition 5.1 by saying that T verifies this equation iff it constitutes a critical point of the real functional J on some infinite-dimensional manifold M . This manifold consists of the tensor measures obtained by transporting T along carriers with velocity fields in $\mathcal{D}^1(X,X')$. The computational use of this remark may be contemplated by discretizing the carrier as a *moving finite element mesh*. The approximate representation of a tensor measure relatively to such a mesh, and of its transport when the mesh deforms, are easily imagined. Starting with a chosen tensor measure T_o , the mesh will be deformed stepwise. If T_o *and therefore all the tensor measures obtained by transporting it* possess the nonnegativity property involved in Proposition 7.2 , each step of mesh displacement will be devised so as to generate a walk toward a *minimal point* of J .

The convergence of the process is naturally related to the boundary conditions one intends to satisfy. All the preceding has been developed with a carrier Λ whose velocity field has compact support in X . The mechanical analog is a continuous medium whose boundary particles are fixed. Determining a placement $p : \Lambda \to X$ which minimizes the real functional J under such a boundary constraint may then be seen as a special problem of *hyperelastic equilibrium*. With the notations of Sections 4 and 5, the corresponding elastic energy is expressed as $\int_\Lambda \gamma_{ij}\, d\Theta^{ij}$; here recall that Θ is a doubly contravariant tensor measure, constant on the manifold Λ . The doubly covariant tensor field γ , i.e. the image of g under p^{-1} , depends *quadratically* on the tangent mapping $\partial p/\partial\lambda$;

in fact, if $p^k(\lambda^1,\ldots,\lambda^n)$ denote the components of $p(\lambda)$ relative to some orthonormal Cartesian frame of X, one has

$$\gamma_{ij}(\lambda) = \frac{\partial p^k}{\partial \lambda^i}\frac{\partial p^k}{\partial \lambda^j} \; .$$

This makes the determination of the equilibrium placement p a boundary value problem concerning a *linear* system of partial differential equations in the λ^i variables, with a priori nonsmooth coefficients. Of course, the existence of solutions can only be expected in a weak sense, involving placements which are no more C^1 mappings of Λ onto X but only elements of some Sobolev spaces. The numerical treatment of this system of partial differential equations requires a mesh in the manifold Λ : this is the same as the moving mesh considered in the preceding.

This throws some light on the structure of the set of the solutions to the equations of Hydrodynamics, as investigated in Section 6. These equations do no actually reduce to the simple form $T^{ij}{}_{,j} = 0$, but the analogy with a problem of hyperelastic equilibrium is not destroyed by the additional terms ; the gravity term $\rho U_{,i}\ell$ in (6.1) only plays the part of a loading (a "dead loading" in the usual case of constant gravity) while the pressure term modifies the density of elastic energy by adding a function of the Jacobian determinant of the placement. The essential nonlinearity of the equations of Hydrodynamics lies in the algebraic structure of the tensor measure T^{ij} : its density, relative to the scalar measure $\rho\ell$, equals the tensor product of the vector field u by itself. Provided that the approximation process is initiated with a tentative solution T_o which meets such a requirement, the transport method handles this condition automatically ; the nonnegativity property of Proposition 7.2 is also secured by itself. Observe that, if the hyperelastic analogy is brought about in numerical procedures, the constraint imposed to the fictitious elastic medium Λ may be relaxed : instead of assuming each boundary particle fixed, one may permit the medium to slide along some part S of the boundary of the region X . If the approximation process is initiated with a fluid velocity field u_o tangential to S - a usual circumstance in hydrodynamical problems-the transport by Λ will preserve this condition.

9. REFERENCES

[1] P.R. Garabedian and D.C. Spencer. Extremal methods in cavitational flow, *J. Rational Mech. and Anal.*, 1(1952), 359-409.

[2] P. Casal. Sur l'énergie cinétique d'un écoulement possédant une surface de discontinuité de vitesse, *C.R. Acad. Sci. Paris,* 234 (1952), 804-806.

[3] J.J. Moreau. Fluid dynamics and the calculus of horizontal variations, *Int. J. Engng. Sci.* 20(1982), 389-411.

[4] J.J. Moreau. Variational properties of stationary inviscid incompressible flows with possible abrupt inhomogeneity or free surface, *Int. J. Engng. Sci.* 23(1985), 461-481.

[5] N. Bourbaki. Integration, Hermann, Paris.

[6] J.J. Moreau. Le transport d'une mesure vectorielle par un fluide et le théorème de Kelvin-Helmholtz, *Rev. Roum. Math. Pures et Appl.* 27(1982), 375-383.

[7] J.E. Marsden and T.J.R. Hughes. Mathematical foundations of elasticity, Prentice-Hall, 1983.

[8] M. Ros. Formes quadratiques positives sur des espaces de gradients, Séminaire d'Analyse Convexe, USTL, Montpellier, 7 (1977), exp. n° 5.

[9] G. De Rham. Variétés différentiables, Hermann, 1955.

Shape Sensitivity Analysis of

Nonsmooth Variational Problems

Jan Sokolowski[(1)]

Systems Research Institute

Polish Academy of Sciences

ul. Newelska 6, 01-447 Warszawa

POLAND

Abstract

This paper is concerned with the shape sensitivity analysis of solutions of variational inequalities of the second kind. The method of sensitivity analysis proposed in [18] is exploited throughout. The Euler and the Lagrange derivatives of the solution of variational inequality in the direction of a vector field are given in the form of solutions to the auxiliary variational problems.

Key words. shape sensitivity analysis, Euler derivative. Lagrange derivative, variational inequality

(1) This work was completed while the author was visiting the Mathematics Department, University of Florida, Gainesville, Florida.

1. INTRODUCTION.

The paper is devoted to the shape sensitivity analysis of the variational inequalities of the second kind. We use the method of sensitivity analysis proposed by Sokolowski and Zolesio [17,18], combined with the Mignot results [10] on the differential stability of solutions to the variational inequalities. The results presented in this paper can be extended to the case of contact problems with adhesive friction [13,20].

We refer the reader to [1,2,8,11,12,15-21] for the related results on the sensitivity analysis of variational inequalities. The standard notation is used [3] throughout the paper.

2. Differential Stability Analysis.

Let $\Omega \subset R^n$ be a given domain with smooth boundary $\Gamma = \partial\Omega$. Let there be given an element $f \in H^1(R^n)$.

We will consider the following problem

Problem (P)

Find an element $u \in H^1(\Omega)$ which minimizes the functional

$$J(\phi) = \frac{1}{2}a(\phi,\phi) - (f,\phi) + j(\phi) \tag{2.1}$$

$$= \frac{1}{2}\int_\Omega \{|\nabla\phi(x)|^2 + |\phi(x)|^2\}dx$$

$$- \int_\Omega f(x)\phi(x)dx$$

$$+ \int_{\partial\Omega} |\phi(x)|d\Gamma$$

over the space $H^1(\Omega)$.

It is useful for our purposes to observe that the convex, nonsmooth functional

$$j(\phi) = \int_{\partial\Omega} |\phi(x)| d\Gamma, \quad \phi \in L^2(\partial\Omega) \tag{2.2}$$

can be defined as follows

$$j(\phi) = \max\{\int_{\partial\Omega} \mu(x)\phi(x) d\Gamma \,|\, -1 \leq \mu(x) \leq 1 \text{ for a.e. } x \in \partial\Omega\} \tag{2.3}$$

We denote

$$\Lambda = \{\mu \in L^\infty(\partial\Omega) \,|\, -1 \leq \mu(x) \leq 1, \text{ for a.e. } x \in \partial\Omega\} \tag{2.4}$$

hence

$$j(\phi) = \max\{\langle\mu,\phi\rangle \,|\, \phi \in \Lambda\} \tag{2.5}$$

here we denote

$$\langle\mu,\phi\rangle = \int_{\partial\Omega} \mu(x)\phi(x)\, d\Gamma, \ \forall \mu,\phi \in L^2(\partial\Omega) \tag{2.6}$$

It is clear that the functional

$$J(\phi) = \frac{1}{2}a(\phi,\phi) - (f,\phi) + \max\{\langle\mu,\phi\rangle \,|\, \mu \in \Lambda\} \tag{2.7}$$

is nondifferentiable on the space $H^1(\Omega)$. The unique solution $u \in H^1(\Omega)$ of the problem (P) verifies the following variational inequality of the second kind

$$a(u,\phi - u) + (f,\phi - u) + j(\phi) - j(u) \geq 0 \quad \forall \phi \in H^1(\Omega) \tag{2.8}$$

We will show that the solution $u \in H^1(\Omega)$ of the variational inequality (2.8) is directionally differentiable, with respect to the element $f \in (H^1(\Omega))'$ as well as to the perturbations of the boundary of the domain of integration Ω.

First let us observe that in view of (2.5) the element $u \in H^1(\Omega)$ can be obtained by solving the following problem

Problem ($\tilde{P}$):

Find $(u,\lambda) \in H^1(\Omega) \times \Lambda$ such that

$$\mathscr{L}(u,\mu) \leq \mathscr{L}(u,\lambda) \leq \mathscr{L}(\phi,\lambda) \qquad (2.9)$$
$$\forall \mu \in \Lambda, \quad \forall \phi \in H^1(\Omega)$$

where

$$\mathscr{L}(\phi,\mu) = \frac{1}{2}a(\phi,\phi) - (f,\phi) + \langle \mu,\phi \rangle \qquad (2.10)$$
$$\forall \phi \in H^1(\Omega), \ \forall \mu \in H^{-1/2}(\partial\Omega)$$

here $\langle \cdot,\cdot \rangle$ denotes the duality pairing between $H^{-1/2}(\partial\Omega)$, $H^{1/2}(\partial\Omega)$; $H^{-1/2}(\partial\Omega) = (H^{1/2}(\partial\Omega))'$ denotes the dual space.

The right inequality in Eq. (2.9) is equivalent to the following variational equation

$$a(u,\phi) = (f,\phi) - \langle \lambda,\phi \rangle, \ \forall \phi \in H^1(\Omega) \qquad (2.11)$$

therefore

$$u = -z + w \qquad (2.12)$$

where the elements $z = z(\lambda)$, $w=w(f) \in H^1(\Omega)$ are given by the unique solutions of the following equations

$$a(z(\lambda),\phi) = \langle\lambda,\phi\rangle, \ \forall\phi \in H^1(\Omega) \tag{2.13}$$

$$a(w(f),\phi) = (f,\phi), \ \forall\phi \in H^1(\Omega) \tag{2.14}$$

On the other hand the left inequality in Eq. (2.9) leads to the following inequality

$$\lambda \in \Lambda: \quad \langle\mu,u\rangle \leq \langle\lambda,u\rangle, \ \forall\mu \in \Lambda \tag{2.15}$$

whence, in view of (2.12), we obtain the following variational inequality

$$\lambda \in \Lambda: \quad \langle z(\lambda), \mu-\lambda\rangle \geq \langle w(f), \mu-\lambda\rangle \quad \forall\mu \in \Lambda \tag{2.16}$$

We denote by $b(\cdot,\cdot): H^{-1/2}(\partial\Omega) \times H^{-1/2}(\partial\Omega) \to R$ the symmetric bilinear form

$$b(\mu,\eta) = \langle z(\mu), \eta\rangle, \ \forall\mu,\eta \in H^{-1/2}(\partial\Omega) \tag{2.17}$$

It can be shown [13] that the bilinear form (2.17) is coercive i.e.,

$$b(\mu,\mu) \geq \alpha||\mu||^2_{H^{-1/2}(\partial\Omega)}, \quad \alpha > 0, \ \forall\mu \in H^{-1/2}(\partial\Omega) \tag{2.18}$$

Let us consider the differential stability of solutions to the variational inequality (2.16) which takes the form

$$\lambda \in \Lambda \subset H^{-1/2}(\partial\Omega)$$

$$b(\lambda,\mu-\lambda) \geq \langle w,\mu-\lambda\rangle, \tag{2.19}$$

$$\forall \mu \in \Lambda$$

It can be verified that the set (2.4) is a closed, convex subset of the Sobolov space $H^{-1/2}(\partial\Omega)$. We will use the following notation.

$$\Xi^{\pm} = \{x \in \partial\Omega|\ \lambda(x) = \mp 1\} \tag{2.20}$$

$$\Xi_0 = \{x \in \Xi^+ \cup \Xi^-|\ u(x) = 0\} \tag{2.21}$$

We will assume that

$$\text{meas}(\Xi^{\pm} \setminus \text{int } \Xi^{\pm}) = 0 \tag{2.22}$$

$$\text{meas}(\Xi_0 \setminus \text{int } \Xi_0) = 0 \tag{2.23}$$

here meas(Ξ) is the (n-1)-dimensional measure of a set $\Xi \subset \partial\Omega$. Furthermore we assume that the sets Ξ^+, Ξ^-, Ξ_0 are sufficiently regular so that the closure in $H^{-1/2}(\partial\Omega)$ of the following sets:

$$K_1 = \{\phi \in L^2(\partial\Omega)|\ \phi(x) \geq 0 \quad \text{a.e. on } \Xi^+, \tag{2.24}$$
$$\phi(x) \leq 0 \quad \text{a.e. on } \Xi^-\}$$

$$K_2 = \{\phi \in K_1|\ \phi(x) = 0 \quad \text{a.e. on } \Xi_0\} \tag{2.25}$$

takes the form

$$\overline{K}_1 = \{\phi \in H^{-1/2}|\int_{\partial\Omega} \phi(x)\eta(x)d\Gamma \geq 0, \tag{2.26}$$

for all $\eta \in C_0(\partial\Omega)$ such that

supp $\eta \subset \Xi^+$ and $\eta(x) \geq 0$ on Ξ^+

supp $\eta \subset \Xi^-$ and $\eta(x) \leq 0$ on $\Xi^{-'}$ }

$$\overline{K}_2 = \{\phi \in \overline{K}_1 | \int_{\partial\Omega} \phi(x)\eta(x)d\Gamma = 0, \quad \forall\eta \in C_0(\partial\Omega), \text{ supp } \eta \in \Xi_0\} \tag{2.27}$$

We denote by $C_\lambda(\Lambda) \subset H^{-1/2}(\partial\Omega)$ the tangent cone

$$C_\lambda(\Lambda) = \{\mu \in H^{-1/2}(\partial\Omega) | \exists\tau > 0 \text{ such that } \lambda + \tau\mu \in \Lambda\} \tag{2.28}$$

It is obvious, in view of (2.4), that

$$C_\lambda(\Lambda) \subset L^\infty(\partial\Omega) \tag{2.29}$$

Finally we denote

$$M = \{\mu \in H^{-1/2}(\partial\Omega) | \int_{\partial\Omega} \mu(x)u(x)d\Gamma = 0\} \tag{2.30}$$

M is the linear, closed subspace of the space $H^{-1/2}(\partial\Omega)$.

We have the following result proved in [13].

<u>Lemma 2.1</u>

Assume that the sets Ξ^+, Ξ^-, Ξ_0 are sufficiently regular then

$$S_\lambda(\Lambda) = S = \overline{C_\lambda(\Lambda)} \cap M = \overline{C_\lambda(\Lambda) \cap M} = \tag{2.31}$$

$$= \{\mu \in H^{1/2}(\partial\Omega) \mid \int_{\partial\Omega} \mu(x)\eta(x)d\Gamma \geq 0$$

for all $\eta \in C_0(\partial\Omega)$ such that

supp $\eta \subset \Xi^+$ and $\eta \geq 0$, or supp $\eta \subset \Xi^-$ and $\eta \leq 0$,

$$\int_{\partial\Omega} \mu(x)\phi(x)d\Gamma = 0,\ \forall\phi \in C_0(\partial\Omega),\ \text{supp}\ \phi \subset \Xi_0\}$$

By Lemma 2.1 it follows that the set $\Lambda \subset H^{-1/2}(\partial\Omega)$ is polyhedric in the sense of Mignot [10] therefore the metric projection in the space $H^{-1/2}(\partial\Omega)$ with respect to the norm $||\phi||_{-1/2,\Gamma} = (b(\phi,\phi))^{1/2}$, onto the set $\Lambda \subset H^{-1/2}(\partial\Omega)$ is conically differentiable [10] and we have the following result.

<u>Theorem 2.1.</u>

Let

$$f_\varepsilon = f + \varepsilon f' + o(\varepsilon),\ \text{in}\ (H^1(\Omega))' \tag{2.32}$$

and $\lambda_\varepsilon \in \Lambda$ denotes the unique solution of the variational inequality

$$\lambda_\varepsilon \in \Lambda$$

$$b(\lambda_\varepsilon,\ \mu-\lambda_\varepsilon) \geq \langle w(f_\varepsilon),\ \mu - \lambda_\varepsilon\rangle \quad \forall\mu \in \Lambda$$

For $\varepsilon > 0$, ε small enough

$$\lambda_\varepsilon = \lambda + \varepsilon\lambda' + o(\varepsilon),\ \text{in}\ H^{-1/2}(\partial\Omega) \tag{2.34}$$

where $||o(\varepsilon)||_{H^{-1/2}(\partial\Omega)}/\varepsilon \to 0$ with $\varepsilon \downarrow 0$.

The element $\lambda' \in H^{-1/2}(\partial\Omega)$ is given by the unique solution of the following variational inequality

$$\lambda' \in S$$
$$b(\lambda', \mu - \lambda') \geq \langle w(f'), \mu - \lambda' \rangle \qquad \forall \mu \in S \tag{2.35}$$

The proof of theorem 1 is given in [13].

3. Shape Sensitivity Analysis.

Let us consider the differential stability of solutions to the problem (P) with respect to the perturbations of the boundary of domain Ω.

We define a family of domains $\{\Omega_\varepsilon\} \subset R^n$ depending on the parameter $\varepsilon \in [0,\delta)$.

Let there be given a vector field

$$V(\cdot,\cdot) \in C^1(0,\delta;\ C^2(R^n;\ R^n)) \tag{3.1}$$

The family $\{\Omega_\varepsilon\}$ is defined as follows

$$\Omega_\varepsilon = T_\varepsilon(V)(\Omega) \tag{3.2}$$

where the mapping $T_\varepsilon(V)\colon R^n \to R^n$, $\varepsilon \in [0,\delta)$, takes the form

$$x(\varepsilon) = T_\varepsilon(V)(X), \quad X \in R^n \tag{3.3}$$

here $x(\cdot)\colon[0,\delta) \to R^n$ denotes the unique solution of the following ordinary differential equation

$$\frac{dx}{dt}(t) = V(t,x(t)),\ t \in (0,\delta) \tag{3.4}$$

$$x(0) = X$$

We denote by $DT_\varepsilon(X)$ the Jacobian of the mapping $T_\varepsilon = T_\varepsilon(V)$ evaluated at the point $X \in R^n$, $DT_\varepsilon^{-1}(X)$ is the inverse of $DT_\varepsilon(x)$, $^*DT_\varepsilon^{-1}(X)$ denotes the adjoint of $DT_\varepsilon^{-1}(X)$.

Let us consider the problem (P) defined in the domain Ω_ε for $\varepsilon \in [0,\delta)$.

<u>Problem (P_ε):</u>

Find an element $u_\varepsilon \in H^1(\Omega_\varepsilon)$ which minimizes the functional

$$J_\varepsilon(\phi) = \frac{1}{2}\int_{\Omega_\varepsilon} \{|\nabla\phi(x)|^2 + |\phi(x)|^2\}dx \tag{3.5}$$

$$-\int_{\Omega_\varepsilon} f(x)\phi(x)dx + \int_{\partial\Omega_\varepsilon} |\phi(x)|d\Gamma$$

over the space $H^1(\Omega_\varepsilon)$

It can be verified that there exists the unique solution $u_\varepsilon \in H^1(\Omega_\varepsilon)$ of the problem (P_ε) for any $\varepsilon \in (0,\delta)$; for $\varepsilon = 0$ the problem (P_ε) coincides with the problem (P).

In order to derive the form of the so-called Euler derivative $\dot{u} \in H^1(\Omega)$ of the solution of the problem (P) in the direction of a vector field $V(\cdot,\cdot)$ we denote

$$u^{\varepsilon} = u_{\varepsilon} \circ T_{\varepsilon} \in H^1(\Omega), \; \varepsilon \in [0,\delta) \tag{3.6}$$

and we define the Euler derivative

$$u = \lim_{\varepsilon \downarrow 0}(u^{\varepsilon} - u^{o})/\varepsilon, \text{ in } H^1(\Omega) \tag{3.7}$$

The element $u^{\varepsilon} \in H^1(\Omega)$, $\varepsilon \in [0,\delta)$ is given by the unique solution of an auxiliary problem (P^{ε}). We derive the form of problem (P^{ε}). To this end we transport the functional $J_{\varepsilon}(\phi)$, $\phi \in H^1(\Omega_{\varepsilon})$ to the fixed domain, using the mapping $T_{\varepsilon}: \Omega \to \Omega_{\varepsilon}$, the resulting functional is denoted $J^{\varepsilon}(\cdot)$. We define

$$J^{\varepsilon}(\phi o T_{\varepsilon}) = J_{\varepsilon}(\phi), \; \forall \phi \in H^1(\Omega_{\varepsilon}) \tag{3.8}$$

Simple computations show [22] that

$$\begin{aligned} \forall \phi \in H^1(\Omega)&: J^{\varepsilon}(\phi) = \tfrac{1}{2}a^{\varepsilon}(\phi,\phi) - (f^{\varepsilon},\phi) + j_{\varepsilon}(\phi) = \\ &= \frac{1}{2}\int_{\Omega}\{<A_{\varepsilon}(x).\nabla\phi(x), \nabla\phi(x)>_{R^n} + |\phi(x)|^2 \beta_{\varepsilon}(x)\}dx \\ &-\int_{\Omega} f^{\varepsilon}(x)\phi(x)dx + \int_{\partial\Omega}|\phi(x)|\sigma_{\varepsilon}(x)d\Gamma \end{aligned} \tag{3.9}$$

where we denote

$$A_{\varepsilon}(x) = \det(DT_{\varepsilon}(x)) \, DT_{\varepsilon}^{-1}(x) \cdot {}^{*}DT_{\varepsilon}^{-1}(x), \; x \in \Omega \tag{3.10}$$

$$\beta_{\varepsilon}(x) = \det(DT_{\varepsilon}(x)), \; x \in \Omega \tag{3.11}$$

$$\sigma_{\varepsilon}(x) = ||\det(DT_{\varepsilon}(x)) {}^{*}DT_{\varepsilon}^{-1}(x).n(x)||_{R^n}, \; x \in \partial\Omega \tag{3.12}$$

$n(x)$, $x \in \partial\Omega$ is unit, outward normal vector on $\partial\Omega$,

$$f^{\varepsilon}(x) = (f o T_{\varepsilon})(x)\beta_{\varepsilon}(x), \; x \in \Omega \tag{3.13}$$

We introduce the problem defined on fixed domain Ω.

Problem (P^ε):

Find an element $u^\varepsilon \in H^1(\Omega)$ which minimizes the functional $J^\varepsilon(\phi)$ over the space $H^1(\Omega)$.

We will proceed in exactly the same way as in the previous section.
We denote

$$u^\varepsilon = -z^\varepsilon(\lambda^\varepsilon) + w^\varepsilon(f^\varepsilon) \tag{3.14}$$
$$= -z^\varepsilon + w^\varepsilon$$

where

$$a^\varepsilon(w^\varepsilon, \phi) = (f^\varepsilon, \phi), \ \forall \phi \in H^1(\Omega) \tag{3.15}$$
$$z^\varepsilon = z^\varepsilon(\lambda^\varepsilon) : a^\varepsilon(z^\varepsilon, \phi) = \langle \lambda^\varepsilon, \phi \rangle, \ \forall \phi \in H^1(\Omega) \tag{3.16}$$

The element $\lambda^\varepsilon \in L^\infty(\partial\Omega) \subset H^{-1/2}(\partial\Omega)$ is given by the unique solution of the following variational inequality

$$\lambda^\varepsilon \in \Lambda$$
$$b^\varepsilon(\lambda^\varepsilon, \mu - \lambda^\varepsilon) \geq \langle \sigma_\varepsilon w^\varepsilon, \mu - \lambda^\varepsilon \rangle \tag{3.17}$$
$$\forall \mu \in \Lambda$$

where the bilinear form

$$b^{\varepsilon}(\cdot,\cdot): H^{-1/2}(\partial\Omega) \times H^{-1/2}(\partial\Omega) \to R \tag{3.18}$$

is defined by

$$b^{\varepsilon}(\xi,\mu) = \int_{\partial\Omega} z^{\varepsilon}(\xi)(x)\mu(x)d\Gamma, \quad \forall \mu,\xi \in H^{-1/2}(\partial\Omega) \tag{3.19}$$

where for any $\xi \in H^{-1/2}(\partial\Omega)$ the element $z^{\varepsilon}(\xi)$ solves the equation

$$a^{\varepsilon}(z^{\varepsilon}(\xi),\phi) = \int_{\partial\Omega} \xi(x)\phi(x)d\Gamma, \quad \forall\phi \in H^{1}(\Omega) \tag{3.20}$$

In view of (3.12) it follows that the Lagrange derivative $\dot{u}$ takes the form

$$\dot{u} = - z'(\lambda) - z(\dot{\lambda}) + \dot{w} \tag{3.21}$$

here the elements $z'(\lambda)$, $\dot{w} \in H^{1}(\Omega)$ can be determined, in view of (3.15), (3.16) and (3.20), by solving the following equations

$$z' = z'(\lambda) \in H^{1}(\Omega)$$

$$a(z',\phi) = \int_{\partial\Omega} \lambda(x)\phi(x)d\Gamma - a'(z,\phi), \ \forall\phi \in H^{1}(\Omega) \tag{3.22}$$

$$\dot{w} \in H^{1}(\Omega)$$

$$a(\dot{w},\phi) = \int_{\Omega} \dot{f}\phi dx - a'(w,\phi), \ \forall \phi \in H^{1}(\Omega) \tag{3.23}$$

where we denote $\lambda = \lambda^{\circ}$, $z = z^{\circ}$, $w = w^{\circ}$,

$$a'(y,\phi) = \int_\Omega \{\langle A'(x)\cdot\nabla y(x), \nabla\phi(x)\rangle_{R^n} + \tag{3.24}$$
$$+ \beta'(x)y(x)\phi(x)\}dx, \ \forall y, \ \phi \in H^1(\Omega)$$

$$\beta'(x) = \mathrm{div}(V(0,x)) \tag{3.25}$$

$$A'(x) = \mathrm{div}\ V(0,x)I - DV(0,x) - {}^*DV(0,x) \tag{3.26}$$

Finally we characterize the element $\dot\lambda \in H^{-1/2}(\partial\Omega)$ which is the Euler derivative of the solution of the problem (2.19) in the direction of a vector field $V(\cdot,\cdot)$.

Lemma 3.1

For $\varepsilon > 0$, ε small enough

$$\lambda^\varepsilon = \lambda + \varepsilon\dot\lambda + o(\varepsilon), \text{ in } H^{-1/2}(\partial\Omega) \tag{3.27}$$

where $||o(\varepsilon)||_{H^{-1/2}(\partial\Omega)}/\varepsilon \to 0$ with $\varepsilon \downarrow 0$.

The element $\dot\lambda \in H^{-1/2}(\partial\Omega)$ is given by the unique solution of the following variational inequality

$$\dot\lambda \in S_\lambda(\Lambda)$$
$$(\dot\lambda, \mu - \dot\lambda) \geq \langle\dot w + \sigma', \mu - \dot\lambda\rangle - b'(\lambda, \mu - \dot\lambda) \tag{3.28}$$
$$\forall\mu \in S_\lambda(\Lambda)$$

Here we denote

$$b'(\lambda,\mu) = \int_{\partial\Omega} \{z'(\lambda)(x)\mu(x) + z(\lambda)(x)\mu(x)\sigma'(x)\}d\Gamma \tag{3.29}$$
$$\forall\mu \in H^{-1/2}(\partial\Omega)$$

$$\sigma'(x) = \text{div } V(0,x) - \langle DV(0,x).n(x), n(x)\rangle_{R^n} \tag{3.30}$$

The proof of Lemma 3.1 is based on the abstract results presented in [18] and it is omitted here.

Let us observe that in view of (2.11), (3.9) the element $u^\varepsilon \in H^1(\Omega)$ solves the equation

$$a^\varepsilon(u^\varepsilon,\phi) = (f^\varepsilon,\phi) - \langle\sigma_\varepsilon\lambda^\varepsilon,\phi\rangle, \quad \forall \varepsilon \in H^1(\Omega) \tag{3.31}$$

therefore

$$a(\dot{u},\phi) + a'(u,\phi) = (\dot{f},\phi) - \langle\dot{\lambda} + \sigma'\lambda,\phi\rangle \quad \forall \phi \in H'(\Omega) \tag{3.32}$$

here we denote $u = u^o$, $\lambda = \lambda^o$,

$$\dot{f}(x) = \nabla(f(x).V(0,x)), \quad x \in \Omega \tag{3.33}$$

From (3.31) it follows that the element $\dot{u} \in H^1(\Omega)$ minimizes the quadratic functional

$$I(\phi) = \frac{1}{2}a(\phi,\phi) + a'(u,\phi) - (\dot{f},\phi) + \langle\dot{\lambda} + \sigma'\lambda, \phi\rangle \tag{3.34}$$

over the space $H^1(\Omega)$.

On the other hand from (3.28) it follows that

$$\langle \dot{w} + \sigma' w, \dot{\lambda} \rangle - (\dot{\lambda}, \dot{\lambda}) - b'(\lambda, \dot{\lambda}) \geq$$

$$\langle \dot{w} + \sigma' w, \mu \rangle - (\dot{\lambda}, \mu) - b'(\lambda, \mu), \ \forall \mu \in \Lambda \tag{3.35}$$

hence, in view of (3.21) - 3.23), 3.29), we obtain

$$\int_{\partial\Omega} (\dot{u} + \sigma' u)\dot{\lambda} d\Gamma \geq \int_{\partial\Omega} (\dot{u} + \sigma' u)\mu d\Gamma \tag{3.36}$$

$$\forall \mu \in S_\lambda(\Lambda)$$

so we can replace the term $\langle \dot{\lambda}, \phi \rangle$ in Eq. (3.34) by the term

$$\max\{ \int_{\partial\Omega} (\dot{u} + \sigma' u)\mu \, d\Gamma | \mu \in S_\lambda(\Lambda)\} =$$

$$= \max\{\langle \mu, \dot{u} + \sigma' u \rangle | \mu \in S_\lambda(\Lambda)\} \tag{3.37}$$

It leads to the following result.

<u>Theorem 3.1</u>

For $\varepsilon > 0$, ε small enough

$$u^\varepsilon = u + \varepsilon \dot{u} + o(\varepsilon), \text{ in } H^1(\Omega) \tag{3.38}$$

where $||o(\varepsilon)||_{H^1(\Omega)} / \varepsilon \to 0$ with $\varepsilon \downarrow 0$.

The element $\dot{u} \in H^1(\Omega)$ minimizes the functional

$$I(\phi) = \frac{1}{2} a(\phi, \phi) + a'(u, \phi) - (\dot{f}, \phi) + \tag{3.39}$$

$$+ \langle \sigma' \lambda, \phi \rangle + \max\{\langle \mu, \phi + \sigma' u \rangle | \mu \in S_\lambda(\Lambda)\}.$$

over the space $H^1(\Omega)$.

We derive the form of the so-called Lagrange derivative $u' \in H^1(\Omega)$ of the solution of problem (P) in the direction of vector field $V(\cdot,\cdot)$. We have [22]

$$u' = \dot{u} - \nabla u.V(0) \tag{3.40}$$

therefore the Lagrange derivative $u' \in H^1(\Omega)$ is well defined provided the following condition is satisfied

$$\nabla u.V(0) \in H^1(\Omega) \tag{3.41}$$

It is known [22] that for any vector field $V(\cdot,\cdot) \in C^1(0,\delta;C^2(R^n;\ R^n))$ such that

$$v_n(x) = \langle V(0,x),n(x)\rangle_{R^n} = 0 \tag{3.42}$$

it follows that

$$u' = 0,\ \text{i.e. } \dot{u} = \nabla u.V(0) \tag{3.43}$$

furthermore $\dot{\lambda} = v_\tau \partial\lambda/\partial\tau$ where $\tau(x)$, $x \in \partial\Omega$ denote the unit tangent vector on $\partial\Omega$.

Let us assume that the condition (3.42) is verifed for a vector field $V(\cdot,\cdot)$, then from (3.32) in view of (3.43) we obtain

$$\begin{aligned} a(\nabla u.V(0)),\ \phi) + a'(u,\phi) = (\dot{f},\phi) \\ -\langle v_\tau \frac{\partial\lambda}{\partial\tau} + \sigma'\lambda,\phi\rangle,\ \forall\phi \in H^1(\Omega) \end{aligned} \tag{3.44}$$

for such a vector field $V(\cdot,\cdot)$. Therefore [22] there exists the distribution $g_n(\phi) \in \mathcal{D}'(\partial\Omega)$, $\phi \in H^1(\Omega)$ such that for any vector field $V(\cdot,\cdot) \in C^1(0,\delta;\ C^2(R^n R^n))$ from (3.32), (3.40), in view of (3.44), we have

$$\begin{aligned} & u' \in H^1(\Omega) \\ & a(u,\phi) = \langle\langle g_n(\phi),\ v_n \rangle\rangle_{\mathcal{D}_1'(\partial\Omega) \times \mathcal{D}_1(\partial\Omega)} \\ & + \langle \sigma'\lambda, \phi\rangle + \langle \lambda', \phi \rangle \\ & \qquad \forall \phi \in H^1(\Omega) \end{aligned} \tag{3.45}$$

Here λ' denotes the Lagrange derivative of the element λ in the direction of the vector field $V(\cdot,\cdot)$,

$$\begin{aligned} & \lambda' = \dot{\lambda} - v_\tau \frac{\partial \lambda}{\partial \tau} \\ & v_\tau(x) = \langle V(0,x),\ \tau(x)\rangle_{R^n},\ x \in \partial\Omega \end{aligned} \tag{3.46}$$

The element $\lambda' \in H^{-1/2}(\partial\Omega)$ is given by the unique solution of the following variational inequality

$$\begin{aligned} & \lambda' \in S_\lambda(\Lambda) \\ & b(\lambda', \mu - \lambda') \geq \langle u v_n H + v_n \frac{\partial u}{\partial n} + w' - z'(\lambda),\ \mu - \lambda' \rangle \\ & \qquad \forall \mu \in S_\lambda(\Lambda) \end{aligned} \tag{3.47}$$

here H denotes the mean curvatuire of the boundary $\partial\Omega$.

We can determine the Lagrange derivative $u' \in H^1(\Omega)$ as the unique solution of the variational problem.

Theorem 3.2

Let us assume that the condition (3.41) is verified. The Lagrange derivative $u' \in H^1(\Omega)$ minimizes the functional

$$I(\phi) = \frac{1}{2} a(\phi,\phi) - \langle\langle g_n(\phi), v_n \rangle\rangle_{\mathcal{D}_1' \times \mathcal{D}_1} + \max\{\langle \mu, \phi + v_n uH + v_n \frac{\partial u}{\partial n} \rangle \mid \mu \in S_\lambda(\Lambda)\} \qquad (3.48)$$

over the space $H^1(\Omega)$.

Remark:

The distribution $g_n(\phi) \in \mathcal{D}_1'(\partial\Omega)$ has the following representation

$$\langle\langle g_n(\phi), v_n \rangle\rangle_{\mathcal{D}_1' \mathcal{D}_1} = \int_{\partial\Omega} v_n [\frac{\partial u}{\partial \tau} \frac{\partial \phi}{\partial \tau} - f\phi + H\lambda\phi] d\Gamma$$

for any element $\phi \in H^2(\Omega)$.

References

[1] Bendsøe M. P., Olhoff N., Sokolowski J. (1985) Sensitivity analysis of problems of elasticity with unilateral constraints. J. Struct. Mech. 13(2), 201-222.

[2] Benedict R., Sokolowski J., Zolesio J. P. (1984) Shape optimization for contact problems. In: Thoft-Chrsitensen P. (ed) System modelling and optimization, LNCIS vol. 59, Springer Verlag, 790-799.

[3] Duvaut G., Lions J. L. (1972) Les inequations en mechanique et en physique. Dunod, Paris.

[4] Fichera G. (1972) Boundary value problems of elasticity with unilateral constraints, In: Handbuch der Physik, Band 6a/2, Springer Verlag.

[5] Friedman A. (1982) Variational principles and free boundary problems, J. Wiley and Sons, New York.

[6] Hanouzet B., Joly J. L. (1979) Méthodes d'ordre dans l'interprétation de certaines inéquations variationnelles et applications. J. Functional Analysis, 34: 216-249.

[7] Haraux A. (1977) How to differentiate the projection on a convex set in Hilbert space. Some applications to variational inequalities. J. Math. Soc. Japan, 29(4) 615-631.

[8] Haslinger J., Horak V., Neittaanmaki P. (1985) Shape optimization in contact problems with friction, Report 10/1985, Lappeenranta University of Technology, Department of Physics and Mathematics, Finland.

[9] Lions J. L., Magenes E. (1968) Problémes aux limites non homogénes et applications. Vol. 1, Dunod, Paris.

[10] Mignot F. (1976) Controle dans les inequations variationelles elliptiques. J. Functional Analysis, 22, 130-185.

[11] Neittaanmaki P., Sokolowski J., Zolesio J. P. Optimization of the domain in elliptic variational inequalities, Appl. Math. & Optim. (to appear).

[12] Sokolowski J. (1981) Sensitivity analysis for a class of variational inequalities. In: Haug E. J., Cea J. (eds), Optimization of distributed parameter structures, vol. 2, Sijthoff & Noordhoff, Alphen aan den Rijn, The Netherlands, 1600-1609.

[13] Sokolowski J., Sensitivity analysis of contact problems with adhesive friction, (to appear).

[14] Sokolowski J. (1983) Optimal control in coefficients of boundary value problems with unilateral constraints. Bulletin of the Polish Academy of Sciences, Technical Sciences, vol. 31 (1-12): 71-81.

[15] Sokolowski J. (1984) Sensitivity analysis of Signorini variational inequality. In: Bojarski B. (ed) Banach Center Publications, Polish Scientific Publisher, Warsaw. (to appear)

[16] Sokolowski J. (1985) Differential stability of solutions to constrained optimization problems. Appl. Math. Optim. 13: 97-115.

[17] Sokolowski J., Zolesio J. P. (1985), Dérivéé par rapport au domaine de la solution d'un problème unilatéral. C. R. Acad. Sc. Paris, t. 301, Série I, n°4: 103-106.

[18] Sokolowski J., Zolesio J. P. (1985) Shape sensitivity analysis of unilateral problems. Publication Mathematiques No. 67, Université de Nice.

[19] Sokolowski J., Zolesio J. P. (1985) Shape sensitivity analysis of an elastic-plastic torsion problem. Bulletin of the Polish Academy of Sciences, Technical Sciences 33:579-586.

[20] Sokolowski, J., Zolesio J. P. Shape sensitivity analysis of contact problems with adhesive friction (in preparation).

[21] Zolesio J. P. (1984) Shape controlability for free boundaries. In: Thoft-Christensen P. (ed) System modelling and optimization, LNCIS vol. 59, Springer Verlag, 354-361.

[22] Zolesio J. P. (1979) Identification de domaines par deformation. Thése d'Etat, Université de Nice.

SHAPE NEWTON METHOD IN NAVAL HYDRODYNAMIC

M. SOULI
Département de Mathématiques
Université de Nice
Parc Valrose
O6034 NICE CEDEX
France

We develop here a Newtoon method for shape optimization. In some examples (see A. Bern - J.L. Chenot - Y. Demay - J.P. Zolesio , to appear) a "quasi-Newton" method can easily be derived by neglecting the adjoint equations contributions in the calculation of the newton matrix. Here in the non linear water wave problem we cannot neglet these terms then the calculation is very heavy but we effectively improve the results obtained by the gradient method in particular for the difficult situation when the Froude number is less then one

I. Introduction
II. Problem with obstacle at the bottom
III. Problem with obstacle piercing the surface
iV. Numerical results.

I. INTRODUCTION.

We consider now the same free boundary problem using a Newton method to solve the shape problem. We condider a weak formulation for the non linear Bernoulli condition on the free surface. We discretize by finite elements this condition and for the n nodes lying on the free boundary, we calculate the derivative with respect to the nodes coordinate as in [2], introducing n adjoint problems who have the same matrix, and we solve the location of boundary nodes by Newton method.

This technique converges more quickly than all the gradient methods presented in [3].

Also for the same little values of the Froude number F, the Newton method converges while the point fixed method fails. This was encouraging enough to turn to the realistic situation where the obstacle is piercing the free surface.

II. PROBLEM WITH AN OBSTACLE AT THE BOTTOM.

The problem is that of the stationary flow around an obstacle resting at the bottom. The flow is assumed to be uniform at the upstream infinity, for a detailled study of the mouvement equations, we can refer to Cahouet's thesis [1], where we reduce to a bounded domain by the localized finite elements method.

Howewer the limit conditions given on the lateral boundary of the domain allow the consideration of limit conditions infinity.

In the problem, the flow is characterized by the number froude $F = \frac{c}{\sqrt{gH}}$

C : Celerity of the flow at upstream infinity

H : the depth

g : the gravity.

if $F > 1$ the flow is uniform downstream

$F < 1$ creation of a wave downstream.

Inthis work, we only consider the case $F < 1$.

The stream function ψ is solution of :

$$(1.1)\quad \begin{cases} 1 & \Delta\psi=0 & D \\ 2 & \psi=0 & B \\ 3 & \psi=1 & S \\ 4 & \dfrac{\partial\psi}{\partial n} = \dfrac{\partial\psi_k}{\partial\eta} & S_k \quad k=1,2 \\ 5 & \displaystyle\int_{S_k} (\psi-\psi_k) g_i \, dy=0 & k=1,2 \quad i=1,N \end{cases}$$

The Dirichlet conditions on B and S express the fact that B and S are stream-lines. The functions ψ_1 and ψ_2 are respectively defined on D_1 and D_2

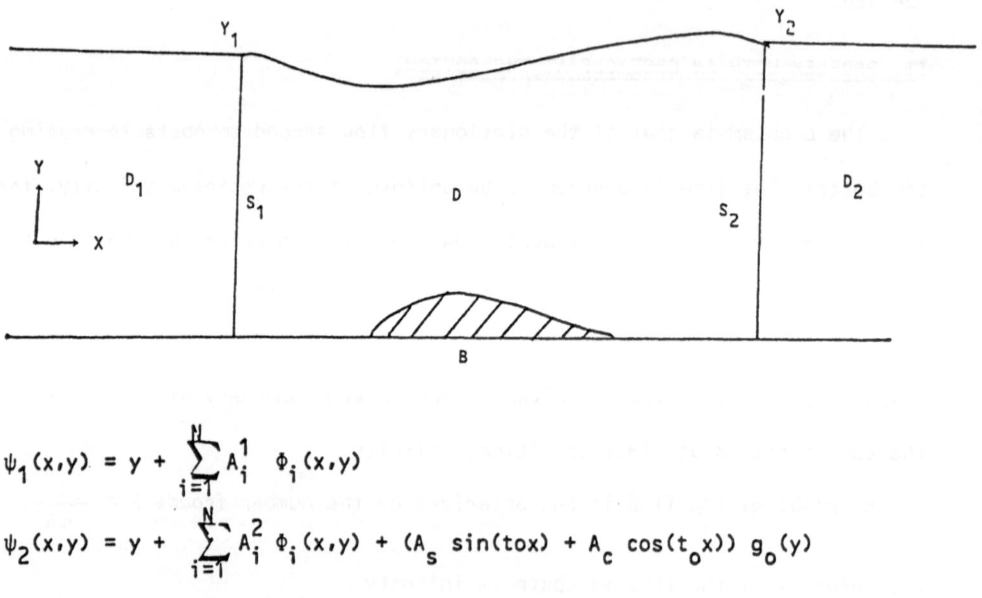

$$(1.2)\quad \psi_1(x,y) = y + \sum_{i=1}^{N} A_i^1 \ \phi_i(x,y)$$

$$\psi_2(x,y) = y + \sum_{i=1}^{N} A_i^2 \ \phi_i(x,y) + (A_s \ \sin(tox) + A_c \ \cos(t_o x)) \ g_o(y)$$

$$\phi_i^k(x,y) = e^{(3-2k)tix} \ g_i(y)$$

$$g_1(y) = \sin(tiy) \quad g_o(y) = sh(toy)$$

ti are the pulsation solution of :

$$tg(ti\ y_k) = F^2\ y_k \qquad k=1,2$$

$$th(to\ y_2) = F^2\ y_2$$

the coefficient A_i^k, A_2, A_3 are determined by 5 in (1.1) in order to ascertain a C^1 link of the stream function ψ with ψ_k across the lateral boundary S_k (k=1,2)

In (1.2) the term of g_o (exists only for $F < 1$) express that the flow is not uniform downstream, it is the wave term.

In (1.1), the boundary S is an unknown of the problem. We must verify a condition expressing the continuity of the pressure across the free surface S, and this by applying Bernoulli's law

$$(1.3) \qquad \frac{\partial\psi}{\partial n} = G(y) \text{ with } G(y) = (1 - \frac{2}{F^2}\ (y-1))^{1/2}$$

of a numerical point of view, the condition (1.3) is sensitive to discretize ; we give a weak condition of (1.3) by applying Green's formula

$$\forall\ v \in V = \{v \in H^1(D) \quad v=0 \text{ on } B\}$$

$$(1.4) \qquad \int_S \frac{\partial\psi}{\partial n}\ v\ ds = \int_D \nabla\psi\ \nabla v\ dx \ - \sum_{k=1}^{2} \int_{S_k} \frac{\partial\psi}{\partial n}\ v\ ds$$

then the condition (1.3) becomes $\forall\ v \in V$

$$(1.4) \qquad \int_D \nabla\psi\nabla v\ dv - \sum_{k=1}^{2} \int_{S_k} \frac{\partial\psi}{\partial n}\ v\ ds - \int_S G(y)v\ ds = 0$$

2. NEWTON'S METHOD.

The equation (1.4) is at the form F(D)=0 where D and then S is the unknown of the problem, to apply Newton's method, we must confine ourselves to a problem given in finite dimension, where the unknowns are the coordinates of the nodes. In all the problem, we assume the x-coordinates fixed.

REMARK.

With intent to reduce the dimension of the Newton matrix, we only consider the nodes situated on the free boundary. In [3] , we have remarked that the gradient with respect to the internal nodes is neglected compared with the one of the nodes on the free boundary.

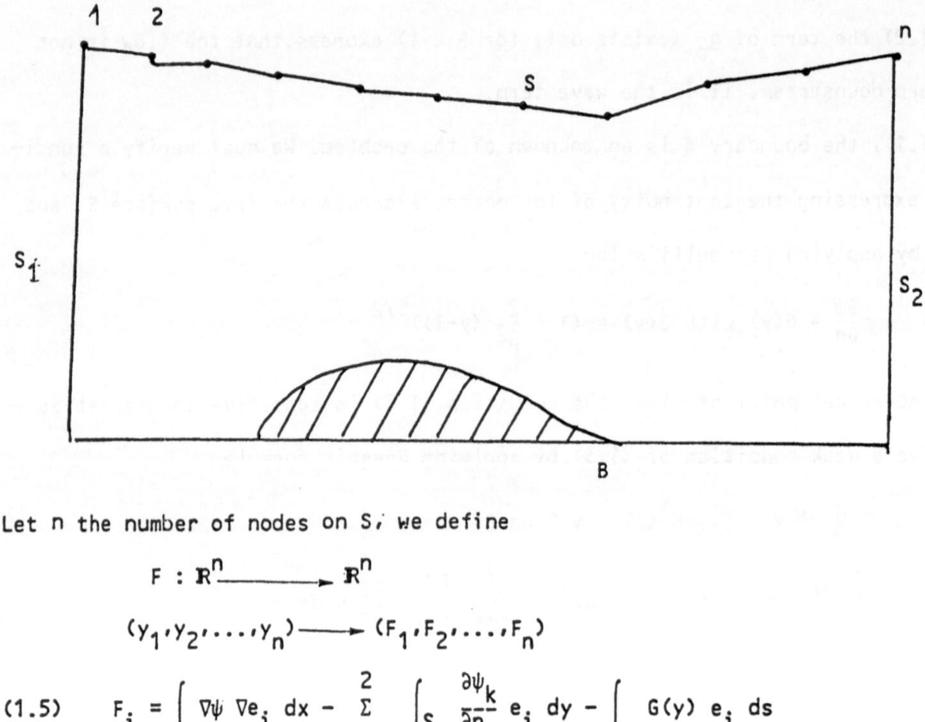

Let n the number of nodes on S, we define

$$F : \mathbb{R}^n \longrightarrow \mathbb{R}^n$$

$$(y_1, y_2, \ldots, y_n) \longrightarrow (F_1, F_2, \ldots, F_n)$$

(1.5) $$F_i = \int_D \nabla\psi \, \nabla e_i \, dx - \sum_{k=1}^{2} \int_{S_k} \frac{\partial \psi_k}{\partial n} e_i \, dy - \int_S G(y) \, e_i \, ds$$

e_i : the basis function at the node i

ψ_k : the prolongation of z on D_k, given by (1.2)

ψ : the discrete solution of the variational problem :

$$\psi \in V_h = \left\{ u \in H^1(D),\ u/k \in Q_1 \quad \begin{array}{l} u=0 \text{ on } B \\ u=1 \text{ on } S \end{array} \right\}$$

$$\forall\, u \in V_{h_o} \left\{ u \quad (D),\; u_{/k} \quad Q_1 \quad \begin{array}{l} u=0 \text{ on } B \\ u=0 \text{ on } S \end{array} \right\}$$

$$(1.6) \quad \int_D \nabla\psi\,\nabla u\, dx - \sum_k \int_{S_k} \frac{\partial\psi_k}{\partial n}\, u\, dy = 0$$

$$\int_{S_k} (\psi-\psi_k) gi\, dy=0 \quad k=1,2 \quad i=1,N$$

3. CALCULATION OF DF.

The calculation of $\frac{\partial F_i}{\partial y_j}$ necessitates the derivative of the equation (1.5) and that of the variational formulation (1.6) discretized by an isoparametric finite elements method Q_1. We use the technique of the calculation of the gradient with respect to the coordinates of the nodes given in | 2 | and | 4 | where the derivatives of (1.5) and (1.6) are simplified by the fact that the derivative of the basis function e_i is nil.

This technique is a particular case of the general theory for the calculation of the gradient developed in | 5 | where we take as a field of velocity the basis functions $\{e_i\}_1^n$

The derivative expression of Fi with respect to the y-coordinate of the node j is :

$$(1.7) \qquad \frac{\partial F_i}{\partial y_j} = \int_D \langle A_j\, \nabla_\psi . \nabla e_i \rangle\, dx + D(\psi,e_i) - \int_S \left(\frac{dG}{dy} + G(y)\,\beta_j\right) e_i\, ds$$

$$+ \int_D \langle A_j\, \nabla\psi .\ \nabla P \rangle\, dx + D(\psi,P)$$

P is the discrete adjoint state, $P \in V_{h_o}$ such that $\forall\, u \in V_{h_o}$

$$\int_D \nabla P\, \nabla u\, dx - \sum_{k=1}^{2} \int_{S_k} \frac{\partial P_k}{\partial n}\, u\, dy = \int_D \nabla e_i\, \nabla u\, dx - \sum_{\substack{k=1\\k=2}} \sum_{j=1}^{N} T_j^k(e_i,u)$$

$$\int_{S_k} (P-P_k)\, g_i\, dy = 0 \quad k=1,2 \quad i=1,N$$

P_k is the prolongation of P on D_k, as for ψ_k in (1.2)

$$P_1(x,y) = \sum_{j=1}^{N} B_j^1 \, \Phi_i(x,y) + (B_s \sin(\text{to } x) + B_c \cos(t_o n))\varrho_o(y)$$

$$P_2(x,y) = \sum_{j=1}^{N} B_j^2 \, \Phi_i(x,y)$$

$$A_j = - \begin{pmatrix} -\partial_2 e_j & \partial_1 e_j \\ \partial_1 e_j & \partial_2 e_j \end{pmatrix}$$

$$\beta_j = n_1^2 \, \partial_2 \; e_j - n_1 \, n_2 \, \partial_1 \, e_j$$

$\vec{n} = (n_1, n_2)$ the outward unit normal at $x \in S$

$D_1(\psi, \rho)$, $T_j^k(e_i, u)$ result from the coupling terms (annex)

III. PROBLEM WITH OBSTACLE PIERCING THE FREE SURFACE S.

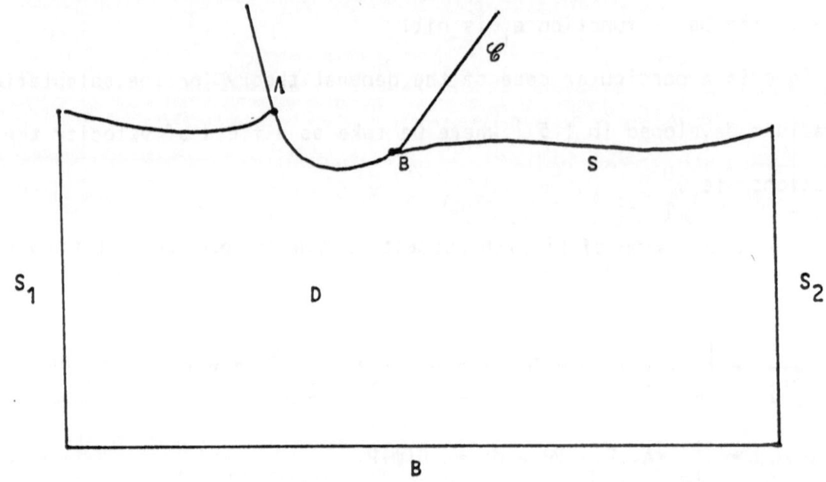

The problem is identical to the first, only the bernoulli condition is changed we assume the hull $\mathscr{C}$ fixed, we applied on $S/\mathscr{C}$ the Bernoulli equation : $S/\mathscr{C}$ is the free part of the boundary S.

$$\frac{\partial \psi}{\partial n} = G(y) \qquad S_L = S/\mathscr{C}$$

The points A and B move remaining on the hull $\mathscr{C}$. In fact A is fixed by physical condition at the highest position in the Bernoulli condition corresponding to velocity equal to zero, B is determined by the fact that S_L is tangent to $\mathscr{C}$ at B.

IV. NUMERICAL RESULTS.

For the calculation of the gradient, we have only taken into account the nodes on the free surface S, because the gradient of the internal nodes is relatively negligeable.

In order to test the validity of the gradient, we have compared our results with the gradient calculated by finite difference :

$$\frac{\partial F_i}{\partial y_j} = \frac{F_i(y_1,\dots,y_j+\ ,\dots,y_N) - F_i(y_1,\dots,y_N)}{\varepsilon} \tag{4.1}$$

The calculation is done in I.B.M 3081 in double precision for different values of ε. The values of the derivative become stabilized for $\varepsilon < 10^{-4}$ (this test is done for the obstacle piercing the free surfaces)

a) Obstacle at the bottom.

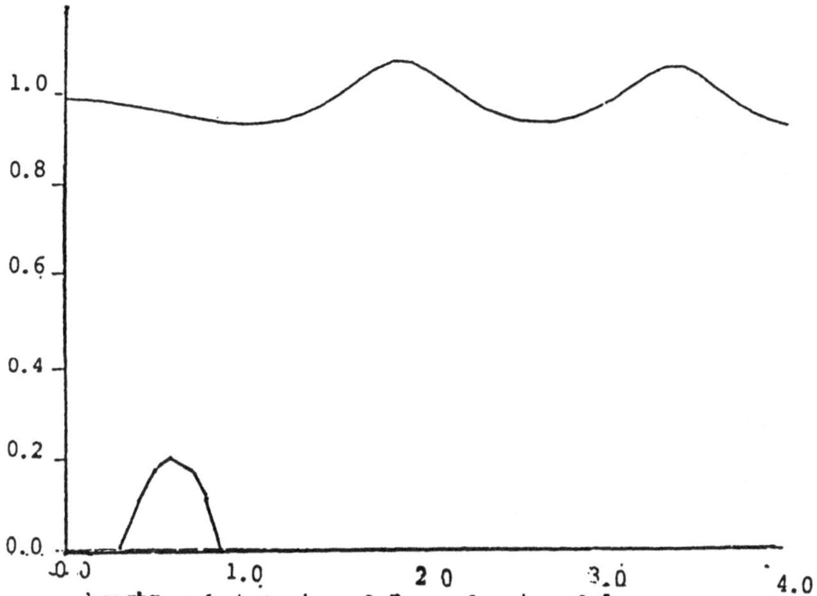

height of obstacle = 0.2   Froude = 0.5

b) Obstacle piercing S.

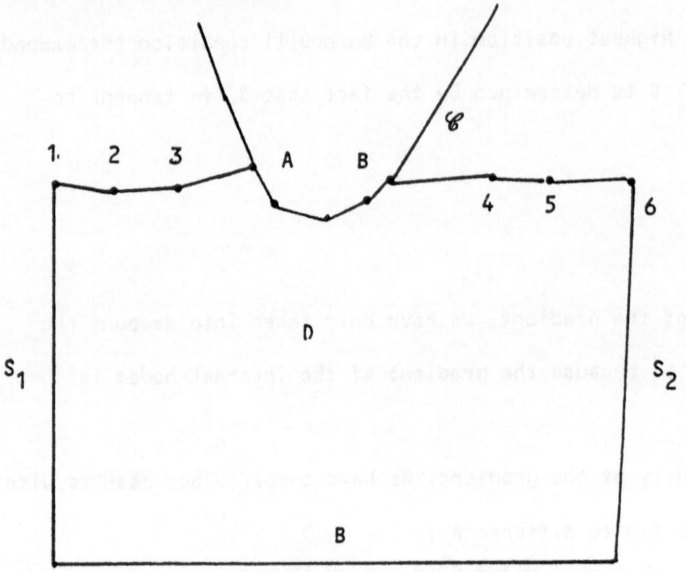

F=0.5 height =0.05 ℓ=0.2

Matrix $(DF)_{i,j)} = \dfrac{\partial F_i'}{\partial y_j}$ calculated by (1.7)

-.2347E+00	-.1908E-01	0.5711E-01	0.1380E-02	0.3483E-03	0.1366E-03
0.1705E+00	-.9985E+00	0.4322E+00	0.4604E-02	0.1063E-02	0.4143E-03
-.2345E+00	-.3077E+00	-.8278E+00	0.1567E-01	0.3226E+00	0.1260E-02
-.2667E+01	-.1470E+01	-.1064E+01	-.1309E+01	0.3809E+00	0.1508E+00
-.7520E+01	-.4169E+01	-.3071E+00	-.2711E+00	-.1800E+01	0.6339E+00
-.7352E+01	-.4078E+01	-.3010E+01	-.5426E+00	-.7836E+00	-.5179E+00

Matrix $(DF)_{i,j} = \dfrac{\partial F_i}{\partial y_j}$ calculated by finite difference (4.1)

-.2347E+00	-.1921E-01	0.5711E-01	0.1379E-02	0.3481E-03	0.1365E-03
0.1705E+00	-.9985E+00	0.4321E+00	0.4602E-02	0.1062E-02	0.4138E-03
-.2345E+00	0.3077E+00	-.8278E+00	0.1567E-01	0.3223E-02	0.1259E-02
-.2667E+01	-.1470E+01	-.1064E+01	-.1309E+01	0.3806E+00	0.1507E+00
-.7420E+01	-.4169E+01	-.3070E+01	-.2709E+00	-.1799E+01	0.6331E+00
-.7352E+01	-.4078E+01	-.3009E+01	-.5425E+00	-.7826E+00	-.5175E+00

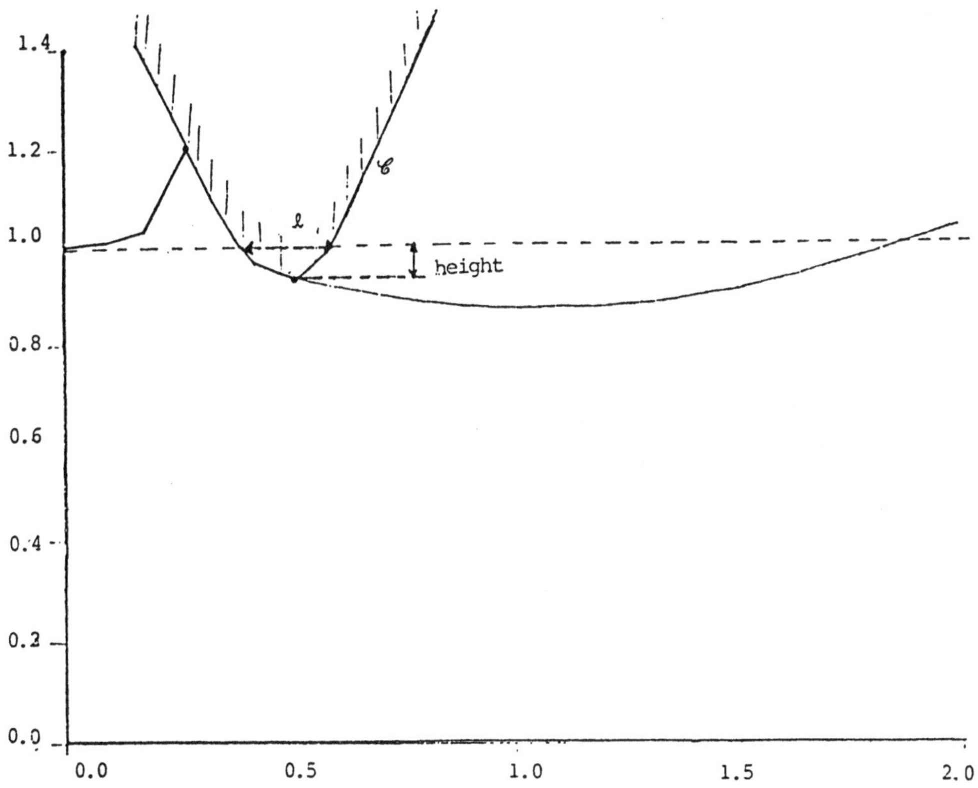

Froude = 0.64 height of obstacle = 0.05 ℓ=0.2

For height =0.05 we have convergence of the Newton method for $0.4 < F < 0.9$ the fixed point method converges only for $0.6 < F < 0.72$

Annex

$$D(\Psi,p)=\sum_{\substack{k=1\\ k=3}}\sum_{i=1}^{N} l_i^k + A_S.L_1 + B_S.L_2$$

$$l_i^k = \mu_i A_i^k.B_i^k |g_i|^2{}_{L^2(S_k)} + t_i^k.B_i^k[\mu_i \int_{S_k} y\cos(t_{iy})(\Psi-y)dy$$

$$-A_i^k(2\mu_i^k \int_{S_k} g_i.y.\cos(t_{iy})dy + g_i{}^2(A_k)) + (\Psi-y)g_i(A_k)] + t_i^k.A_i^k(\mu_i \int_{S_k} y\cos(t_{iy}).Pdy + g_i P(A_k))$$

$L_1 = -\mu_o \int_{S_2} y.ch(t_{oy}).Pdy - sh(t_o y_n)P(A_n)$

$L_2 = -sh(t_{oy})(\Psi-y)(A_1)$

Where :

A_i^k, A_s and B_i^k, B_s are the Lagrangien parameters defined by (1.2) and (2.3).

μ_i and μ_o verify $\frac{d}{dy_k}t_i=\mu_i$ and $\frac{d}{dy_2}t_o=\mu_o$.

REFERENCES :

[1] J. CAHOUET Etude numérique et expérimentale du Problème de la résistance de vague non linéaire. Thesis Paris 84.

[2] M. SOULI, J.P. ZOLESIO.

[3] M. SOULI, J.P. ZOLESIO. Discrete and semi-discrete gradient with respect to the domain in wave problems. IFIP Nice Juin 1986.

[4] O. PIRONNEAU. Optimal shape design for elliptic systems. Springer-Verlag.

[5] J.P. ZOLESIO. Identification de Domaine par déformation. Thesis Nice 79.

SEMI-DISCRETE AND DISCRETE GRADIENT FOR NON LINEAR WATER WAVE PROBLEMS

M. SOULI
Département Mathématique
Parc Valrose

J.P. ZOLESIO
C.N.R.S
U.S.T.L. , Place E. Bataillon
34060 Montpellier Cedex

I - INTRODUCTION

We are concerned with numerical computation of water in a two dimensionel flow, see stoker [7] . The water is assimiled to a perfect, incompressible fluid ; the velocity $\bar{v}$ has the following form $\bar{v} = \nabla \Phi$ where Φ is an harmonic potential .

In two dimensions flow, the velocity is expressed in terms of the stream function ψ by

$$\bar{v} = (u,v) = \left(\frac{\partial \psi}{\partial y}, \frac{\partial \psi}{\partial x}\right) .$$

The domain D which is occupied by the stationary fluid, is bounded on its upper part by a "free boundary" S on which two conditions must be verified :

a) S is a streamline that is a Dirichlet condition for ψ .

b) The energy conservation law formulated by the non linear Bernoulli condition $1/2\,|v|^2 + e$ is constant an each streamline (e being the internal energy where here is neglected the surface tension of S .

Many works have been done for this free boundary problem by using a linearized Bernoulli condition, recents experiments in [2] prove that for large waves, the linearized theory fails completly.

A first work taking into account the non linear condition has been done by J. Cahouet and M. Lenoir [3] , in this paper , we propose two shape formulations of the problem.

An important parameter is the Froude number, the situation $F > 1$ coresponds to uniform downstream flow generated by height velocity C of the upstream infinity flow, this is the best one in a mathematical an numerical point of view.

We focus here on the worst situation $f < 1$, which corresponds to a downstream undamped wave.

To use a finite element method we reduce the domain to a bounded one. The obstacle generating the perturbation of the flow lies at the bottom. For this, we introduce two vertical boundaries S1 (upstream) and S2 (downstream).

The harmonic stream function in D is C^o liked across $S_k(k = 1,2)$ using the explicit solution for the linearized problem out of D , that is far enough of the obstacle. This classicaly leads to the boundary conditions given in (1.1) following [2].

We try also in this work to improve the results given by Cahoues [2] where he uses the fixed point method.

The free boundary problem is formulated as a shape optimization problem and we are driven to the minimization of a cost functional. We numericaly compute the gradient (the so called shape derivative) by two differents methods and we compare the results. It is impossible to give in a short paper more then just a flaviour of all the complexity concerning the numerical treatement of this problem ; the state-of-art being now to systematicaly avoid the use of the continuous gradient of the problem, as it was formely done, but to make use of semi-discrete gradient and derivatives with respect to the nodes of the finite element model.

1 - Position of the problem

We forsee in this work the search for the position of the free surface of the bidimensional wave by the shape optimal control method.

The problem studied is that of the stationary flow arround an obstacle placed at the bottom. The flow is assumed uniform upstream. The equations of the flow are detailled in [1], see also M. Souli [5] .

The stream function ψ is solution of :

$$\Delta\psi = 0 \quad D$$

$$\psi = 0 \quad B$$

$$\frac{\partial\psi}{\partial n} = v(y)(\psi-y)+1$$

(1.1)
$$\frac{\partial\psi}{\partial n} = \frac{\partial\psi_k}{\partial n} \quad S_k$$

$$\int_{S_t} (\psi-\psi_k)\, g_i dy = 0 \qquad k=1,2$$

ψ_k is the prolongation of ψ on D_k

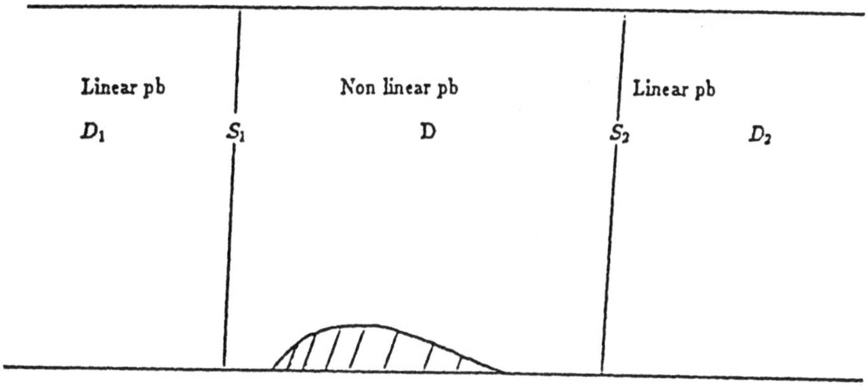

The limit condition are defined to assure a C1 link of ψ across the boundary S_k :

We have the explicit expressions :

(1.2)
$$\psi_1(x,y) = u + \sum_{i=1}^{N} A_i^1\, \Phi\,(x,y)$$

$$\psi_2(x,y) = y + \sum_{i=1}^{N} A_i^2\, \Phi_i(x,u) + A_S \sin(t_0x) + A_c \operatorname{cost}(t_0x))\, g_0(y)$$

A_i^k, A. and A_c are the unknowns of the problem determined by the condition 1.1 . The ψ_i are obtained by separated variables in the linearized problem formulated in the strip D_1 and D_2 .

$$\Phi_i(x,y) = e^{(3-2k)t_ix}\, g_i(y)$$

$$g_i(y) = \sin(t_i y)$$

t_i are the pulsations solution of the classical dispersion equation

$$\tan(t_i y_k) = F2t_i$$

$$\tanh(t_o y_2) = F2t_o$$

where F is the Froude number defined by

$$F = \frac{c}{\sqrt{gH}}$$

C : Celerity of the flow at upstream

H : The depth

g : The gravity

And finaly in condition (3) , the function ν is given by :

$$\nu(y) = \frac{2}{F^2}\,(1 + (1 - \frac{2}{F^2}(y-1)^{-1}$$

Remark

The presence of the term in g_o in the ψ_2 expression (only for $F < 1$) takes into account the oscillating behavior of ψ_2 dowstream (Wave term). Upstream, we suppose the flow uniform and thus $\lim \psi_1(x,y) = y$ if $x \longrightarrow -\infty$.

For $F > 1$, we show by a simple calculation that the wave term is nil .

The boundary S being an unknown of the problem, the stream function must verity an additional condition on S expressing the fact that S is a streamline :

(1.2) $\psi = 1$ on S

2 - Application of the optimal control

We formulate the problem (1.1) and (1.2) under the form of minimization of a functional cost J defined on D .

(1.3) $$\text{Find } D^* \in \theta_{ad}$$

$$D^* = \text{Arg Min } J(D)$$

$$J(D) = \frac{1}{2} \int_s (\psi - 1)^2 ds$$

θ_{ad} set of admissible domains .

ψ solution of (1.1) .

The calculation of the derivative dJ with respect to the domain is done with the help of the formalism introduced in [8] . It is defined by :

(1.4) $$dJ(D,\bar{v}) = \lim_{t \to 0} \frac{J(D_t) - J(D)}{t} \quad v \in V$$

V a field of velocity constructing the virtual deformation of D . For a smooth domain ans a field of velocity of C^1 class at least, the derivative dJ is :

$$dJ(D,\bar{v}) = \int_s g.v(0).\bar{n}ds$$

Calculation of the gradient

We make explicit in this work two methods for the calculationof the gradient dJ .

a- We assume the boundary of the domain polygonal, pieces smooth and the field of velocity smooth in a neighborhood of D . The gradient dJ will then allow the intervention of terms coming out of the singularities of the boundary, it is the semi-discrete gradient (continuous graient on a discrete geometry).

b- We discretize the domain D, and we consider a discrete field of velocity. In this case the gradient is nothing but the one with respect to the nodes that we meet in [4]and [6].

II Semi-discrete gradient

The calculation of the continuous gradient serves us on a basis for calculating the semi-discrete gradient. Thus we recall some results concerning the calculation of the continuous gradient.

Théorème [9]

L et us suppose :

(H_1) D_1 open bounded in R^2 , with boundary S_1 of C^2 class

(H_2) V field smooth velocity

(H_2) f of class C^0

If $g(t) = \int_{S_1} f(t,x)\, ds$ then the Eulerian derivative of g is given by :

$$g^1(t) = \int_{S_1} [\frac{\partial}{\partial t} f(t,x) + (\frac{\partial}{\partial n} f(t,x) + H.f(t,x)) \langle \bar{v}.\bar{n} \rangle]\, ds$$

H the mean curvative of S_1 .

$\bar{n}$ the outward unit normal .

Remark

In the continuous gradient, the derivative of J allows the intervention of the curvature H. from a numerical point of view, the calculation of H is delicate . To get rid of this difficulty, we develop a new method for the calculation of the gradient.

The method consists in considering a boundary S non smooth in the following way

H_1 1 : it exists singularities at the points $A_1, \dots, A_n$ such that S is pieces smooth .

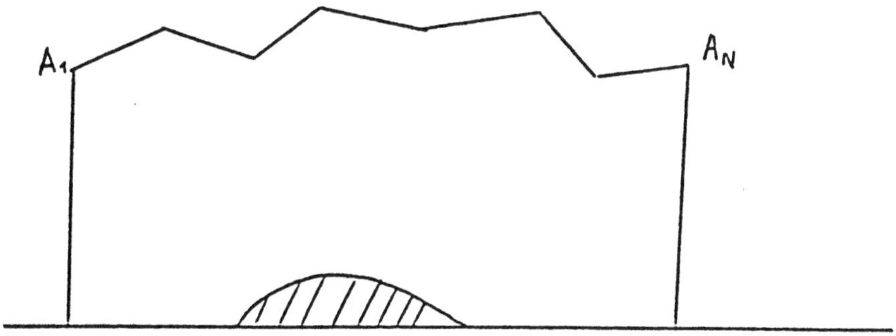

Theorem 2 [9]

Under the hypothesis H_1 1 , H2 , H3 we have

$$g^1(t) = \int_{S_1} [\frac{af(t,x)}{\partial t} + \frac{\partial f(t,x)}{\partial n} \langle v \}n \rangle]ds + \sum_{i=1}^{n} \bar{T}.\bar{V}(A_k).f(t,A_k)$$

$\bar{T} = t_1 - t_2$ is the difference between the tangent vectors at A_k .
For the functional cost defined by (1.3) , theorem 2 gives :

$$dJ(D,V) = \int_S - \nabla_T \psi \nabla_T P.vdx + \int_S \frac{\partial}{n}(\frac{\partial\psi}{\partial n}).Pvds + \int_{S_0} (\psi-1)\frac{\partial\psi}{\partial n} vds$$

$$+ \sum_{i=1}^{n} T.V(A_i).(\nu(y_i)(\psi_i - y_i) + 1) P_i + \frac{1}{2}(\psi-1)^2 + \emptyset$$

D (ψ,P) is coming out of the coupling terms (annex) $v = \bar{V}.\bar{n}$
P is the adjoint state solution of the problem.

$$\nabla P = 0 \quad \text{in } D$$

$$\frac{\partial P}{\partial n} = \psi - 1 \quad \text{on } S$$

$$P = 0 \quad \text{on } B$$

$$\frac{\partial P}{\partial n} = \frac{\partial P_k}{\partial n} \quad \text{on } S_k \quad k=1,2$$

$$\int_{S_k} (P-P_k)\, g_i dy = 0$$

P_k the prolongation of P on D_k defined by :

$$P_1(x,y) = \sum_{i=1}^{n} B_i^1 \Phi_i(x,y) + [B_s \sin(t_o y) + B_c \cos(t_o y)]\, g_o(y)$$

$$P_2(x,y) = \sum_{i=1}^{n} B_i^2 \Phi_i(x,y)$$

B_i^k, B_s and B_c and the unknowns determined by the condition in (2.2) , Φ_i the basic function defined in (2.2) .

3 - Discrete gradient

We establish in this chapter similar results, when we consider a discrete field of velocity on a discrete domain. We have shown in [6] that when we consider the velocity $\bar{v}=(e_i,e_i)$ in (1.4) where ei is the basis function at the node i , the gradient in the $(e_i,0)$ (resp $(0,e_i)$) direction is nothing else but the gradient with respect to the coordinate x'x (resp y'y) of the node i .

As for the semi discrete gradient, we introduce an adjoint state P solution of the discrete formulation coming out of the problem (2.2) . The discrete gradient in the $\bar{v}$ direction is :

$$dJ(D,v) = -\int_d \langle A \nabla\psi \,,\, \nabla P \rangle dx + \int_S (\nu(y)(\psi-y)+1)\, Bvds$$

$$+ \int_S (\nu'(y)(\psi-y) - \nu(y))\, Pv_2 ds + \frac{1}{2}\int_2 (\psi-1)^2\, Bvds + D(\psi,P)$$

For $\bar{v} = (e_i,0)$ (resp $\bar{v} = (0,e_i)$) the matrix A has the form :

$$A = - \begin{matrix} \partial_1 e_i & \partial_2 e_i \\ \partial_2 e_i & -\partial_1 e_i \end{matrix}$$

$$\left(\text{resp } A = - \begin{matrix} \partial_2 e_i & \partial_1 e_i \\ \partial_1 e_i & \partial_2 e_i \end{matrix} \right)$$

$B\bar{v} = \operatorname{div}\bar{v} - (D\bar{v}.\bar{n},\bar{n})$

As for the semi discrete gradient, $D(\psi;P)$ is coming out of the coupling terms (Annex)

ψ and P are discrete solutions of (1.1) , (2.2) .

Remark

We recognize the gradient with respect to the x_i and Y_i coordinate of the node (X_i,Y_i) .

$$\frac{\partial J}{\partial X_i} = dJ\,(D\,,(e_i\,,0))$$

$$\frac{\partial J}{\partial Y_i} <= dJ\,(D\,,(0\,,e_i))$$

Algorithm of resolution

The domain D being discretized, the free boundary S is determined by the position of the nodes situated on this boundary. In the search for the new boundary, we assume the abscisses of the nodes fixed, thus

$$\bar{v} = (o, e_i)$$

The ordinates Yi are determined by the algorithm given by . A. Buckley [1] using an approximation of the Hessian matrix to search for the optimal slope. This algorithm is implemented in I.B.M. 3081 in Harwell library.

In the Wave problem, the crirical parameters are the heitht of the obstacle, and the Froude number when it approximates the value 1.

With the gradient method, we have convergence for the heights of the obstacle and the Froude number for which the fixed point method [2] fails.

Numerical results

a) To test the validity of the two gradients, we have compared them to the gradient obtained by finite difference.

$$DJ_i = \frac{J(Y_1, \dots, Y_i + \varepsilon, \dots, Y_n) - J(Y_1, \dots, Y_n)}{\varepsilon}$$

The calculation of dJ is in double precision on I.B.M. 3081 for different values of ε. The values of dJ_i become stable for $\varepsilon \leqslant 10^{-4}$.

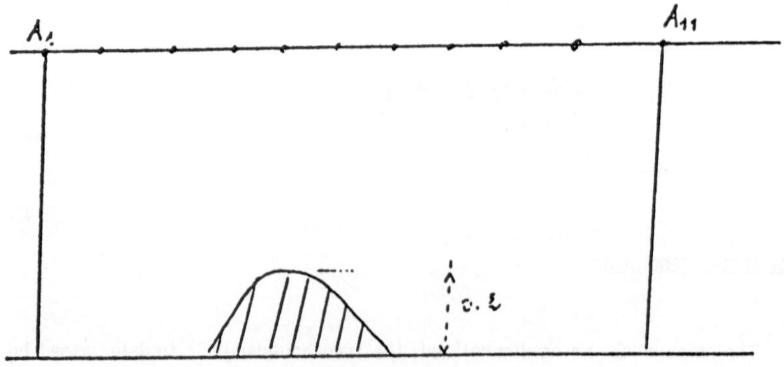

$F = 0.5$ height $= 0.2$

A	B	C
0.7458E+00	0.7458E+00	0.7210E+00
-.1129E+00	-.1130E+00	-.1463E+00
-.2367E+00	-.2368E+00	-.2575E+00
-.2835E+00	-.2836E+00	-.2898E+00
-.2577E+00	-.2573E+00	-.2541E+00
-.1892E+00	-.1693E+00	-.1820E+00
-.1105E+00	-.1106E+00	-.1046E+00
-.4493E-01	-.4498E-01	-.4194E-01
-.3621E-02	-.3634E-01	-.3662E-01
0.1433E-01	0.1434E-01	0.1512E-01
0.1116E-01	0.1120E-01	0.1048E-01

a : Gradient by finite difference at the nodes $i=1$ to 11 , $\varepsilon=10^{-5}$

b : Discrete gradient

C : Semi discrete gradient

2) In fig. 1 we heve convergence after 40 iterations initial cost $j_o = 0.2701$, initial cost $J_{40} = 0.14.10^{-2}$ for a height of the obstacle equal 0.4 (The hight of the flow upstream equal 1) . The fixed point method fails for each $F < 1$.

To check the robustness of these gradients we did an experiment involving a very large obstacle. We make that when the Froude number F approaches 1 , oscillations are generated by this algorithm.

Fig 2

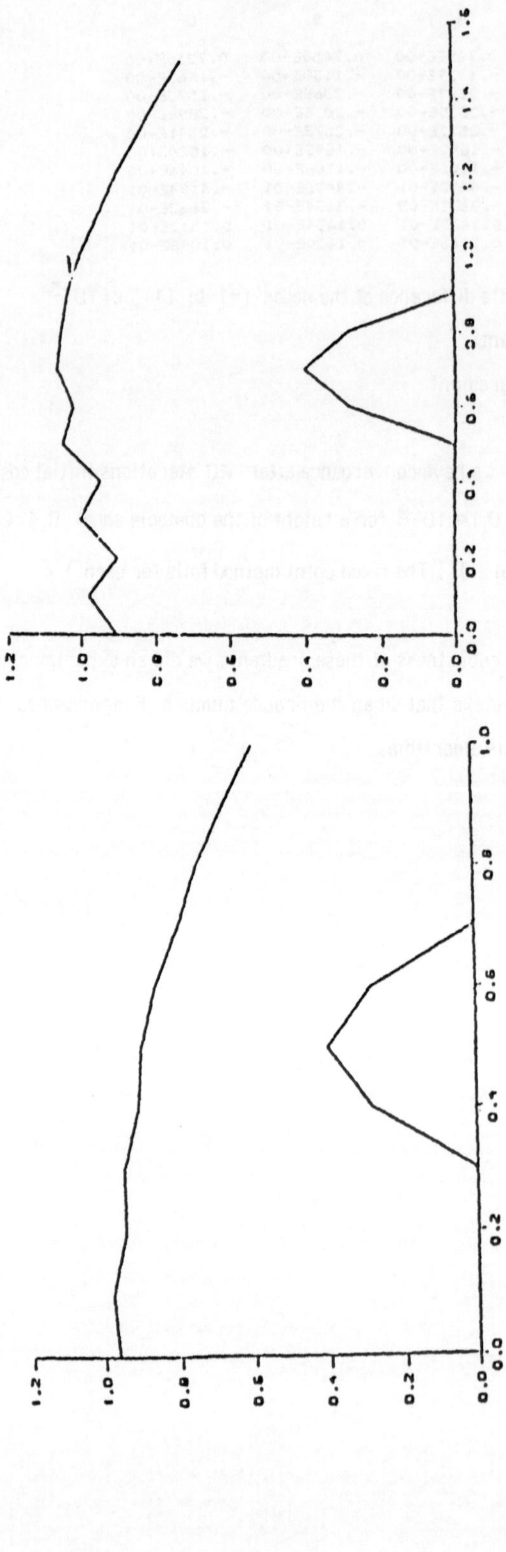

height obst = 0.4 $F = 0.7$

height obst = 0.4 $F = 0.5$

Annex

$$D(\psi,p) = \sum_{\substack{k=1\\k=2}} \sum_{i=1}^{N} 1_i^k + A_S.L_1 + B_S.L_2$$

$$L_i^k = \mu_i A_i^k.B_i^k |g_i|^2 \; L^2(S_k) \; + \; t_i^k.B_i^k |\mu_i \int_{S_k} y \cos t\,(t_{iy})\,(\psi - y)dy$$

$$- A_i^k(2\mu_i^k \int_{S_k} g_i.y.\cos t(t_{iy})dy + g_i^2(A_k)) + (\psi - y)g_i(A_k)) + t_i^k A_i^k(\mu_i \int_{S_k} y \cos t(t_{iy}).Pdy + g_i P(A_k))$$

$$L_1 = \psi_0 \int_{S_1} y.ch(t_{oy})\,.\,Pdy - sh(t_{oyn})\,P\,(A_n)$$

$$L_2 = sh(t_{oy})\,(\psi - y)\,(A_1)$$

where :

A_i^k, A_S and B_i^k, B_c are the largrangien parameters defined by (1.2) and (2.3) .

μ_i and μ_0 verify $\frac{d}{dy_t} t_i = \mu_i$ and $\frac{d}{dy_2} t_0 = \mu_0$.

REFERENCES

[1] A. Buckleley : An alternate implementation of Goldfarb's minimization algorithm. Mathematical programming 8, 1975. p 207-231

[2] J. Cahouet : Etude numérique et expérimentale du problème bidimensionnel de la résistance de vague non linéaire. Thesis Paris 1984 .

[3] J. Cahouet, M. Lenoir : Résolution numérique du problème non linéaire de la résistance de vague bidimensionnelle. C.R.A.S. 297, 1983 .

[4] O. Pironneau : Optimal shape design for elliptic systems. Springer. Verlag.

[5] M. Souli, J.P. Zolésio : Shape derivative of discretized problems. To appear .

[6] M. Souli : Shape Newton method on naval hydrodynamic. WG7 I.F.I.P. Nice 1986 .

[7] J.J. Stoker : Water wave . The mathematical theory with applications.

[8] J.P.Zolésio : Identification de domaines par déformation. Thesis Nice 1979.

[9] J.P. Zolésio : Gradient des coûts gouvernés par des problèmes de Neumann posés sur des ouverts anguleux en optimisation de domaine. Cours I.N.R.I.A. Nice 1983, ANS Rapport CRM nº 1021 Montréal, 1982 .

[10] J.P. Zolésio : Les Dérivées par rapport aux noeuds des triangularisations et leur utilisation en identification de domaine, rapport CRM n_0 Montréal 1982 and Ann. Sc. Math. Quebec 8 (1984) 97-120.

Gradient with respect to nodes for non-isoparametric finite elements

TIMO TIIHONEN*
JEAN-PAUL ZOLESIO

Abstract. We consider the problem of controlling the solution of a finite element model using the nodal co-ordinates as control variables. The main emphasis is on the study of the applicability of the domain deformation method for different element types. The results are applied to a simple problem of finite element grid optimization.

1. INTRODUCTION

In this work we study the problem of controlling the solution of the finite element discretization of a partial differential equation. More precicely, we are interested in determining the derivative of the finite element solution with respect to the nodal co-ordinates of the finite element mesh. The principal applications of these derivatives can be found in the fields of the finite element grid optimization, the optimal shape design and the numerical simulation of big deformations of elastic materials.

For the sake of notational simplicity we have chosen a very simple model example, namely the Poisson equation in $\mathbf{R}^n$. In chapter 2 we recall the basic results of Zolésio [2] on the so-called speed (or velocity) method for domain deformations. These results are then applied to our model problem and also to its Galerkin approximation. We then proceed by studying formally the problem of finding the 'optimal' Galerkin approximation scheme. Using the formal results thus obtained we recall and explicit the results of Zolésio [3] and Souli and Zolésio [1] in the case of (iso)parametric lagrangian elements. A case study is then made for a non-isoparametric finite element model.

2. DERIVATIVE WITH RESPECT TO DOMAIN DEFORMATIONS

We shall use the so-called speed method [2] to modelize the domain deformations. If $V(t,x)$ stands for a sufficiently regular vector field defined in the neighbourhood of a domain $\Omega \subset \mathbf{R}^n$ (typically $V \in C^0([0,t_0], C^k(\mathbf{R}^n))$ i.e. V is continuous in time and C^k in space), we note by $T_t : x \to X(t,x)$ the mapping from Ω to $\mathbf{R}^n$ defined by

$$\begin{cases} \frac{d}{dt}X(t,x) &= V(t,X(t,x)), \\ X(0,x) &= x. \end{cases} \tag{1}$$

*This work was initiated when the first author was staying at the University of Nice. The financial support of the French governement is gratefully acknowledged.

We note by $\Omega_t := T_t(\Omega)$ the domain transported by the velocity field V after the time t. It is well known [2] that if T_t and T_t^{-1} are mappings of C^k, then

$$\phi \in H^s(\Omega) \iff \phi^t := \phi \circ T_t^{-1} \in H^s(\Omega_t)$$

as soon as $k \geq s$.

Let $y(\Omega_t)$ be a regular function defined on Ω_t. The material (or speed) derivative of $y(\Omega_t)$ at $t = 0$, if it exists, is defined by

$$\dot{y}(x) = \lim_{t \to 0} \frac{y(\Omega_t)(T_t x) - y(\Omega)(x)}{t}. \tag{2}$$

If the functions $y(\Omega_t)$ are not sufficiently regular for the pointwise definition to make sense, the derivative can be defined in a Sobolev space setting; $\dot{y} \in H^m(\Omega)$ s.t.

$$\lim_{t \to 0} \| \frac{y(\Omega_t) \circ T_t - y(\Omega)}{t} - \dot{y} \|_{H^m(\Omega)} = 0. \tag{2'}$$

From [2] we know that if $\dot{y}$ depends continuously on V, then it depends only on $V(0)$.

Let us now introduce some basic formulae for $\dot{y}$, cf. [2]:

$$\widehat{(\nabla y)} = -DV^* \nabla y + \nabla \dot{y} \tag{3}$$

where DV is the Jacobian of $V(0)$ and $*$ denotes the transpose.

$$\frac{d}{dt} \int_{\Omega_t} y(\Omega_t) dx \Big|_{t=0} = \int_\Omega \dot{y}(\Omega) dx + \int_\Omega y(\Omega) \mathrm{div} V(0) dx. \tag{4}$$

For $f : \mathbf{R}^n \to \mathbf{R}$ which does not depend on t we have

$$\dot{f} = \nabla \cdot V(0), \tag{5}$$

$$\frac{d}{dt} \int_{\Omega_t} f dx = \int_\Omega \mathrm{div}(fV) dx. \tag{6}$$

We are now ready to study our model problem:

$$\begin{gathered} \text{Find } y \in H_0^1(\Omega) \text{ s.t.} \\ \int_\Omega \nabla y \nabla \phi dx = \int_\Omega f \phi dx \qquad \forall \phi \in H_0^1(\Omega). \end{gathered} \tag{7}$$

As we shall be interested mainly in the finite element applications we choose for simplicity the velocity field V in such a way that $V = 0$ on the boundary, which is often the case in the grid optimization applications. That means that we do not deform the shape of the boundary. The perturbed state problem reads:

$$\text{Find } y_t \in H_0^1(\Omega_t) \text{ s.t.}$$
$$\int_\Omega (\nabla y_t)\circ T_t(\nabla\phi_t)\circ T_t \det(DT_t)dx = \int_\Omega f\circ T_t\phi\circ T_t\det(DT_t)dx \qquad \forall\phi_t\in H_0^1(\Omega_t). \tag{7'}$$

Differentation on the both sides of (7') yields an equation for the material derivative of the solution, $\dot{y}\in H_0^1(\Omega)$

$$\int_\Omega \nabla\dot{y}\nabla\phi dx = \int_\Omega (-\mathrm{div}V+DV^*+DV)\nabla y\nabla\phi+\mathrm{div}(fV)\phi+\nabla y\nabla\dot{\phi}+f\dot{\phi}dx, \quad \forall\phi\in H_0^1(\Omega). \tag{8}$$

We note that we can take $\phi_t = \phi\circ T_t^{-1}$ for all ϕ so that $\dot{\phi}=0$. Thus the two last terms on the right hand side of the equation vanish.

3. Galerkin approximation

Let us study, instead of (7), the following problem:

$$\text{Find } \hat{y}\in W \text{ s.t.}$$
$$\int_\Omega \nabla\hat{y}\nabla\phi dx = \int_\Omega f\phi dx \qquad \forall\phi\in W \tag{9}$$

where $W\subset H_0^1(\Omega)$ is some finite dimensional subspace.

Assume that we have fixed a basis $\{\phi_j\}$, $j=1,\dots,N$ of W and a basis $\{\phi_j^t\}$, $j=1,\dots,N$ of W_t. If ϕ^t:s are chosen so that the limit $\lim_{t\to 0}(\phi_j^t(T_tx)-\phi_j(x))/t =: \dot{\phi}_j$ exists a.e. in Ω for all j, then we can, at least formally, derive an analogue for (8);

$$\int_\Omega \nabla\dot{\hat{y}}\nabla\phi_j dx = \int_\Omega (-\mathrm{div}V+DV+DV^*)\nabla\hat{y}\nabla\phi_j + \mathrm{div}(fV)\phi_j - \nabla\hat{y}\nabla\dot{\phi}_j + f\dot{\phi}_j dx \quad \forall j. \tag{10}$$

We can not, however, prove that $\dot{\hat{y}}\in W$. In fact, if we write $\hat{y}=\sum_i q_i\phi_i$ we have $\dot{\hat{y}} = \sum_i \dot{q}_i\phi_i + \sum_i q_i\dot{\phi}_i$. Replacing this expression in (10) we get an equation for $\dot{q}_i$:s

$$\int_\Omega \sum_i \dot{q}_i\nabla\phi_i\nabla\phi_j dx = \int_\Omega (-\mathrm{div}V+DV^*+DV)\nabla\hat{y}\nabla\phi_j + \mathrm{div}(fV)\phi_j$$
$$-\sum_i q_i(\nabla\phi_i\nabla\dot{\phi}_j + \nabla\dot{\phi}_i\nabla\phi_j) + f\dot{\phi}_j dx \quad \forall j. \tag{11}$$

One possible application for formulas like (11) is the construction of in some sense optimal Galerkin approximations. Namely, if we denote by y the solution of (7) and by $\hat{y}$ that of (9) we have that

$$\begin{aligned}|y-\hat{y}|_{1,\Omega}^2 &= \int_\Omega \nabla y \nabla y dx - 2\int_\Omega \nabla y \nabla \hat{y} dx + \int_\Omega \nabla \hat{y} \nabla \hat{y} dx \\ &= \int_\Omega \nabla y \nabla y dx - 2\int_\Omega f\hat{y} dx + \int_\Omega \nabla \hat{y} \nabla \hat{y} dx \\ &= -\int_\Omega f\hat{y} dx + \int_\Omega \nabla y \nabla y dx.\end{aligned}$$

Thus, smaller $J(\hat{y}) := -\int_\Omega f\hat{y}dx$ gets, the better is the approximation $\hat{y}$.

Assume now that we have been given a deformation field V. Using the formulas of chapter 2 we can differentiate $J(\hat{y})$ with respect to V:

$$\dot{J}(\hat{y}) = -\int_\Omega \mathrm{div}(fV)\hat{y}dx - \int_\Omega f\dot{\hat{y}}dx. \tag{12}$$

To evaluate the second term on the right we can decompose

$$\int_\Omega f\dot{\hat{y}}dx = \int_\Omega f(\sum_i \dot{q}_i\phi_i + \sum_i q_i\dot{\phi}_i)dx. \tag{13}$$

Now, on the other hand from (11)

$$\begin{aligned}&\sum_i \dot{q}_i \int_\Omega f\phi_i dx = \int_\Omega \sum_i \dot{q}_i \nabla\phi_i \nabla\hat{y} dx \\ &= \int_\Omega (-\mathrm{div}V + DV^* + DV)\nabla\hat{y}\nabla\hat{y} + \mathrm{div}(fV)\hat{y} - 2\sum_i q_i \nabla\dot{\phi}_i \nabla\hat{y} + f\sum_i q_i\dot{\phi}_i dx.\end{aligned} \tag{14}$$

Thus, cōmbining (12), (13) and (14) we have

$$\dot{J}(\hat{y}) = -\int_\Omega (-\mathrm{div}V + DV^* + DV)\nabla\hat{y}\nabla\hat{y}dx + 2\int_\Omega -\mathrm{div}(fV)\hat{y} + \sum_i q_i\nabla\dot{\phi}_i\nabla\hat{y} + \sum_i q_i\dot{\phi}_i f dx. \tag{15}$$

4. The finite element case

Isoparametric elements. For notational convenience we shall speak only of the quadrilateral isoparametric element in $\mathbf{R}^2$ although the results can be applied to all isoparametric elements.

Let $\mathcal{T}_h$ be a regular finite element partitioning of Ω by quadrilateral elements. Given this 'triangulation' we can find deformation fields V in such a manner that also the deformed grid $T_t \circ \mathcal{T}_h$ will be a regular finite element partitioning of the deformed domain Ω_t. In fact Souli and Zolésio [1] have shown that if the components of the vector field V are chosen from the space of isoparametric finite element functions, then the corresponding mapping preserves the elements (i.e.

the image of a quadrilateral element is also a quadrilateral element). Moreover this choice convects also the isoparametric finite element basis functions. In other words, if we denote by ϕ_i a basis function (Q_1-isoparametric) on the grid $\mathcal{T}_h$ and if we choose the field V to be spanned by vectors of the type $(\phi_j, 0)$ and $(0, \phi_k)$ then $\phi_i^t = \phi_i \circ T_t^{-1}$ will be a basis function on the grid $\mathcal{T}_h^t$. This, on the other hand means that $\dot{\phi}_i = 0$.

The crucial point here was not the fact that the element was isoparametric. In fact it is not difficult to see that the basis functions are preserved also for the subparametric elements provided that the deformation field V is generated by the same basis functions that are used to define the element geometries. The essential is that the deformation field preserves for each point the coordinates in the reference element.

The fact that $\dot{\phi}_i = 0$ for all i permits us to write the equation for the material derivative $\dot{\hat{y}}$ in the form

$$\int_\Omega \nabla\dot{\hat{y}}\nabla\phi_j dx = \int_\Omega (-\mathrm{div}V + DV^* + DV)\nabla\hat{y}\nabla\phi_j + \mathrm{div}(fV)\phi_j dx \qquad \forall j. \quad (16)$$

Similarily, the equation (15) for the directional derivative of J can now be written as

$$\dot{J}(\hat{y}) = -\int_\Omega (-\mathrm{div}V + DV^* + DV)\nabla\hat{y}\nabla\hat{y}dx - 2\int_\Omega \mathrm{div}(fV)\hat{y}dx. \qquad (17)$$

It must be remarked, however, that while the domain deformation approach is succesful in giving elegant derivative expressions it has a limited range of applicability as it relies heavily on the fact that all the integrals are to be evaluated exaxtly. This means that (16) or (17) can be made exact only for some simple elements and simple force functions f.

The non-isoparametric case. Let us now consider the situation where the elements are not defined using some reference element but are constructed directly in the physical co-ordinates. The most simple example of the elements of this genre is the non-isoparametric Q_1-element. As another example we might consider the Argyris element with its degrees of freedom which depend on the normals of the element sides.

As before we denote by $\mathcal{T}_h$ a regular, quadrilateral subdivision of Ω. The non-isoparametric Q_1 finite element functions are continuous functions which are piecewise bilinear, i.e. they can be expressed locally in the form $\psi_i(x,y) = a + bx + cy + dxy$. The i:th basis function is chosen to be the function which equals 1 at the i:th node and vanishes at other nodes. Thus, if we fix an element Q and denote its nodal co-ordinates by (x_i, y_i), $i = 1\ldots4$, the basis function ψ_i related to the i:th node can be expressed locally in the form

$$\psi_i(x,y) = \frac{|A_i(x,y)|}{|A|}$$

where

$$A = \begin{pmatrix} 1 & x_1 & y_1 & x_1y_1 \\ 1 & x_2 & y_2 & x_2y_2 \\ 1 & x_3 & y_3 & x_3y_3 \\ 1 & x_4 & y_4 & x_4y_4 \end{pmatrix}$$

and A_i is obtained from A by replacing the i:th line with the vector $(1, x, y, xy)$.

Let us now analyze what happens as the element Q is deformed by some mapping $T_t = I + tV$. If we denote by ψ_i^t the i:th basis function on the element $Q^t := T_t(Q)$, then the material derivative of ψ_i is by definition

$$\dot{\psi}_i(x) = \lim_{t \to 0} \frac{\psi_i^t(T_t(x)) - \psi_i(x)}{t}.$$

This can also be expressed in another form

$$\dot{\psi}_i(x) = \frac{\partial}{\partial t}\Psi_i(t,x) + \nabla\psi_i(x) \cdot V(x)$$

where $\Psi_i(t,x) = \psi_i^t(x)$. Now, if we denote by "$'$" the partial derivative with respect to t, we have that

$$\frac{\partial}{\partial t}\Psi_i(t,x) = \psi_i'(x) = \frac{|A_i'(x)|}{|A|} - \frac{|A_i(x)||A'|}{|A|^2}.$$

It can be seen that although ψ_i' is locally in the space of bilinear functions, it is not continuous between the elements in general. Thus ψ_i' does not belong to the finite element space. On the other hand the term $\nabla\psi_i \cdot V$ involves the deformation velocity V which is in the space of isoparametric finite element functions. Thus it is not even a polynomial in the global co-ordinates (x, y) and, consequently, is not contained in the finite element space. Hence we can conclude that the material derivative $\dot{\psi}_i$ of the basis function ψ does not vanish. Neither does it belong to the finite element space.

An immediate consequence of this is that the material derivatives of the basis functions have to be explicited when evaluating the formulas like (11) or (15). This means that, for example, the grid optimization is more complicated for the non-parametric elements, at least when the domain deformation method is used.

References

1. M. Souli, J.-P. Zolésio, in preparation.
2. J.-P. Zolésio, "Identification de domaines par déformations," Thèse d'état, Université de Nice, 1979.
3. J.-P. Zolésio, *Les derivées par raport aux nœuds des triangulation et leur utilization en identification de domaine*, Ann. Sc. Mat. Québec 8 (1984), 97–120.

University of Jyväskylä, Department of Mathematics, SF–40100 Jyväskylä, Finland
CEMEF, Sophia Antipolis, F-06560 Valbonne, France

Exact controllability for wave equation with Neumann boundary control *

R. Triggiani
Department of Mathematics
University of Florida
Gainesville, Florida 32611

and

Department of Applied Mathematics, Thornton Hall
University of Virginia
Charlottesville, Virginia 22903

the author presented a summary of very recent results on exact controllability for the wave equation under boundary control exercised either in the Dirichlet or else in the Neumann boundary conditions. For lack of space, the present paper deals exclusively with the Neumann case, while for the Dirichlet case reference is made to [T.1].

The results presented in this paper were obtained jointly with I. Lasiecka and J. L. Lions during the period February - July 1987 and are part of a more complete and more comprehensive joint work by I. Lasiecka, J. L. Lions, and R. Triggiani presently in progress.

The work of the first two authors was sponsored by AFOSR under Grant 84-0365A and by NSF under Grant DMS-8301668 whose financial support is gratefully acknowledged.

1. Introduction, statement of main results, literature.

1.1 Statement of problem and assumptions

Let Ω be an open bounded domain in R^n $(n > 2)$ with sufficiently smooth boundary $\partial\Omega = \Gamma$. We assume that Γ consists of two parts: Γ_o and Γ_1, $\Gamma_o \cup \Gamma_1 = \Gamma$, with Γ_o possibly empty and Γ_1 non-empty and relatively open in Γ. We consider the exact controllability problem for the solution $y(x,t)$ of the wave equation

$$(1.1)\quad \begin{cases} \text{a)} & y_{tt} = \Delta y & \text{in } Q = \Omega x(0,T) \\ \text{b)} & y(x,0) = y^o(x),\ y_t(x,0) = y^1(x) & \text{in } \Omega \\ \text{c)} & y = 0 & \text{in } \Sigma_o = \Gamma_o x(0,T) \\ \text{d)} & \dfrac{\partial y}{\partial \nu} = v & \text{in } \Sigma_1 = \Gamma_1 x(0,T) \end{cases}$$

where Δ is the Laplacian acting on the n-dimensional space variable x and ν is the unit normal of Γ pointing toward the exterior of Ω. We likewise set $\Sigma = \Gamma x(0,T)$. Qualitatively this means: given $\{\Omega,\Gamma_o,\Gamma_1\}$ we ask whether there is some $T_m > 0$ (depending on the geometry of the triplet) such that if $T > T_m$, the following steering property of (1.1) holds true: for all initial data y^0, y^1 in some preassigned space $Z = Z_1 x Z_2$ based on Ω, there exists a suitable control function v on some preassigned space V_{Σ_1} based on Γ_1 and $[0,T]$, whose corresponding solution of (1.1) satisifies $y(\cdot,T) = y(T) = y_t(\cdot,T) = y_t(T) \equiv 0$. We then say that the dynamics (1.1) is exactly controllable on the space Z over the interval $[0,T]$, by means of control functions $v \in V_{\Sigma_1}$. We shall consider various choices of pairs $[Z,V_{\Sigma_1}]$ of spaces.

In the course of our study, we shall need to invoke (a subset of the following) four hypotheses on the geometry of the triplet $\{\Omega,\Gamma_o,\Gamma_1\}$, which we find convenient to list here at the outset for easy future reference.

Let the triplet $\{\Omega,\Gamma_o,\Gamma_1\}$ possess a vector field $h(x) = [h_1(x),\ldots,h_n(x)] \in C^2(\overline{\Omega})$ such that:

(H.1):

$$(1.1)\quad \begin{cases} h\cdot\nu < 0 & \text{on } \Gamma_o \\ h\cdot\nu > 0 & \text{on } \Gamma_1 \end{cases}, \quad \nu = \text{outward unit normal;}$$

(H.2):

$$(1.2)\quad \int_\Omega H(x)v(x)\cdot v(x)d\Omega > \rho \int_\Omega |v(x)|^2_{R^n} d\Omega, \text{ for some constant } \rho > 0 \quad \forall v \in [L_2(\Omega)]^n$$

where

$$(1.3)\quad H(x) = \begin{vmatrix} \frac{\partial h_1}{\partial x_1}, & \cdots, & \frac{\partial h_1}{\partial x_n} \\ \vdots & & \\ \frac{\partial h_n}{\partial x_1}, & \cdots, & \frac{\partial h_n}{\partial x_n} \end{vmatrix}$$

[A sufficient checkable condition for (1.2) to hold is that $H(x) + H^*(x)$ be uniformly positive definite in $\bar{\Omega}$];

(H.3):

$$(1.4)\qquad \rho > 2\, G_h C_p$$

where:

$$(1.5)\qquad 4G_h \equiv \max_{\bar{\Omega}} |\nabla(\operatorname{div} h)|;$$

$$(1.6)\qquad \int_\Omega \phi^2 d\Omega < C_p^2 \int_\Omega |\nabla\phi|^2 d\Omega,\ \forall\ \phi \in H^1_{\Gamma_o}(\Omega),\ \text{defined in (1.9) below}$$

$0 < C_p$ = Poincare constant.

Remark 1.1 Assumptions (H.1) through (H.3) apply, in particular to (suitably smooth) triplets $\{\Omega, \Gamma_o, \Gamma_1\}$ which are "star-complemented - star-shaped" [C-1 [R.1]. This means that there exists a point $x_o \in R^n$ such that

$$(1.7)\qquad \begin{cases} \text{a)}\ (x - x_o)\cdot\nu < 0 & \text{on } \Gamma_o \quad (\Gamma_o \text{ is star complemented w.r.t. } x_o) \\ \text{b)}\ (x - x_o)\cdot\nu > 0 & \text{on } \Gamma_1 \quad (\Gamma_1 \text{ is star shaped w.r.t. } x_o) \end{cases}$$

and we can take then $h(x) = x - x_o$ (radial field) in the statement of assumptions (H.1) through (H.3). Indeed, in this case, we have

$$H(x) \equiv \text{identity matrix}; \qquad \operatorname{div} h \equiv n = \dim \Omega; \qquad G_h = 0$$

and assumptions (H.2), (H.3) are automatically satisfied with $\rho = 1$ □

Remark 1.2 It is possible to construct smooth domains Ω (say, in the plane R^2), such that hypotheses (H.1) through (H.3) for a corresponding triplet $\{\Omega, \Gamma_o, \Gamma_1\}$ are satisfied by a suitable smooth vector field $h(x)$, while condition (1.7) fails for any radial field

$x-x_o$, $x_o \in R^n$ $(n=2)$. More precisely, for such Ω, use of radial vector fields yields an active portion Γ_1' of the boundary Γ strictly larger than the active portion Γ_1 obtained via a suitable (non radial) vector field $h(x)$ satisfying (H.1) through (H.3). See [T.1] □

In some statements below, we shall replace the somewhat unnatural assumption (H.3) with another hypotheses (H.4) of different nature. This is a uniqueness property[1] of the set Γ_1 over the time interval $[0,T_u]$ $0 < T_u < \infty$, for a corresponding homogeneous problem:

(H.4)

(1.8)

$$\left\{\begin{array}{l}\left\{\begin{array}{ll}\phi_{tt} = \Delta\phi & \text{on } Q = \Omega x(0,T)\\ \phi|_\Sigma = 0 & \text{in } \Sigma = \Gamma x(0,T)\\ \frac{\partial\phi}{\partial\nu}\Big|_{\Sigma_1} = 0 & \text{in } \Sigma_1 = \Gamma_1 x(0,T)\end{array}\right.\\ T > T_u\\ \text{implies}\\ \phi \equiv 0 \text{ in } Q.\end{array}\right.$$

Remark 1.3 If $\Gamma_1 = \Gamma$, i.e. $\Gamma_o = \emptyset$ (empty) (Neumann problem), then property (1.8) always holds true (classical Holmgren - F. John theorem [J.1]) or else [R.1]. If instead, $\Gamma_o \neq \emptyset$, we understand that ongoing work by W. Littman [L.9] should provide the required nontrival extension of Holmgren - F. John's uniqueness theorem.

1.2 Exact controllability on $H^1_{\Gamma_o}(\Omega)xL^2(\Omega)$ with controls $v \in L^2(0,T; L^2(\Gamma_1)) \equiv L^2(\Sigma_1)$ and $\mathcal{M}(y^0) = 0$ if $\Gamma_o = \emptyset$.

We shall first consider $v \in L^2(\Sigma_1)$ in (1.1d) and study the corresponding exact controllability problem on the space $H^1_{\Gamma_o}(\Omega)xL^2(\Omega)$, where

(1.9) $$H^k_{\Gamma_o}(\Omega) = \{f \in H^k(\Omega): f|_{\Gamma_o} = 0\}, \ k = 1,2,3...$$

even though recent studies on regularity theory for problem (1.1) with, say $\Gamma_o = \emptyset$ i.e. $\Gamma_1 = \Gamma$ and $y^0 = y^1 = 0$, show that the map $v \to [y,y_t]$ is not continuous $L^2(\Sigma) \to C([0,T]; H^1(\Omega)xL^2(\Omega))$, $0 < T < \infty$. Indeed, for $v \in L^2(\Sigma)$ and $y^0 = y^1 = 0$, the

---

This corresponds, in modern control theoretic terminology, to an 'observablity' property (as 'distinct from' continuous 'observability' [R.1])

solution $[y(T), y_t(T)]$ of (1.1) lies in a space strictly larger than $H^1(\Omega)xL^2(\Omega)$, which appears to depend on the domain Ω [L-T.3] (Example: $H^{2/3}(\Omega)xH^{-1/3}(\Omega)$ for Ω a sphere; $H^{3/4-\varepsilon}(\Omega)xH^{-1/4-\varepsilon}(\Omega)$ for Ω a parallelepiped [L-T.1]). Nevertheless the space $H^1_{\Gamma_0}(\Omega)xL^2(\Omega)$ is of physical interest ("energy space") and this justifies our study of exact controllability here. As a result, even if we shall find say for $\Gamma_o \neq \emptyset$ that any pre-assigned initial pair $(y^0, y^1) \in H^1_{\Gamma_o}(\Omega)xL^2(\Omega)$ can be steered to rest: $y(T) = y_t(T) = 0$ for all $T >$ some (universal) time $T_m > 0$ by some suitable control $v \in L^2(0,T;L^2(\Gamma_1))$, there is no guarantee that during the transfer $0 < t < T$, the solution $[y(\cdot,t), y_t(\cdot,t)]$ remains in $H^1_{\Gamma_o}(\Omega)xL^2(\Omega)$.

Our main results are as follows; in particular, in contrast with the Dirichlet case [H.1], [L.3-6], [T.1], they require geometrical conditions even when $\Gamma_1 = \Gamma$.

Theorem 1.1 Let either $\Gamma_o \neq \emptyset$ or else $\Gamma_o = \emptyset$. Let the triplet $\{\Omega,\Gamma_o,\Gamma_1\}$ satisfy hypotheses (H.1) = (1.1), (H.2) = (1.2), and (H.3) = (1.4).

Then there exists $T_m > 0$ (to be specified below) such that if $T > T_m$ then: for any $(y^0, y^1) \in H^1_{\Gamma_o}(\Omega)xL^2(\Omega)$, subject to the further requirements that $\mathcal{M}(y^o) \equiv \int_\Omega y^o d\Omega = 0$ and $\mathcal{M}(y^1) = \int_\Omega y^1 d\Omega = 0$ if $\Gamma_o = \emptyset$, there is suitable $v \in L^2(0,T;L^2(\Gamma_1))$ such that the corresponding solution of problem (1.1) satisfies $y(T) = y_t(T) = 0$. For T_m we may take

$$T_m = \frac{2(D_h C_p + M_h)}{\rho - 2G_h C_p} + C_p \tag{1.10}$$

$$M_h \equiv \max_{\overline{\Omega}} |h|; \; 2D_h = \max_{\overline{\Omega}} |\text{div } h|; \; G_h \text{ and } C_p \text{ as in (1.5), (1.6)} \quad \Box \tag{1.11}$$

The next result replaces assumption (H.3) with assumption (H.4) and, consequently, provides a different T_m; moreover, it avoids the assumption $\mathcal{M}(y^1) = 0$ when $\Gamma_o = \emptyset$.

Theorem 1.2 Let either $\Gamma_o \neq \emptyset$ or else $\Gamma_o = \emptyset$. Let the triplet $\{\Omega,\Gamma_o\Gamma_1\}$ satisfy assumptions (H.1) = (1.1), (H.2) = (1.2), and (H.4) = (1.8) for $T > T_u$. ((H.4) is automatically true if $\Gamma_o = \emptyset$). Then there exists $T'_m > 0$, (see below) such that if $T > T'_m$ then: for any $(y^0, y^1) \in H^1_{\Gamma_o}(\Omega)xL^2(\Omega)$, subject to the further requirement that $\mathcal{M}(y^o) = \int_\Omega y^o d\Omega = 0$ if $\Gamma_o = \emptyset$, there is a suitable $v \in L^2(0,T;L^2(\Gamma_1))$ such that the corresponding solution of problem (1.1) satisfies $y(T) = y_t(T) = 0$.
For T'_m we can take the following values

For $\Gamma_o \neq \emptyset$:

(1.12) $$T'_m > \max\{2\frac{C_p D_h + M_h}{\rho} + C_p, T_u\}$$

For $\Gamma_o = \emptyset$

(1.13) $$T'_m > \max\{\frac{2M_h}{\rho}, T_u\}$$ □

1.3 Exact controllability on $L^2(\Omega)\times[H^1_{\Gamma_o}(\Omega)]'$ with controls $v \in H^{-1}(0,T;L^2(\Gamma_1))$.

In this case we have

Theorem 1.3 Let either $\Gamma_o \neq \emptyset$ or else $\Gamma_o = \emptyset$. Let the triplet $\{\Omega,\Gamma_o,\Gamma_1\}$ satisfy assumptions (H.1) = (1.1), (H.2) = (1.2), and (H.4) = (1.8) for $T > T_u$ [(H.4) is automatically satisfied if $\Gamma_o = \emptyset$].

Then, there exists $T'_m > 0$ (see below) such that if $T > T'_m$ then: for any $(y^0, y^1) \in L^2(\Omega)\times[H^1_{\Gamma_o}(\Omega)]'$, there is a suitable $v \in H^{-1}(0,T;L^2(\Gamma_1))$ such that the corresponding solution of problem (1.1) satisfies $y(T) = y_t(T) = 0$.

For T'_m we can take the same T'_m stated in Theorem 1.2, see (1.12) and (1.13) for $\Gamma_o \neq \emptyset$ and $\Gamma_o = \emptyset$, respectively □

1.4 Exact controllability on $L^2(\Omega)\times[H^1(\Omega)]'$ in the Neumann case ($\Gamma_o = \emptyset$) in the absence of geometrical conditions on Ω with a special class of controls v

The preceding results, in particular Theorems 1.2 and 1.3, yields exact controllability for problem (1.1) with $\Gamma_o = \emptyset$ (Neumann case) on the spaces $H^1(\Omega)\times L^2(\Omega)$ and $L^2(\Omega)\times[H^1(\Omega)]'$ respectively, subject at least to the geometrical conditions (H.1) and (H.2) on Ω. It turns out, however, that the same analysis which culminates with these results permits to show - modulo modifications that introduce no essential additional difficulties - an exact controllability result on $L^2(\Omega)\times[H^1(\Omega)]'$ for the Neumann problem (1.1), i.e. $\Gamma_o = \emptyset$, with no requirement of geometrical conditions on Ω. Elimination of the geometrical conditions does not come without a price. The price consists in the class of admissible controls v in (1.1d) with $\Gamma_o = \emptyset$, which is considerable larger than $L^2(\Sigma)$ and, moreover, must have a special structure, as described below. Given Ω, we fix a point x_o in R^n and set with $\nu(x)$ the unit outward normal at $x \in \Gamma$:

$$(1.14)\qquad \begin{aligned} &a) \quad \Gamma(x^o) = \{x \in \Gamma : (x - x_o)\cdot\nu(x) > 0\} \\ &b) \quad \Gamma_*(x^o) = \Gamma \setminus \Gamma(x^o) \end{aligned}$$

We then define the following class of controls v in (1.1d) with $\Gamma_o = \emptyset$:

$$(1.15)\qquad v = \begin{cases} v_1 + \dfrac{\partial v_2}{\partial t}, & v_1, v_2 \in L^2(0,T;\ \Gamma(x^o)) \\ v_3, & v_3 \in L^2(0,T;\ H^{-1}(\Gamma_*(x^o))) \end{cases}$$

We then have

<u>Theorem</u> 1.4 Let $\Gamma_o = \emptyset$. There exists $T_m > 0$ (see below) such that if $T > T_m$ then: for any $(y^0, y^1) \in L^2(\Omega)\times[H^1(\Omega)]'$, there exists a suitable control v in the class described by (1.15) such that the corresponding solution of problem (1.1) with $\Gamma_o = \emptyset$ satisfies $y(T) = y_t(T) = 0$.

For T_m we can take

$$(1.16)\qquad T_m = \max\{T_u,\quad 2\max_{\overline{\Omega}}|x - x_o|\}$$

where T_u is the time defined by the uniqueness property (1.8) with $\Gamma_o = \emptyset$, see Remark 1.3 □

2. A-priori inequalities

2.1 Preliminaries. Case $\Gamma_0 \neq \emptyset$ versus case $\Gamma_0 = \emptyset$.

Let $A: L^2(\Omega) \supset \mathcal{D}(A) \to L^2(\Omega)$ be the operator defined by

$$(2.1) \qquad Af = -\Delta f, \ \mathcal{D}(A) = \{f \in H^2(\Omega): f|_{\Gamma_0} = \frac{\partial f}{\partial \nu}\Big|_{\Gamma_1} = 0\}.$$

Then, A is a non-negative self-adjoint operator with compact resolvent $R(\lambda,A)$. We distinguish two cases.

Case a) Let $\Gamma_0 \neq \emptyset$. Then A is actually positive, since the problem

$$(2.2) \qquad \begin{aligned} &Af = 0, \ f \in \mathcal{D}(A), \text{ i.e.} \\ &\Delta f = 0 \text{ in } \Omega, \ f|_{\Gamma_0} = 0, \ \frac{\partial f}{\partial \nu}\Big|_{\Gamma_1} = 0 \qquad \text{implies } f = 0 \end{aligned}$$

by Green theorem applied to $0 = (\Delta f, f)_{L^2(\Omega)}$. Thus, in this case

$$(2.3) \qquad A^{-1} \in \mathcal{L}(L^2(\Omega)).$$

Consider the space $\mathcal{D}(A^{1/2})$, domain of the positive operator $A^{1/2}$, topologized as usual by (1)

$$(2.4) \qquad \|z\|^2_{\mathcal{D}(A^{1/2})} = \|A^{1/2}z\|^2_{L^2(\Omega)} = (Az,z)_{L^2(\Omega)}, \qquad z \in \mathcal{D}(A^{1/2}).$$

We have

$$(2.5a) \qquad \begin{cases} \mathcal{D}(A^{1/2}) = H^1_{\Gamma_0}(\Omega), \text{ with equivalent norms, } ^{(1)} \\ \|z\|^2_{\mathcal{D}(A^{1/2})} = \|A^{1/2}z\|^2_{L^2(\Omega)} = \int_\Omega |\nabla z|^2 d\Omega \end{cases}$$

In fact for $z \in \mathcal{D}(A)$, one has by (2.1) and Green theorem

$$(2.6) \qquad (Az,z)_{L^2(\Omega)} = \int_\Omega |\nabla z|^2 d\Omega$$

(1) The graph-norm of $\mathcal{D}(A^{1/2})$ is the same as the $H^1_{\Gamma_0}(\Omega)$- norm.

The above identity can be extended to all $z \in H^1_{\Gamma_0}(\Omega)$ defined in (1.), yielding $H^1_{\Gamma_0}(\Omega) \subset \mathcal{D}(A^{1/2})$. On the other hand, $\mathcal{D}(A) \subset H^2_{\Gamma_0}(\Omega)$ and by interpolation

$$\mathcal{D}(A^{1/2}) = [\mathcal{D}(A), L^2(\Omega)]_{1/2} \subset [H^2_{\Gamma_0}(\Omega), L^2(\Omega)]_{1/2} = H^1_{\Gamma_0}(\Omega)$$

Moreover, writing $z = A^{-1/2}A^{1/2}z$, for $z \in \mathcal{D}(A^{1/2})$ we obtain the generalized Poincare inequality

$$\text{(2.7a)} \quad \begin{cases} \int_\Omega z^2 d\Omega = \|z\|^2_{L^2(\Omega)} \le c_p^2 \|A^{1/2}z\|^2_{L^2(\Omega)} = c_p^2 \int_\Omega |\nabla z|^2 d\Omega \\ \text{for all } z \in \mathcal{D}(A^{1/2}) = H^1_{\Gamma_0}(\Omega),\ c_p = \|A^{-1/2}\| \end{cases}$$

to be often invoked

Case b) Let $\Gamma_0 = \emptyset$. (Neumann problem for (1.1)). Then A in (2.1) is not invertible on $L^2(\Omega)$, but is invertible (with bounded inverse) in the space

$$\text{(2.8)} \qquad L^2_0(\Omega) = L^2(\Omega)/\mathcal{N}(A) = \{f \in L^2(\Omega): \int_\Omega f d\Omega = 0\}$$

where $\mathcal{N}(A)$ is the null space of A, spanned by the normalized constant function in Ω. We have

$$\text{(2.9)} \quad \begin{cases} L^2(\Omega) = L^2_0(\Omega) + \mathcal{N}(A) & \text{(orthogonal sum)} \\ z = \bar{z} + \bar{c} & \bar{z} \in L^2_0(\Omega),\ \bar{c} = \text{const} \in \mathcal{N}(A) \end{cases}$$

We then introduce the quantity

$$\text{(2.10)} \qquad \mathcal{M}(z) = \frac{1}{\text{meas } \Omega} \int_\Omega z d\Omega = \frac{1}{\text{meas } \Omega} (z,1)_\Omega$$

so that, in the notation of (2.9), we have $(\bar{z},1)_\Omega = 0$ and

$$\text{(2.11)} \qquad \bar{c} = \mathcal{M}(z) \quad \text{and} \quad z \in L^2_0(\Omega) \Longleftrightarrow \{z \in L^2(\Omega) \text{ and } \mathcal{M}(z) = 0\}$$

The counterparts of (2.5a) - (2.7a) are now

$$\text{(2.5b)} \quad \begin{cases} \mathcal{D}(A^{1/2}) = \{f \in H^1(\Omega): \mathcal{M}(f) = 0\} & \text{with equivalent norms} \\ \|f\|^2_{\mathcal{D}(A^{1/2})} \equiv \|A^{1/2}f\|^2_{L^2(\Omega)} = \int_\Omega |\nabla f|^2 d\Omega,\ \mathcal{M}(f) = 0 \end{cases}$$

$$(2.5c)\quad \begin{cases} \mathcal{D}((A+I)^{1/2}) = H^1(\Omega), \text{ with equal norms} \\ \|f\|^2_{H^1(\Omega)} = \|f\|^2_{\mathcal{D}((A+I)^{1/2})} = ((A+I)f,f)_{L^2(\Omega)} = \int_\Omega |f|^2 + |\nabla f|^2 d\Omega \end{cases}$$

$$(2.7b)\quad \|f\|^2_{L^2(\Omega)} \le C_p^2 \|A^{1/2} f\|^2_{L^2(\Omega)} = C_p^2 \int_\Omega |\nabla f|^2 d\Omega, \ \mathcal{M}(f) = 0$$

2.2 Preliminary a-priori inequalities

We shall see in section 3 that exact controllability for problem (1.1) on the space $H^1_{\Gamma_0}(\Omega)xL^2(\Omega)$, or else on the space $L^2(\Omega)xH^{-1}_{\Gamma_0}(\Omega)$, etc. is equivalent to certain inequalities for the associated homogeneous problem

$$(2.12)\quad \begin{cases} \text{a)} & \phi_{tt} = \Delta\phi & \text{in } Q \\ \text{b)} & \phi(x,0) = \phi^o, \ \phi_t(x,0) = \phi^1 & \text{in } \Omega \\ \text{c)} & \phi \equiv 0 & \text{in } \Sigma_0 \\ \text{d)} & \frac{\partial\phi}{\partial\nu} \equiv 0 & \text{in } \Sigma_1 \end{cases}$$

whose solution with $\phi^o, \phi^1 \in H^1_{\Gamma_0}(\Omega)xL^2(\Omega)$ is given by

$$(2.12e)\quad \begin{aligned} \phi(t) &= C(t)\phi^o + S(t)\phi^1 = \\ &= C(t)\overline{\phi^o} + S(t)\overline{\phi^1} + \mathcal{M}(\phi^o) + t\mathcal{M}(\phi^1) \in C([0,T]; H^1_{\Gamma_0}(\Omega)) \end{aligned}$$

$$(2.12f)\quad \phi_t(t) = C(t)\phi^1 - AS(t)\overline{\phi^o} \in C([0,T]; L^2(\Omega))$$

where according to (2.9) we write

$$(2.12g)\quad \phi^o = \overline{\phi^o} + \mathcal{M}(\phi^o), \ \phi^1 = \overline{\phi^1} + \mathcal{M}(\phi^1)$$

and where $C(t)$ is the strongly continuous cosine operator generated by $-A$ and $S(t) = \int_0^t C(\tau)d\tau$, $t \in R$. Note that $\mathcal{M}(\phi^o) = \mathcal{M}(\phi^1) = 0$ implies $\mathcal{M}(\phi(t)) = 0$ from (2.12e).

Accordingly, we find it convenient to assemble these inequalities in the present section, for easy future reference, by showing under what conditions (typically on the geometry of the triplet $\{\Omega, \Gamma_0, \Gamma_1\}$) they hold true.

The main goal of this subsection is to prove the following three results.

Theorem 2.1 Let either $\Gamma_0 \neq \emptyset$ or else $\Gamma_0 = \emptyset$. If $\Gamma_0 = \emptyset$, assume further that $\mathcal{M}(\phi^o) = \mathcal{M}(\phi^1) = 0$ (cf (2.10)). Let the triplet $\{\Omega, \Gamma_0, \Gamma_1\}$ satisfy the geometrical hypotheses (H.1) = (1.1), (H.2) = (1.2), and (H.3) = (1.4), in terms of a suitable vector field $h(x) \in C^2(\overline{\Omega})$, as in the statement of Theorem 1.1.

Then, there exists $T_m > 0$, to be specified below, such that for $T > T_m$ the following inequality holds true for problem (2.12a-d):

(2.13) $$\int_{\Sigma_1} \phi_t^2 d\Sigma_1 \geq c_{h,\rho}(T-T_m)E(0)$$

(2.14) $$E(0) = \int_\Omega |\nabla\phi^o|^2 + |\phi^1|^2 d\Omega,$$ equal to the $\mathcal{D}(A^{1/2})xL^2(\Omega)$ - norm of $\{\phi^o, \phi^1\}$ (cf (2.5a)), in turn equivalent to its $H^1_{\Gamma_0}(\Omega)xL^2(\Omega)$ - norm;

for all ϕ^o, ϕ^1 in $H^1_{\Gamma_0}(\Omega)xL^2(\Omega)$ for which the left hand side of (2.13a) is finite. Moreover, for T_m and $c_{h,\rho}$ in (2.13a) we can take

(2.15a) $$T_m = \frac{2K_h}{\rho-2G_hC_p} + C_p; \qquad K_h = D_hC_p + M_h$$

(2.15b) $$c_{h,\rho} = \frac{\rho-2G_hC_p}{\sup_{\Gamma_1}|h|},$$

(2.15c) $$M_h \equiv \max_{\overline{\Omega}} |h|; \quad 2D_h = \max_{\overline{\Omega}} |\text{div } h|; \quad 4G_h \equiv \max_{\overline{\Omega}} |\nabla(\text{div } h)| \square$$

Theorem 2.1 will follow from the following

Lemma 2.2 Let $\{\Omega, \Gamma_0, \Gamma_1\}$ satisfy the geometrical assumptions (H.1) = (1.1) and (H.2) = (1.2) in terms of a suitable vector field $h(x) \in C^2(\overline{\Omega})$. Then, the following inequality holds true for the solution of problem (2.12a-d):

(2.16) $\frac{1}{2}\int_{\Sigma_1} \phi_t^2 \, h\cdot\nu d\Sigma_1 > \rho \int_Q |\nabla\phi|^2 dQ + \frac{1}{2}\int_Q \phi\nabla(\mathrm{div}\, h)\cdot\nabla\phi dQ + \beta_{0,T}$

where $\beta_{0,T}$ (boundary term at t = 0 and t = T) is given by

(2.17a) $\beta_{0,T} = \frac{1}{2}[(\phi_t, \phi \,\mathrm{div}\, h)_\Omega]_0^T + [(\phi_t, h\cdot\nabla\phi)_\Omega]_0^T.$

Moreover, if $\Gamma_0 \neq \emptyset$, or else if $\Gamma_0 = \emptyset$ but $\mathcal{M}(\phi^o) = \mathcal{M}(\phi^1) = 0$ then (see (2.7a-b), (2.15c))

(2.18b) $|\beta_{0,T}| < [D_h C_p + M_h] \, E(0) = K_h E(0), \; \Gamma_0 \neq \emptyset$; or $\Gamma_0 = \emptyset$ and $\mathcal{M}(\phi^o) = \mathcal{M}(\phi^1) = 0$

Proof of Lemma 2.2 We use a multiplier technique as in [ ]

Step 1 With h(x) the assumed vector field, we multiply both sides of (2.12a) by $h\cdot\nabla\phi$. Proceeding as in [ ], we obtain the following identity:

$$\int_\Sigma \frac{\partial\phi}{\partial\nu}(h\cdot\nabla\phi)d\Sigma + \frac{1}{2}\int_\Sigma \phi_t^2 h\cdot\nu d\Sigma - \frac{1}{2}\int_\Sigma |\nabla\phi|^2 h\cdot\nu d\Sigma =$$

$$= \int_Q H\nabla\phi\cdot\nabla\phi dQ + \frac{1}{2}\int_Q [\phi_t^2 - |\nabla\phi|^2] \,\mathrm{div}\, h \, dQ$$

$$+ [(\phi_t, h\cdot\nabla\phi)_{L^2(\Omega)}]_0^T \tag{2.18}$$

which we write here without using boundary conditions for easy future reference to various cases. Here, H = H(x) is the matrix defined in (1.). To make the present paper self-contained, we provide in Appendix 1 a derivation of (2.18).

Step 2. To estimate the second integral on the right of (2.18), we multiply both sides of (2.12a) by ϕ div h and integrate by parts. Using the identity

(2.19) $\nabla\phi\cdot\nabla(\phi \,\mathrm{div}\, h) = \phi\nabla(\mathrm{div}\, h)\cdot\nabla\phi + |\nabla\phi|^2 \mathrm{div}\, h$

we obtain (details in Appendix 2)

$$\int_Q [\phi_t^2 - |\nabla\phi|^2] \,\mathrm{div}\, h dQ = \int_Q \phi\nabla(\mathrm{div}\, h)\cdot\nabla\phi dQ$$

$$- \int_\Sigma \frac{\partial\phi}{\partial\nu}\phi \,\mathrm{div}\, h d\Sigma + [(\phi_t, \phi \,\mathrm{div}\, h)_{L^2(\Omega)}]_0^T \tag{2.20}$$

which we again write without using boundary conditions. For future reference, we note that the above argument yielding identity (2.20) holds for any smooth vector field, not only the postulated h. Specializing to div h $\equiv 1$ (i.e. multiplying (2.12a) simply by ϕ) gives the identity

$$(2.21) \qquad \int_Q \phi_t^2 - |\nabla\phi|^2 \, dQ = [(\phi_t, \phi)_{L^2(\Omega)}]_0^T$$

to be invoked below.

Step 3 We now use the boundary conditions (2.12c-d).
Thus

$$(2.22) \qquad \text{on } \Sigma_0: \quad \phi \equiv \phi_t \equiv 0; \; |\nabla\phi| = \left|\frac{\partial\phi}{\partial\nu}\right|; \; h\cdot\nabla\phi = (h\cdot\nu)\frac{\partial\phi}{\partial\nu}$$

Hence, combining (2.18) and (2.20) and using (2.12d) and (2.22) yields

$$\tfrac{1}{2}\int_{\Sigma_0} \left(\frac{\partial\phi}{\partial\nu}\right)^2 h.\nu d\Sigma_0 + \tfrac{1}{2}\int_{\Sigma_1} \phi_t^2 \, h\cdot\nu d\Sigma_1 - \tfrac{1}{2}\int_{\Sigma_1} |\nabla\phi|^2 h.\nu d\Sigma_1 =$$

$$(2.23) \qquad = \int_Q H\nabla\phi\cdot\nabla\phi dQ + \tfrac{1}{2}\int_Q \phi\nabla(\operatorname{div} h)\cdot\nabla\phi dQ + \beta_{0,T}$$

with $\beta_{0,T}$ defined by (2.17a). Using assumption (H.2) = (1.2) in (2.23) yields (2.16) as desired. It remains to prove estimate (2.17b) when $\Gamma_0 \neq \emptyset$. First, we recall the standard result that

$$(2.24) \qquad E(t) \equiv \int_\Omega |\nabla\phi(t)|^2 + \phi_t^2(t) d\Omega \equiv E(0)$$

for the conservative problem (2.12). (Multiply (2.12a) by ϕ_t and integrate by parts). To handle the first term in (2.17a), we use Schwarz inequality and Poincare inequality (2.7a,b) (the latter justified since $\Gamma_0 \neq \emptyset$ or else $\Gamma_0 = \emptyset$ but $\mathcal{M}(\phi^o) = \mathcal{M}(\phi^1) = 0$, so that $\mathcal{M}(\phi) = 0$) and we obtain

$$(2.25a) \qquad |[(\phi_t, \phi \operatorname{div} h)_\Omega]_0^T| \le 2D_h C_p E(0), \quad \Gamma_0 \neq \emptyset; \text{ or } \Gamma_0 = \emptyset, \text{ but } \mathcal{M}(\phi^o) = \mathcal{M}(\phi^1) = 0$$

see (2.15c), (2.14), and (2.24). Applying now Schwarz inequality to the second term of (2.17a), we obtain the estimate (see (2.15c), (2.14), and (2.24)):

$$(2.25b) \qquad |[(\phi_t, h\cdot\nabla\phi)_\Omega]_0^T| \le M_h E(0)$$

which is valid in both cases $\Gamma_0 \neq \emptyset$ and $\Gamma_0 = \emptyset$. Then, (2.25a-b) yields (2.17b) □

<u>Proof of Theorem 2.1</u> <u>Step</u> 1. We return to inequality (2.16). With reference to the second integral on the right hand side of (2.16), we have

(2.26) $$\tfrac{1}{2}\int_\Omega \phi\nabla(\operatorname{div} h)\cdot\nabla\phi d\Omega \geq -2\, G_h \int_\Omega |\phi||\nabla\phi| d\Omega = -2\, G_h\, (|\phi|, |\nabla\phi|)_\Omega$$

recalling (2.15c). Since $\Gamma_0 \neq \emptyset$, the generalized Poincare inequality (2.7a) holds and yields

(2.27) $$-2\, G_h(|\phi|, |\nabla\phi|)_\Omega \geq -2\, G_h \|\phi\|\, \||\nabla\phi|\| \geq -2G_h C_p \int_\Omega |\nabla\phi|^2 d\Omega$$

Thus, using first (2.27) in (2.26) and then (2.26) in (2.16), we obtain

(2.28) $$\tfrac{1}{2}\int_{\Sigma_1} \phi_t^2 h.\nu d\Sigma_1 \geq \text{R.H.S. of (2.23)} \geq (\rho-2G_h C_p)\int_Q |\nabla\phi|^2 dQ - K_h\, E(0)$$

<u>Step</u> 2 We now recall identity (2.21). Since $\Gamma_0 \neq \emptyset$ or else $\Gamma_0 = \emptyset$ but $\mathcal{M}(\phi^o) = \mathcal{M}(\phi^1) = 0$ so that $\mathcal{M}(\phi) = 0$, we obtain by Poincare inequality (2.7a,b) and (2.24) [as in (2.25a)]

(2.29a) $$\left|\int_Q \phi_t^2 - |\nabla\phi|^2 dQ\right| = \left|[(\phi_t,\phi)_\Omega]_0^T\right| \leq C_p\, E(0),\quad \Gamma_0 \neq \emptyset;\ \Gamma_0 = \emptyset \text{ and } \mathcal{M}(\phi^o) = \mathcal{M}(\phi^1) = 0$$

Hence

(2.29b) $$\int_Q |\nabla\phi|^2\, dQ \geq \int_Q \phi_t^2 dQ - C_p E(0),\quad \Gamma_0 \neq \emptyset;\ \text{or } \Gamma_0 = \emptyset \text{ and } \mathcal{M}(\phi^o) = \mathcal{M}(\phi^1) = 0$$

and recalling (2.24)

$$\int_Q |\nabla\phi|^2\, dQ \geq \tfrac{1}{2}\int_Q |\nabla\phi|^2\, dQ + \tfrac{1}{2}\int_Q \phi_t^2\, dQ - \frac{C_p}{2} E(0)$$

(2.30) $$= \frac{T-C_p}{2} E(0) \qquad \Gamma_0 \neq \emptyset;\ \text{or } \Gamma_0 = \emptyset \text{ and } \mathcal{M}(\phi^o) = \mathcal{M}(\phi^1) = 0$$

Under assumption (H.3) = (1.4), then (2.30) inserted in (2.28) gives

(2.31) $$\int_{\Sigma_1} \phi_t^2\, h\cdot\nu d\Sigma_1 \geq [(\rho-2G_h C_p)(T-C_p) - 2K_h]\, E(0)$$

$$= (\rho - 2G_h C_p)\left[T - \left(\frac{2K_h}{\rho-2G_h C_p} + C_p\right)\right] E(0)$$

Then (2.31) yields the desired conclusion (2.13) - (2.15) of Theorem 2.1 □

An important variation of Lemma 2.2 - to be crucially used in the developments of subsection 2.3 - is the following Lemma which is valid for both cases $\Gamma_0 \neq \emptyset$ and $\Gamma_0 = \emptyset$. Its impact is that it will allow us to dispense with assumption (H.3) = (1.4) and replace it with assumption (H.4) = (1.8). On the other hand, in the case $\Gamma_0 = \emptyset$, Lemma 2.3 does not provide the constant C_T' as a linear function of T.

<u>Lemma</u> 2.3 Let $\{\Omega, \Gamma_0, \Gamma_1\}$ satisfy the geometrical assumptions (H.1) = (1.1) and (H.2) = (1.2), in terms of a suitable vector field $h(x) \in C^2(\overline{\Omega})$.

a) Let $\Gamma_0 \neq \emptyset$. Then, there is $T_m > 0$ (to be specified below) such that if $T > T_m$, then the following inequality holds true for problem (2.12a-d)

$$(2.32) \qquad \int_{\Sigma_1} \phi_t^2 \, d\Sigma_1 + \int_Q \phi^2 dQ \geq C_T \, E(0)$$

$$(2.33a) \qquad C_T = \frac{(\rho-\varepsilon G_h)}{\mu_{h,\varepsilon}} [T - (\frac{2K_h}{\rho-\varepsilon G_h} + C_p)]$$

$$(2.33b) \qquad \mu_{h,\varepsilon} \equiv \max \{\max_{\overline{\Gamma}_1} |h|, 2\frac{G_h}{\varepsilon}\}, \; K_h = D_h C_p + M_h \text{ as in (2.15a)}$$

$$(2.33c) \qquad T_m = 2\frac{K_h}{\rho} + C_p$$

for any $0 < \varepsilon < \rho/G_h$. (Compare T_m in (2.33c) with T_m in (2.15a))

b) Let now $\Gamma_0 = \emptyset$. There is $T_m' > 0$ (specified below) such that if $T > T_m'$ then the following inequality holds true for problem (2.12a-d)

$$(2.34) \qquad \int_{\Sigma} \phi_t^2 d\Sigma + \|\phi\|^2_{C([0,T];\, L^2(\Omega))} \geq C_T' \, E(0)$$

$$(2.35a) \qquad C_T' = \frac{(\rho-\varepsilon G_h)}{\mu'_{h,\varepsilon,\varepsilon_1}(T)} \; [T - (\varepsilon_1 + 2\,\frac{\varepsilon_1 D_h + M_h}{\rho - \varepsilon G_h})]$$

$$(2.35b) \qquad \mu'_{h,\varepsilon,\varepsilon_1}(T) \equiv \max \{\max_{\overline{\Gamma}}|h|, \; 2\,\frac{G_h}{\varepsilon}\, T + \frac{2D_h}{\varepsilon_1} + (\rho-\varepsilon G_h)\,\frac{1}{\varepsilon_1}\}$$

$$(2.35c) \qquad T'_m = \frac{2M_h}{\rho}$$

With T_m given by (2.33c) [resp. T'_m given by (2.35c)], for any $T > T_m$ [resp. T'_m] we select $\varepsilon > 0$ [resp. $\varepsilon_1 > 0$ and $\varepsilon > 0$] small enough, as to make the expression in the square bracket in (2.33a) [resp. in (2.35a)] positive.

Proof of Lemma 2.3. Step 1. We return to identity (2.23) (which was derived without making use of either assumption (H.1) = (1.1) or assumption (H.2) = (1.2)). We estimate the second integral on the right of (2.23) in a way different from (2.26) - (2.27) (the latter equation requiring $\Gamma_0 \neq \emptyset$):

$$(2.36) \qquad 1/2 \int_\Omega \phi\nabla(\operatorname{div} h)\cdot\nabla\phi d\Omega \geq -2G_h(|\phi|,|\nabla\phi|)_\Omega \geq -a_\varepsilon \int_\Omega|\phi|^2 d\Omega - b_\varepsilon \int_\Omega|\nabla\phi|^2 d\Omega$$

for any $\varepsilon > 0$ where

$$(2.37) \qquad a_\varepsilon = \frac{G_h}{\varepsilon} \qquad\qquad b_\varepsilon = \varepsilon G_h$$

Invoking now assumption (H.2) = (1.2), we then obtain for the right hand side (R.H.S.) of (2.23) via (2.36)

$$(2.38) \qquad \text{R.H.S. of (2.23)} \geq (\rho-b_\varepsilon) \int_Q|\nabla\phi|^2 dQ - a_\varepsilon \int_Q \phi^2 dQ + \beta_{0,T}$$

for ε sufficiently small. We note explicitly that (2.38) is valid in both cases $\Gamma_0 \neq \emptyset$ and $\Gamma_0 = \emptyset$.

Step 2 To estimate $\beta_{0,T}$ we recall that its second term in (2.17a) is likewise upper bounded in both cases $\Gamma_0 \neq \emptyset$ and $\Gamma_0 = \emptyset$ in (2.25b). It is at the level of estimating the first term in (2.17a) that we must distinguish between $\Gamma_0 \neq \emptyset$ and $\Gamma_0 = \emptyset$: The estimate for $\Gamma_0 \neq \emptyset$ was already obtained in (2.25a), using Poincare inequality. Instead, for $\Gamma_0 = \emptyset$, we write by Schwarz inequality

$$(2.39) \qquad |[(\phi_t, \phi \operatorname{div} h)_\Omega]_0^T| \leq D_h\{\varepsilon_1(\|\phi_t(T)\|^2 + \|\phi^1\|^2) + \frac{1}{\varepsilon_1}(\|\phi(T)\|^2 + \|\phi^0\|^2)\}$$

$$\leq 2D_h[\varepsilon_1 E(0) + \frac{1}{\varepsilon_1}\|\phi\|^2_{C([0,T];L^2(\Omega))}]$$

in place of (2.25a). Thus, putting together (2.17a), (2.25b) and (2.39), we obtain for $\Gamma_0 = \emptyset$ (but also for $\Gamma_0 \neq \emptyset$)

(2.40) $$|\beta_{0,T}| < [\varepsilon_1 D_h + M_h]E(0) + \frac{1}{\varepsilon_1} D_h \|\phi\|^2_{C([0,T];L^2(\Omega))},$$

in place of (2.17b). Thus, using (2.38) and either (2.17b) or else (2.40), we obtain the following estimates for ε sufficiently small: for $\Gamma_0 \neq \emptyset$:

(2.41) $$\text{R.H.S. of (2.23)} > (\rho - b_\varepsilon) \int_Q |\nabla\phi|^2 \, dQ - a_\varepsilon \int_Q \phi^2 dQ - K_h E(0);$$

for $\Gamma_0 = \emptyset$:

(2.42) $$\text{R.H.S. of (2.23)} > (\rho - b_\varepsilon) \int_Q |\nabla\phi|^2 dQ - a_\varepsilon \int_Q \phi^2 dQ - [\varepsilon_1 D_h + M_h]\, E(0)$$
$$- \frac{1}{\varepsilon_1} D_h \|\phi\|^2_{C([0,T];L^2(\Omega))}$$

<u>Step</u> 3 Let first $\Gamma_0 \neq \emptyset$. We use (2.30) in (2.41). Recalling the top line of (2.28) we obtain

$$\tfrac{1}{2}\int_{\Sigma_1} \phi_t^2 \, h\cdot\nu d\Sigma_1 > \text{R.H.S. of (2.23)} >$$

(2.43) $$> (\rho - b_\varepsilon)\left(\frac{T}{2} - \frac{C_p}{2}\right)E(0) - K_h E(0) - a_\varepsilon \int_Q \phi^2 dQ$$

Instead, let $\Gamma_0 = \emptyset$. We return to identity (2.21) and we estimate as in (2.39)

(2.44) $$\left|\int_Q \phi_t^2 - |\nabla\phi|^2 dQ\right| = \left|[(\phi_t, \phi)_\Omega]_0^T\right| < \varepsilon_1 E(0) + \frac{1}{\varepsilon_1} \|\phi\|^2_{C([0,T];L^2(\Omega))}$$

in place of (2.29a). Hence

(2.45) $$\int_Q |\nabla\phi|^2 dQ > \int_Q \phi_t^2 dQ - \varepsilon_1 E(0) - \frac{1}{\varepsilon_1} \|\phi\|^2_{C([0,T];L^2(\Omega))}, \text{ for } \Gamma_0 = \emptyset$$

and recalling (2.24)

(2.46) $$\int_Q |\nabla\phi|^2 dQ > \tfrac{1}{2}\int_Q |\nabla\phi|^2 dQ + \tfrac{1}{2}\int_Q \phi_t^2 dQ - \tfrac{1}{2}\varepsilon_1 E(0) - \frac{1}{2\varepsilon_1} \|\phi\|^2_{C([0,T];L^2(\Omega))}$$
$$= \left(\frac{T}{2} - \tfrac{1}{2}\varepsilon_1\right) E(0) - \frac{1}{2\varepsilon_1} \|\phi\|^2_{C([0,T];L^2(\Omega))}, \text{ for } \Gamma_0 = \emptyset$$

Using (2.46) into (2.42) and recalling the top line of (2.28) results into

(2.47) $$\tfrac{1}{2}\int_{\Sigma_1} \phi_t^2 \, h\cdot\nu d\Sigma_1 > \text{R.H.S. of (2.23)} >$$

$$\geq (\rho - b_\varepsilon)(\tfrac{T}{2} - \tfrac{1}{2}\varepsilon_1)\, E(0) - (\varepsilon_1 D_h + M_h)\, E(0)$$

$$(2.48) \qquad - a_\varepsilon \int_Q \phi^2 dQ - [\frac{1}{\varepsilon_1} D_h + (\rho - b_\varepsilon)\frac{1}{2\varepsilon_1}] \, \|\phi\|^2_{C([0,T];L^2(\Omega))}$$

Step 4 For $\Gamma_0 \neq \emptyset$, inequality (2.43) easily implies (2.32) - (2.33a-b-c), via (2.37). Similarly, for $\Gamma_0 = \emptyset$, (2.47) implies

$$(\max_{\overline{\Gamma}} |h|) \int_\Sigma \phi_t^2 \, d\Sigma + [2a_\varepsilon T + \frac{2D_h}{\varepsilon_1} + (\rho - b_\varepsilon)\frac{1}{\varepsilon_1}] \, \|\phi\|^2_{C([0,T];L(\Omega))}$$

$$(2.49) \qquad \geq (\rho - b_\varepsilon)[T - (\varepsilon_1 + 2\frac{\varepsilon_1 D_h + M_h}{\rho - b_\varepsilon})] \, E(0)$$

from which (2.34) - (2.35a-b-c-) follow. The proof of Lemma 2.3 is complete □

2.3 Absorption of lower order term $\|\varphi\|^2_{C([0,T];L^2(\Omega))}$ under uniqueness assumption (H.4). Another a-priori inequality

We recall that with $\Gamma_0 \neq \emptyset$ the last step in the proof of Thoerem 2.1 (from Eq. (2.26) to conclusion) consists in 'absorbing' the interior term $\|\varphi\|$ by the energy term $\||\nabla\varphi|\|$ by use of Poincare inequality (2.7a) in Eq. (2.27): the price paid in this approach is the requirement of the additional (and undesirable) assumption (H.3) = (1.4) at the level of obtaining (2.31).

In the present subsection, our starting point is Lemma 2.3, thereby we assume only hypotheses (H.1) = (1.1) and (H.2) = (1.2) and, moreover, we consider also the case $\Gamma_0 = \emptyset$. Then, Lemma 2.3, part a) and part b) for $\Gamma_0 \neq \emptyset$, or $\Gamma_0 = \emptyset$ (with no assumption $\mathcal{M}(\varphi^o) = 0$ needed), respectively lead to an a-priori inequality like (2.34), which we re-write here for convenience as

$$(2.50)\quad \begin{cases} \int_{\Sigma_1} \varphi_t^2 \, d\Sigma_1 + \|\varphi\|^2_{C([0,T];L^2(\Omega))} \geq C_{1,T} \, E(0), \\ \text{for all } T > \text{some } T_1 > 0, \text{ for either case } \Gamma_0 \neq \emptyset \text{ or } \Gamma_0 = \emptyset. \end{cases}$$

$$T_1 = \begin{cases} 2\,\dfrac{D_h C_p + M_h}{\rho} + C_p = T_m \text{ in (2.33c)} & \text{if } \Gamma_0 \neq \emptyset \\ 2\,\dfrac{M_h}{\rho} = T'_m \text{ in (2.35c)} & \text{if } \Gamma_0 = \emptyset \end{cases}$$

where the positive constant $C_{1,T}$ depends on T but not on φ^o, φ^1. Indeed, $C_{1,T}$ coincides with C'_T given by Eq. (2.35a) in the case $\Gamma_0 = \emptyset$; while $C_{1,T}$ is given by $C_T/\max\{1,T\}$ in the case $\Gamma_0 \neq \emptyset$, where C_T is defined by (2.33a). Moreover, T_1 coincides with T_m given by (2.33c) in the case $\Gamma_0 = \emptyset$, and with T'_m given by (2.35c) in the case $\Gamma_0 = \emptyset$.

We then provide another, more sophisticated approach to the problem of 'absorbing' $\int_Q \varphi^2 dQ$ in (2.36a) or $\|\varphi\|^2_{C([0,T];L^2(\Omega))}$ in (2.50). This approach is based on a compactness argument given below. While this argument manages to dispense entirely with assumption (H.3), it requires however a different type of assumption on the set Γ_1, namely that the set Γ_1 satisfies the uniqueness property (H.4) = (1.8) over some time interval $[0,T_u]$ for the associated homogeneous problem (2.12a-d). For this question, we refer to Remark 1.1.

Lemma 2.4 Assume that the solution φ in (2.12e) of problem (2.12a-d) with φ^o, φ^1 in $H^1_{\Gamma_0}(\Omega)\times L^2(\Omega)$ satisfies inequality (2.50) for $T > T_1 > 0$ [This is guaranteed to hold true under assumptions (H.1) and (H.2) by virtue of Lemma 2.3 in either case

$\Gamma_0 \neq \emptyset$ and $\Gamma_0 = \emptyset$]. Assume further the uniqueness property (H.4) = (1.8) of the set Γ_1 over the time interval $[0,T_u]$.

Then, for all $T > \max\{T_1, T_u\}$, T_1 defined by (2.50), there exists a constant $C'_{1,T} > 0$, depending on T but not on the initial data, such that

$$(2.51) \qquad \|\phi\|^2_{C([0,T];L^2(\Omega))} \le C'_{1,T} \int_{\Sigma_1} \phi^2 + \phi_t^2 \, d\Sigma_1, \quad T > \max\{T_1, T_u\} \;\square$$

For easy reference, we state as a separate result the following immediate corollary.

Theorem 2.5 In both cases $\Gamma_0 \neq \emptyset$ and $\Gamma_0 = \emptyset$, assume hypotheses (H.1) = (1.1), (H.2) = (1.2) on the vector field h (so that inequality (2.50) holds true) as well as assumption (H.4) = (1.8) on the uniqueness property of the set Γ_1 over the time interval $[0,T_u]$, so that inequality (2.51) holds true.

Then, for all $T > \max\{T_1, T_u\}$, T_1 defined by (2.50), the solution Φ given by (2.12e) of problem (2.12a-d) satisfies (inequality (2.51) and hence) the inequality

$$(2.52) \qquad \int_{\Sigma_1} \phi_t^2 + \phi^2 \, d\Sigma_1 \ge \frac{C_{1,T}}{TC'_{1,T} + 1} E(0)$$

where the constants $C_{1,T}$ and $C'_{1,T}$ are the same as in (2.50) and (2.51), respectively $\square$

Proof of Lemma 2.4 The proof is by contradiction.

Step 1 Suppose there exists a sequence $\{\varphi_n(t)\}$ of solutions to problem (2.12a-d), i.e.

$$(2.53) \quad \begin{cases} \text{a)} & \varphi_n'' = \Delta\varphi_n & \text{in } Q \\ \text{b)} & \varphi_n|_{t=0} = \varphi_n^0 \; H^1_{\Gamma_0}(\Omega), \quad \varphi_n'|_{t=0} = \varphi_n^1 \in L^2(\Omega) & \text{in } \Omega \\ \text{c)} & \varphi_n \equiv 0 & \text{in } \Sigma_0 \\ \text{d)} & \dfrac{\partial\varphi_n}{\partial\nu} \equiv 0 & \text{in } \Sigma_1 \end{cases}$$

over $[0,T]$ explicitly given by

$$(2.53e) \qquad \varphi_n(t) = C(t)\,\varphi_n^0 + S(t)\,\varphi_n^1 \in C([0,T];H^1_{\Gamma_0}(\Omega))$$

such that with $\frac{d}{dt} = {}'$:

$$
\begin{cases}
(2.54a) & \|\phi_n\|_{C([0,T];L^2(\Omega))} \equiv 1 \\
(2.54b) & \int_{\Sigma_1} \phi_n^2 + (\phi_n')^2 \, d\Sigma_1 \to 0 \text{ as } n \to \infty
\end{cases}
$$

By assumption, the solutions $\phi_n(t)$ satisfy inequality (2.50) and by (2.54) we have

$$
(2.55) \qquad E_n(0) = \int_\Omega (\phi_n^1)^2 + |\nabla \phi_n^o|^2 + d\Omega < \text{const, uniformly in } n
$$

We can thus extract a subsequence, still subindexed by n, such that $[\nabla\phi_n^o$ converges to some function in $[L^2(\Omega)]^{\dim \Omega}$ weakly, and hence] there are constants c_n for which

$$
(2.56) \quad
\begin{cases}
\text{a)} & \phi_n^o + c_n \to \text{ some function } \phi^o \text{ in } H^1(\Omega) \text{ weakly} \\
\text{b)} & \phi_n^1 \to \text{ some function } \phi^1 \text{ in } L^2(\Omega) \text{ weakly}
\end{cases}
$$

If $\Gamma_0 \neq \emptyset$, the condition $\phi_n^o + c_n \in H^1_{\Gamma_0}(\Omega)$ implies by (2.53b)

$$
(2.56c) \qquad 0 = [\phi_n^o + c_n]|_{\Gamma_0} = \phi_n^o|_{\Gamma_0} + c_n = c_n
$$

If $\Gamma_0 = \emptyset$, we may assume w.l.o.g. that $\phi_n^o \in L_0^2(\Omega)$ i.e. $\mathcal{M}(\phi_n^o) \equiv 0$, see (2.8) and (2.11), i.e. that $\mathcal{M}(\phi_n^o + c_n) = c_n$.

<u>Step</u> 2 (Solutions to problem (2.12a-d) with initial data as in (2.56)).

If $\Gamma_0 = \emptyset$, it is convenient to split quantities in two orthogonal components in $L_0^2(\Omega)$ and in $\mathcal{N}(A)$, as in (2.9).

Thus set

$$
(2.57) \quad
\begin{aligned}
&\text{a)} \quad \phi_n^1 = \overline{\phi_n^1} + \overline{c_n^1} \\
&\text{b)} \quad \phi^1 = \overline{\phi^1} + \overline{c^1} \\
&\text{c)} \quad \phi^0 = \overline{\phi^0} + \overline{c^0}
\end{aligned}
\qquad\qquad
\begin{aligned}
&\overline{\phi_n^1}, \overline{\phi^1}, \overline{\phi^0} \in L_0^2(\Omega) \\
&\overline{c_n^1}, \overline{c^1}, \overline{c^0} \in \mathcal{N}(A)
\end{aligned}
$$

since $L_0^2(\Omega)$ is invariant under C(t) and S(t), we have

$$
(2.58)\quad
\begin{aligned}
&\text{a)}\quad C(t)\,(\phi_n^o + c_n) = C(t)\,\phi_n^o + c_n\\
&\text{b)}\quad S(t)\phi_n^1 = S(t)\overline{\phi_n^1} + \overline{c_n^1}\,t\\
&\text{c)}\quad C(t)\,\phi^o = C(t)\,\overline{\phi^o} + \overline{c^o}\\
&\text{d)}\quad S(t)\phi^1 = S(t)\,\overline{\phi^1} + \overline{c^1}\,t.
\end{aligned}
$$

Thus, the solutions

$\tilde{\phi}_n(t)$ due to initial data $[\phi_n^o + c_n,\ \phi_n^1]$

and

$\phi(t)$ due to initial data $[\phi^o,\ \phi^1]$

of problem (2.12a-d) are given by

$$
\begin{cases}
(2.59)\quad \tilde{\phi}_n(t) = C(t)(\phi_n^o + c_n) + S(t)\phi_n^1 = \begin{cases} \phi_n(t) & \text{for } \Gamma_0 \neq 0 \text{ (see (2.53e), (2.56c))}\\ \phi_n(t) + c_n = C(t)\phi_n^o + S(t)\overline{\phi_n^1} + c_n + \overline{c_n^1}\,t, & \text{for } \Gamma_0 = \emptyset \end{cases}\\
(2.60)\quad \tilde{\phi}_n'(t) = \phi_n'(t) = -AS(t)\phi_n^o + C(t)\phi_n^1, \quad \text{see (2.53e)}
\end{cases}
$$

$$
\begin{cases}
(2.61)\quad \phi(t) = C(t)\phi^o + S(t)\phi^1\\
(2.62)\quad \phi'(t) = -AS(t)\phi^o + C(t)\phi^1
\end{cases}
$$

<u>Step</u> 3 It follows that

$$
\begin{cases}
(2.63)\quad \tilde{\phi}_n(t) \to \phi(t) \text{ in } L^\infty(0,T;H^1_{\Gamma_0}(\Omega)) \text{ weak star}\\
(2.64)\quad \tilde{\phi}_n'(t) = \phi_n'(t) \to \phi'(t) \text{ in } L^\infty(0,T;L^2(\Omega)) \text{ weak star}
\end{cases}
$$

In fact, with reference to

$$\begin{cases} \tilde{\phi}_n(t) - \phi(t) = C(t)(\phi_n^o - \overline{\phi^o}) + S(t)(\overline{\phi_n^1} - \overline{\phi^1}) + (\overline{c_n^1} - \overline{c^1})t \\ \qquad\qquad + (c_n - \bar{c}_0) \end{cases} \quad (2.65)$$

$$\tilde{\phi}_n'(t) - \phi'(t) = -A^{1/2}S(t)A^{1/2}(\phi_n^o - \overline{\phi^o}) + C(t)(\overline{\phi_n^1} - \overline{\phi^1}) + (\overline{c_n^1} - \overline{c^1}) \quad (2.66)$$

in the case $\Gamma_0 = \emptyset$ (in the case $\Gamma_0 \neq \emptyset$, delete the super script "bar" on $\overline{\phi^o}$, $\overline{\phi_n^1}$, $\overline{\phi^1}$, and set $\overline{c_n^1} = \overline{c^1} = c_n = \bar{c}_0 = 0$, see (2.56 c)), if now

$$g_1 \in L^1(0,T;[\mathcal{D}(A^{1/2})]')$$

$$g_2 \in L^1(0,T;L^2(\Omega))$$

then

$$\int_0^T (A^{1/2}[C(t)(\phi_n^o - \overline{\phi^o}) + S(t)(\overline{\phi_n^1} - \overline{\phi^1})],\ A^{-1/2}g_1(t))_{L^2(\Omega)}$$

$$+ (-A^{1/2}S(t)A^{1/2}(\phi_n^o - \overline{\phi^o}) + C(t)(\overline{\phi_n^1} - \overline{\phi^1}),\ g_2(t))_{L^2(\Omega)}\, dt$$

$$= \int_0^T (A^{1/2}(\mathcal{G}_n - \overline{\mathcal{G}}),\ C(t)A^{-1/2}g_1(t) - A^{1/2}S(t)g_2(t))_{L^2(\Omega)}$$

$$+ (\overline{\phi_n^1} - \overline{\phi^1},\ A^{1/2}S(t)A^{-1/2}g_1(t) + C(t)g_2(t))_{L^2(\Omega)}\, dt \to 0 \quad (2.67)$$

by Lebesgue dominated theorem and (2.56), since $\|C(t)\|$, $\|A^{1/2}S(t)\| <$ const, t in [0,T]. Then (2.67) yields (2.63) - (2.64).

Step 4 It follows from (2.63) and trace theory that for a suitable subsequence

$$\tilde{\phi}_n|_{\Sigma_1} = [\phi_n + c_n]_{\Sigma_1} \to \phi|_{\Sigma_1} \text{ in, say, } L^2(\Sigma_1) \quad (2.68)$$

If $\Gamma_0 \neq \emptyset$, we have seen in (2.56c) that $c_n \equiv 0$.

If $\Gamma_0 = \emptyset$, then condition $\phi_n|_{\Sigma_1} \to 0$ in $L_2(\Sigma_1)$ from (2.54b), combined with (2.68), gives $c_n \to$ some constant c. Thus, we have proved: there is a subsequence such that

$$
(2.69)\quad
\begin{cases}
\text{a)} & \phi_n^o \to \text{some } \phi^o & \text{in } H^1(\Omega) \text{ weakly} \\
\text{b)} & \phi_n^1 \to \phi^1 & \text{in } L^2(\Omega) \text{ weakly}
\end{cases} \text{ from (2.56)}
$$

$$
\begin{cases}
\text{c)} & \phi_n(t) \to \text{some function } \phi(t) & \text{in } L^\infty(0,T;H^1_{\Gamma_0}(\Omega)) \text{ weak star} \\
\text{d)} & \phi_n'(t) \to \phi'(t) & \text{in } L^\infty(0,T;L^2(\Omega)) \text{ weak star}
\end{cases} \text{ from (2.63)}
$$

$$
\text{e)}\quad \phi_n|_{\Sigma_1} \to \phi|_{\Sigma_1} \equiv 0 \qquad \text{in } L^2(\Sigma_1), \text{ from (2.54b), (2.69c) and (2.68)}
$$

<u>Step</u> 5 By (2.69c), $\{\phi_n(t)\}$ is uniformly bounded in $L^\infty(0,T;H^1_{\Gamma_0}(\Omega))$ and by compactness there is a subsequence ϕ_n strongly convergent in $L^\infty(0,T;L^2(\Omega))$ to ϕ. Thus

$$
(2.70)\qquad \|\phi_n\|_{L^\infty(0,T;L^2(\Omega))} \to \|\phi\|_{L^\infty(0,T;L^2(\Omega))} \equiv 1
$$

by (2.54a). But, by (2.53e), the limit ϕ satisfies problem (2.12a-d) and, moreover by (2.69e) we have $\phi|_{\Sigma_1} \equiv 0$. Thus, the limit ϕ satisfies the problem

$$
(2.71)\quad
\begin{cases}
\phi'' = \Delta\phi & \text{in } Q \\
\phi|_\Sigma = 0 & \text{in } \Sigma \\
\left.\dfrac{\partial\phi}{\partial\nu}\right|_{\Sigma_1} = 0 & \text{in } \Sigma_1
\end{cases}
\qquad \text{for } T > \max\{T_1, T_u\}
$$

Then, assumption (H.4) applies and we conclude that $\phi \equiv 0$ in Q, a contradiction with (2.70). Lemma 2.4 is proved □

We next present an important improvement of Theorem 2.5 in the case $\Gamma_0 = \emptyset$, when $E(0)$ is not equivalent to

$$
\|\{\phi^o, \phi^1\}\|^2_{H^1(\Omega)\times L^2(\Omega)} \qquad \text{if } \mathcal{M}(\phi^o) \neq 0.
$$

<u>Theorem</u> 2.6 Let $\Gamma_0 = \emptyset$, so that assumption (H.4) is automatically satisfied. Assume further hypotheses (H.1) and (H.2) on the vector field h, so that conclusion (2.52) of Thoerem 2.4 holds true.

Then, for all $T > \max\{T_1, T_u\}$, T_1 defined by (2.50), there is a positive constant k_T, depending on T but not on ϕ^o, ϕ^1, such that the solution ϕ of problem (2.12a-d) satisfies

$$(2.72) \qquad \int_\Sigma \phi^2 + \phi_t^2 \, d\Sigma > k_T \, \|\{\phi^o, \phi^1\}\|^2_{H^1(\Omega)\times L^2(\Omega)}$$

(Note: we are not assuming $\mathcal{M}(\phi^o) = 0$) □

Proof For $T > T_u$, the uniqueness property (H.4) implies that

$$(2.73) \qquad \{\int_\Sigma \phi^2 + \phi_t^2 \, d\Sigma\}^{1/2}$$

is a norm. Let now $T > \max\{T_1, T_u\}$. If $\{\phi_n(t)\}$ is a Cauchy sequence for this norm with $\phi_n(0) = \phi_n^o$ and $\phi_n'(0) = \phi_n^1$, then the conclusion (2.52) of Theorem 2.5 - which holds true under present assumptions - implies that there is a sequence $\{c_n\}$ such that

(2.74)

a) $\phi_n^o + c_n$ converges in $H^1(\Omega)$

b) ϕ_n^1 converges in $L^2(\Omega)$

and by trace theory

(2.74c) $\phi_n^o + c_n$ converges in $L^2(\Gamma)$

On the other hand, since both $\{\phi_n(t)\}$ and $\{\phi_n'(t)\}$ converge in $L^2(\Sigma)$, i.e. $\{\phi_n(t)\}$ converges in $H^1(0,T;L^2(\Gamma))$, then Sobolev imbedding theorem implies that

(2.75) $\phi_n(0) = \phi_n^o$ converges in $L^2(\Gamma)$

Comparing (2.75) with (2.74c) yields that the numerical sequence $\{c_n\}$ converges. Thus, $\{\phi_n(t)\}$ being a Cauchy sequence for the norm (2.73) implies that $\{\phi_n^o\}$ is convergent in $H^1(\Omega)$ and $\{\phi_n^1\}$ is convergent in $L^2(\Omega)$. Thus, inequality (2.72) is proved □

2.4 <u>Absorption of lower order term $\int_{\Sigma_1} \phi^2 d\Sigma_1$ under uniqueness assumption</u> (H.4).

<u>A-priori inequality of Theorem 2.1 revisited</u>

In the present subsection, our starting point is Theorem 2.5 for $\Gamma_0 \neq \emptyset$ or Theorem 2.6 for $\Gamma_0 = \emptyset$. Thus, we assume throughout hypotheses (H.1) and (H.2) and, in the case $\Gamma_0 \neq \emptyset$, the additional uniqueness property (H.4) of the set Γ_1 over the time interval $[0,T_u]$ (This property is automatically satisfied if $\Gamma_0 = \emptyset$). As a result, we obtain that for all $T > \max\{T_1, T_u\}$, the solution of ϕ of problem (2.12a-d) satisfies inequality (2.52) for $\Gamma_0 \neq \emptyset$ or inequality (2.72) for $\Gamma_0 = \emptyset$; i.e. (we re-write these in a combined form):

$$(2.76)\quad \begin{cases} \int_{\Sigma_1} \phi_t^2 + \phi^2 d\Sigma_1 > C_{2,T} \begin{cases} E(0) & \text{for } \Gamma_0 \neq \emptyset \\ \|\{\phi^o, \phi^1\}\|^2_{H^1(\Omega)xL^2(\Omega)} & \text{for } \Gamma_0 = \emptyset \end{cases} \\ E(0) \text{ equivalent to } \|\{\phi^o, \phi^1\}\|^2_{H^1_{\Gamma_0}(\Omega)xL^2(\Omega)}, \qquad \Gamma_0 \neq \emptyset \\ T > \max\{T_1, T_u\}, \quad T_1 \text{ defined in (2.50)} \end{cases}$$

where $C_{2,T}$ is a positive constant depending on T but not on the initial data

Our next step is then to employ an argument patterned after Lemma 2.4 in order to 'absorb' the term

$$\int_{\Sigma_1} \phi^2 d\Sigma_1 \text{ by the term } \int_{\Sigma_1} \phi_t^2 d\Sigma_1.$$

<u>Lemma</u> 2.7 Assume $\phi^o \in H^1_{\Gamma_0}(\Omega)$ and $\phi^1 \in L^2(\Omega)$ and, if $\Gamma_0 = \emptyset$, assume further $\mathcal{M}(\phi^o) = 0$. Consider the corresponding solution ϕ of problem (2.12a-d) and assume that ϕ satisfies inequality (2.76) [By virtue of Theorems 2.5 or 2.6, this is guaranteed to hold true under assumptions (H.1) and (H.2) and, if $\Gamma_0 \neq \emptyset$, the additional uniqueness property (H.4) of the set Γ_1 over the time interval $[0, T_u]$ (a property which is automatically true if $\Gamma_0 = \emptyset$).

Then, for all $T > \max\{T_1, T_u\}$, T_1 defined by (2.50) there is a positive constant $C'_{2,T}$ depending on T but not on the initial data such that

$$(2.77)\quad \int_{\Sigma_1} \phi^2 d\Sigma_1 < C'_{2,T} \int_{\Sigma_1} \phi_t^2 d\Sigma_1$$

For easy future reference we state as a separate result the following immediate corollary

Theorem 2.8 In both cases $\Gamma_0 \neq \emptyset$ and $\Gamma_0 = \emptyset$, let ϕ^o, $\phi^1 \in H^1_{\Gamma_0}(\Omega) x L^2(\Omega)$ and, if $\Gamma_0 = \emptyset$, let further $\mathcal{M}(\phi^o) = 0$. Assume hypotheses (H.1) and (H.2) on the vector field h as well as assumption (H.4) on the uniqueness property of the set Γ_1 over the time interval $[0, T_u]$, so that inequality (2.76) holds true (assumption (H.4) is automatically satisfied if $\Gamma_0 = \emptyset$).

Then, for all $T > \max\{T_1, T_u\}$, the solution ϕ given by (2.12e) of problem (2.12a-d) satisfies (inequality (2.77) and hence) the inequality

$$(2.78)\quad \begin{cases} \int_{\Sigma_1} \phi_t^2 d\Sigma_1 > \dfrac{C_{2,T}}{1+C'_{2,T}} \begin{cases} E(0) & \text{for } \Gamma_0 \neq \emptyset \\ \|\{\phi^o, \phi^1\}\|^2_{H^1(\Omega)xL^2(\Omega)} & \text{for } \Gamma_0 = \emptyset;\ \mathcal{M}(\phi^o) = 0 \end{cases} \\ E(0) \text{ equivalent to } \|\{\phi^o, \phi^1\}\|^2_{H^1_{\Gamma_0}(\Omega)xL^2(\Omega)}, \quad \Gamma_0 \neq \emptyset \end{cases}$$

$$T > \max\{T_1, T_u\}$$

where the constants $C_{2,T}$ and $C'_{2,T}$ are the same as in (2.76) and (2.77), respectively

Proof of Lemma 2.7 Same ideas as in Lemma 2.3.

Step 1 Suppose, by contradiction, that there exists a sequence $\{\phi_n(t)\}$ of solutions as in (2.53a-d) such that

$$(2.79)\quad \begin{cases} \text{a)} & \|\phi_n\|_{L^2(\Sigma_1)} \equiv 1 \\ \text{b)} & \int_{\Sigma_1} (\phi'_n)^2 d\Sigma_1 \to 0 \text{ as } n \to \infty \end{cases}$$

By assumption, the solutions $\phi_n(t)$ satisfy inequality (2.76) and thus the pairs $\{\phi_n^o, \phi_n^1\}$ are uniformly bounded in $H^1_{\Gamma_0}(\Omega)xL^2(\Omega)$, as in Step 1 of Lemma 2.3, in both cases $\Gamma_0 \neq \emptyset$ and $\Gamma_0 = \emptyset$.

It then follows, as in Steps 2, 3, 4 of Lemma 2.3, that for a suitable subsequence we ha

$$(2.80)\quad \begin{cases} \text{a)} & \phi_n(t) \to \text{some } \phi(t) \text{ in } L^\infty(0,T;H^1_{\Gamma_0}(\Omega)) \text{ weak star} \\ \text{b)} & \phi'_n(t) \to \phi'(t) \text{ in } L^\infty(0,T;L^2(\Omega)) \quad \text{weak star} \end{cases}$$

By (2.80), the sequence $\{\phi_n(t)\}$ is uniformly bounded in $L^\infty(0,T; H^1_{\Gamma_0}(\Omega))$. Hence, by trace theory, the sequence $\{\phi_n|_{\Sigma_1}\}$ is uniformly bounded in $L^\infty(0,T; H^{1/2}(\Gamma_1))$ and thus it lies in a compact set of $L^2(\Sigma_1)$. Then, for a suitable subsequence we have

(2.80c) $\quad \phi_n|_{\Sigma_1} \to \phi|_{\Sigma_1}$ in $L^2(\Sigma_1)$ (strongly)

and by (2.79a) we deduce

(2.80d) $\quad \|\phi\|_{L^2(\Sigma_1)} = 1$

On the other hand, (2.79b) implies

(2.80e) $\quad \phi'|_{\Sigma_1} = 0$

But, by (2.53e), the limit function ϕ satisfies the problem

$$
(2.81)\quad
\begin{cases}
\text{a)} & \phi'' = \Delta\phi & \text{in } Q \\
\text{b)} & \phi|_{\Sigma_0} = 0 & \text{in } \Sigma_0 \\
\text{c)} & \dfrac{\partial\phi}{\partial\nu}\Big|_{\Sigma_1} = 0 & \text{in } \Sigma_1
\end{cases}
$$

Differentiating in time (2.81) and using (2.80e), we obtain that $\phi' = \phi_t$ solves

$$
(2.82)\quad
\begin{cases}
(\phi')_{tt} = \Delta(\phi') & \text{in } Q \\
\phi'|_{\Sigma} = 0 & \text{in } \Sigma \\
\dfrac{\partial\phi'}{\partial\nu}\Big|_{\Sigma_1} = 0 & \text{in } \Sigma_1
\end{cases}
\qquad \text{for } T > \max\{T_1, T_u\}
$$

Then, assumption (H.4) applies and we conclude that $\phi' \equiv 0$ in Q. Thus, $\phi \equiv$ const in Q. If $\Gamma_0 \neq \emptyset$, then (2.81b) yields $\phi \equiv 0$ in Q. If $\Gamma_0 = \emptyset$, the further assumption $\mathcal{M}(\phi^0) = 0$ yields likewise $\phi \equiv 0$ in Q.
Thus in any case, the conclusion $\phi \equiv 0$ in Q contradicts (2.80d). Lemma 2.7 is proved $\|$

2.5 An a-priori inequality in the absence of geometrical conditions on Ω for the Neumann problem ($\Gamma_0 = \emptyset$).

Throughout this subsection we take $\Gamma_0 = \emptyset$ (but we do not assume $\mathcal{M}(\phi^o) = 0$) and we specialize the vector field h(x) to a radial field $x - x^o$, for some fixed $x_o \in R^n$. Thus, recalling Remark 1.1 and (2.15c)

$$
(2.83) \quad \begin{cases} \text{a)} \quad h(x) \equiv m(x) \equiv x-x^o \\ \text{b)} \quad H(x) \equiv \text{identity; matrix} \qquad \operatorname{div} h \equiv n = \dim \Omega; \qquad 2D_h = n \\ \text{c)} \quad M_h = \max_{\overline{\Omega}} |x-x^o| = R(x^o) \end{cases}
$$

In the main statement below, we shall need the tangential gradient $\nabla_\sigma \phi$ of a function $\phi \in C^1(\overline{\Omega})$ on Γ (or part thereof; in our case the set $\Gamma_*(x^o)$ defined in (1.14)). At each point of Γ (sufficiently smooth), consider the unit outward normal ν and a, say, orthogonal system of unit vectors $\tau_1, \dots \tau_{n-1}$ on the corresponding tangent plane. We have

$$\nabla\phi = (\nabla\phi\cdot\nu)\nu + \sum_{i=1}^{n-1} (\nabla\phi\cdot\tau_i)\tau_i$$

$$(2.84) \qquad = \frac{\partial\phi}{\partial\nu}\,\nu + \sum_{i=1}^{n-1} \frac{\partial\phi}{\partial\tau_i}\,\tau_i$$

Thus, if $\frac{\partial\phi}{\partial\nu} = 0$ on Γ we set

$$(2.85) \qquad \sigma_i = \text{first order tangential operator on } \Gamma = \frac{\partial}{\partial\tau_i}$$

and we then define in this case

$$(2.86) \qquad \|\nabla\phi\|^2_{L^2(\Gamma)} = \sum_{i=1}^{n-1} |\sigma_i\phi|^2 \equiv \|\nabla_\sigma\phi\|^2_{L^2(\Gamma)} \quad \text{for } \frac{\partial\phi}{\partial\nu}\Big|_\Gamma = 0$$

Our main result in this subsection is the following inequality which requires no geometrical conditions on Ω.

Theorem 2.9 Let $\Gamma_0 = \emptyset$ and recall the sets $\Gamma(x^o)$ and $\Gamma_*(x^o)$ of Γ defined by (1.14). There exists $T_m > 0$ (given explicitly below) such that for all $T > T_m$, the solution ϕ given by (2.12e) of problem (2.12a-d) with $\Gamma_0 = \emptyset$ satisfies the following inequality

$$\begin{cases} \int_0^T \int_{\Gamma(x^o)} \phi_t^2 \, d\Gamma(x^o)dt + \int_0^T \int_{\Gamma_*(x^o)} |\nabla_\sigma \phi|^2 d\Gamma_*(x^o)dt + \int_0^T \int_\Gamma \phi^2 d\Gamma dt \\ \geq C_T \|\{\phi^o, \phi^1\}\|^2_{H^1(\Omega)\times L^2(\Omega)} \\ \text{for all } \{\phi^o, \phi^1\} \in H^1(\Omega)\times L^2(\Omega) \text{ for which the left hand side is finite} \end{cases} \tag{2.87}$$

where C_T is a positive constant depending on T but not on ϕ^o, ϕ^1.

For T_m we can take (see (2.83)):

$$T_m = \max\{T_u, 2R(x^o)\} \tag{2.88}$$

where T_u is the time defined by the uniqueness property (1.8) with $\Gamma_0 = \emptyset$, see Remark 1.3 □

<u>Proof of Theorem</u> 2.9

<u>Step 1</u> (Variation of Lemma 2.3)

<u>Lemma</u> 2.10 For $\Gamma_0 = \emptyset$ we have for any $\varepsilon > 0$ and $m(x) = x-x^o$

$$\int_0^T \int_{\Gamma(x^o)} \phi_t^2 \, m(x)\cdot\nu(x)d\Gamma(x^o)dt + \int_0^T \int_{\Gamma_*(x^o)} |\nabla\phi|^2 |m(x)| d\Gamma_*(x^o)dt + \frac{(1+n)}{\varepsilon} \|\phi\|^2_{C([0,T];\, L^2(\Omega))} \geq (T-T_\varepsilon)E(0) \tag{2.89}$$

$$T_\varepsilon = \varepsilon(1+n) + 2R(x^o), \quad \phi^o, \phi^1 \in H^1(\Omega)\times L^2(\Omega) \tag{2.90}$$

<u>Proof of Lemma</u> 2.10 (specialization and variation of proof of Lemma 2.3). Under specialization (2.83), the fundamental equality (2.23) for problem (2.12a-d) becomes then

$$\tfrac{1}{2}\int_\Sigma \phi_t^2 \, m\cdot\nu d\Sigma - \tfrac{1}{2}\int_\Sigma |\nabla\phi|^2 \, m\cdot\nu d\Sigma = \int_Q |\nabla\phi|^2 dQ + \beta_{0,T} \tag{2.91}$$

where $\beta_{0,T}$ is given by (2.17a) with $h(x) = m(x)$. We obtain from (2.40) and (2.83)

$$\beta_{0,T} \geq -\left[\frac{n}{2}\varepsilon + R(x^o)\right]E(0) - \frac{n}{2\varepsilon}\|\phi\|^2_{C([0,T];L^2(\Omega))} \tag{2.92}$$

Thus, (2.91), (2.92) give

$$(2.93)\quad \tfrac{1}{2}\int_\Sigma \phi_t^2\, m\cdot\nu d\Sigma - \tfrac{1}{2}\int_\Sigma |\nabla\phi|^2\, m\cdot\nu d\Sigma + \frac{n}{2\varepsilon}\, \|\phi\|^2_{C([0,T];L^2(\Omega))}$$

$$\geq \int_Q |\nabla\phi|^2 dQ - [\varepsilon\tfrac{n}{2} + R(x^o)]\; E(0)$$

As to identity (2.21), by proceeding as in (2.39), we obtain for $\Gamma_0 = \emptyset$:

$$(2.94)\quad \left|\int_Q \phi_t^2 - |\nabla\phi|^2 dQ\right| = \left|[(\phi_t,\ \phi)_\Omega]_0^T\right| \leq \varepsilon\; E(0) + \frac{1}{\varepsilon}\, \|\phi\|^2_{C([0,T];L^2(\Omega))}$$

and hence

$$(2.95)\quad \int_Q |\nabla\phi|^2 dQ \geq \int_Q \phi_t^2\; dQ - [\varepsilon E(0) + \frac{1}{\varepsilon}\, \|\phi\|^2_{C([0,T];L^2(\Omega))}]$$

in place of (2.29 b). Thus, by (2.95) and (2.24), proceeding as in (2.30) we obtain for $\Gamma_0 = \emptyset$:

$$(2.96a)\quad \int_Q |\nabla\phi|^2 dQ \geq \tfrac{1}{2}\int_Q |\nabla\phi|^2 dQ + \tfrac{1}{2}\int_Q \phi_t^2 dQ - \tfrac{1}{2}\,[\varepsilon E(0) + \frac{1}{\varepsilon}\, \|\phi\|^2_{C([0,T];L^2(\Omega))}]$$

$$= \frac{(T-\varepsilon)}{2}\; E(0) - \frac{1}{2\varepsilon}\, \|\phi\|^2_{C([0,T];L^2(\Omega))}$$

in place of (2.30). Inserting (2.96) in (2.83) yields

$$(2.96b)\quad \int_\Sigma \phi_t^2\, m\cdot\nu d\Sigma - \int_\Sigma |\nabla\phi|^2 m\cdot\nu d\Sigma + (\frac{m+1}{\varepsilon})\, \|\phi\|^2_{C([0,T];L^2(\Omega))} \geq (T - T_\varepsilon)E(0)$$

with T_ε as in (2.90). We now split $\int_\Gamma = \int_{\Gamma(x^o)} + \int_{\Gamma_*(x^o)}$, recall definition (1.14), drop negative terms, and then (2.96) yields (2.89) □

<u>Step</u> 2 (Absorption of $\|\phi\|^2_{C([0,T];L^2(\Omega))}$ as in Lemma 2.4)

<u>Lemma</u> 2.11 Let $\Gamma_0 = \emptyset$ and let $T > T_m$, see (2.88). Then

$$(2.97)\quad \|\phi\|^2_{C([0,T];L^2(\Omega))} \leq C_T'\, \{\int_0^T\int_{\Gamma(x^o)} \phi_t^2\; d\Gamma(x^o)dt + \int_0^T\int_{\Gamma_*(x^o)} |\nabla\phi|^2 d\Gamma_*(x^o)dt + \int_0^T\int_\Gamma \phi^2 d\Gamma dt$$

Proof of Lemma 2.11 By contradiction, as in the proof of Lemma 2.4, let $\{\phi_n(t)\}$ be a sequence of solutions (2.53) such that

$$(2.98)\quad \begin{cases} \text{a)} & \|\phi_n\|_{C([0,T];L^2(\Omega))} \equiv 1 \\ \text{b)} & \int_0^T \int_{\Gamma(x^o)} (\phi_n')^2 d\Gamma(x^o)dt + \int_0^T \int_{\Gamma_*(x^o)} |\nabla\phi_n|^2 d\Gamma_*(x^o)dt + \int_\Sigma \phi_n^2 d\Sigma \to 0 \end{cases}$$

Each $\phi_n(t)$ satisfies (2.89). By taking ε sufficiently small, we then have from (2.89) that if $T > T_m$ then: $E_n(0) <$ const as in (2.55). Steps 2 through 5 in the proof of Lemma 2.4 then yield that, in the notation of Lemma 2.4, the limit function ϕ satisfies

$$(2.99)\quad 1 \equiv \|\phi\|_{L^\infty(0,T;L^2(\Omega))} \to \|\phi\|_{L^\infty(0,T;L^2(\Omega))} = 1$$

as well as

$$\begin{cases} \phi'' = \Delta\phi & \text{in } Q \\ \dfrac{\partial\phi}{\partial\nu} = 0 & \text{in } \Sigma \end{cases}$$

But the last integral term in (2.98b) gives likewise

$$\phi = 0 \qquad \text{in } \Sigma$$

For $T > T_m$, the uniqueness property then yields $\phi \equiv 0$ in Q, a contradiction of (2.99) □

Step 3 Putting together Lemmas 2.10 and 2.11, we have immediately

Lemma 2.12 For $\Gamma_0 = \emptyset$, $\varepsilon > 0$, and $T > T_\varepsilon$:

$$(2.100)\quad \begin{aligned} &[\tfrac{(1+n)}{\varepsilon}C_T' + r(x^o)]\ \{\int_0^T \int_{\Gamma(x^o)} \phi_t^2\, d\Gamma(x^o)dt + \int_0^T \int_{\Gamma_*(x^o)} |\nabla\phi|^2 d\Gamma_*(x^o)dt\} \\ &\quad + \tfrac{(1+n)}{\varepsilon}C_T' \int_\Sigma \phi^2 d\Sigma > (T - T_\varepsilon)E(0) \end{aligned}$$

where $r(x^o) = \max_\Gamma |x-x^o|$.

Step 4 Since $\frac{\partial\phi}{\partial\nu} = 0$, then $\nabla\phi$ can be replaced by $\nabla_\sigma\phi$, see (2.85) - (2.86). To prove (2.87) it remains to show that $E(0)$ in (2.100) can be replaced by $\|\{\phi^o, \phi^1\}\|^2_{H^1(\Omega)\times L^2(\Omega)}$.

This can be done in the same way as in the proof of Theorem 2.6, since

$$(2.101) \qquad \{\int_0^T \int_{\Gamma(x^o)} \phi_t^2 \, d\Gamma(x^o)dt + \int_0^T \int_{\Gamma_*(x^o)} |\nabla\phi|^2 \, d\Gamma_*(x^o)dt + \int_0^T \int_\Gamma \phi^2 d\Gamma dt\}^{1/2}$$

is a norm for $T > T_m$. Thus, the proof of Theorem 2.6 using this time (2.100) instead of (2.52) shows that if $\{\phi_n(t)\}$ is a Couchy sequence for the norm (2.101), then $\{\phi_n^o\}$ and $\{\phi_n^1\}$ are convergent in $H^1(\Omega)$ and $L^2(\Omega)$, respectively, and inequality (2.87) follows □

3. Exact controllability on $H^1_{\Gamma_0}(\Omega)xL^2(\Omega)$ with controls $v \in L^2(0,T; L^2(\Gamma_1)) \equiv L^2(\Sigma_1)$. Equivalence to a-priori inequalities. Ontoness approach of the solution operator.

For sake of clarity of exposition, we shall treat the two cases $\Gamma_0 \neq \emptyset$ and $\Gamma_0 = \emptyset$ separately. Even though the conceptual approach will be the same in both cases, there are a number of technical differences that arise between them.

3.1 The case $\Gamma_0 \neq \emptyset$.

The goal of this subsection is to prove the following result.

Theorem 3.1 a) Problem (1.1) is exactly controllable on the space $H^1_{\Gamma_0}(\Omega)xL^2(\Omega)$ over the time interval $[0,T]$, $0 < T < \infty$, by means of $L^2(\Sigma_1)$ - controls v if and only if the following inequality holds:

$$(3.1)\quad \begin{cases} \int_0^T\int_{\Gamma_1} \phi_t^2 \, d\Sigma_1 \geq C_T E(0) = C_T \|\{\phi^o, \phi^1\}\|^2_{\mathcal{D}(A^{1/2})xL^2(\Omega)} \\ E(0) \text{ equivalent to } \|\{\phi^o, \phi^1\}\|^2_{H^1_{\Gamma_0}(\Omega)xL^2(\Omega)} \quad (\text{cf}(2.5\text{a})) \\ \text{for all } \{\phi^o, \phi^1\} \in H^1_{\Gamma_0}(\Omega)xL^2(\Omega) \text{ for which the left hand side of the above} \\ \text{inequality is finite,} \end{cases}$$

where C_T is a positive constant depending on T, but not on ϕ^o, ϕ^1, and where ϕ solves the homogeneous backward problem

$$(3.2)\quad \begin{cases} \phi_{tt} = \Delta\phi & \text{on } Q \\ \phi(\cdot,T) = \phi^o, \ \phi_t(\cdot,T) = \phi^1 & \text{in } \Omega \\ \phi \equiv 0 & \text{in } \Sigma_o \\ \dfrac{\partial\phi}{\partial\nu} = 0 & \text{in } \Sigma_1 \end{cases}$$

which is the time reversed version of the forward problem (2.12a-d).

b) Inequality (3.1) is, in turn, equivalent to the following inequality for the same problem (3.2):

(3.3)
$$\begin{cases} \int_0^T \int_{\Gamma_1} \phi^2 d\Sigma_1 > C_T \|\{\phi^o, \phi^1\}\|^2_{L^2(\Omega)\times H^{-1}_{\Gamma_0}(\Omega)} \\ H^{-1}_{\Gamma_0}(\Omega) \text{ - norm equivalent to the } [\mathcal{D}(A^{1/2})]' \text{ - norm given by} \\ \qquad \|z\|_{[\mathcal{D}(A^{1/2})]'} = \|A^{-1/2} z\|_{L^2(\Omega)} \\ \text{for all } \{\phi^o, \phi^1\} \in L^2(\Omega)\times H^{-1}_{\Gamma_0}(\Omega) \text{ for which the left hand side of the above} \\ \text{inequality is finite}\Box \end{cases}$$

Proof of Theorem 3.1 Step 0. We introduce the operator $\tilde{N}$: continuous $L^2(\Gamma_1) \to H^{3/2}(\Omega)$ by setting (recall (2.2)):

(3.4)
$$w = \tilde{N}g \iff \begin{cases} \Delta w = 0 & \text{in } \Omega \\ w|_{\Gamma_0} = 0 & \text{on } \Gamma_0 \\ \frac{\partial w}{\partial \nu}\Big|_{\Gamma_1} = g & \text{on } \Gamma_1 \end{cases}$$

Let $\tilde{N}^*$ denote the adjoint operator of $\tilde{N}$: $(\tilde{N}v,u)_{L^2(\Omega)} = (v,\tilde{N}^*u)_{L^2(\Gamma_1)}$. We shall need the following Lemma, in the style of [ ].

Lemma 3.2. For $f \in \mathcal{D}(A)$, we have

(3.5)
$$\tilde{N}^*Af = \begin{cases} f|_{\Gamma_1} & \text{on } \Gamma_1 \\ 0 & \text{on } \Gamma_0. \end{cases}$$

Proof of Lemma 3.2 We $g \in L^2(\Gamma)$ we complete by Green second theorem (subscripts denote L^2-norms)

(3.6)
$$\begin{aligned} -(\tilde{N}^*Af, g)_\Gamma &= -(Af, \tilde{N}g)_\Omega = (\Delta f, \tilde{N}g)_\Omega = \\ &= (f,\Delta(\tilde{N}g))_\Omega + (\tfrac{\partial f}{\partial \nu}, \tilde{N}g)_\Gamma - (f, \tfrac{\partial(\tilde{N}g)}{\partial \nu})_\Gamma = \\ &= -(f,g)_{\Gamma_1} \end{aligned}$$

since $f|_{\Gamma_0} = \frac{\partial f}{\partial \nu}\Big|_{\Gamma_1} = 0$ by (2.1); $(\tilde{N}g)|_{\Gamma_0} = 0$, $\frac{\partial \tilde{N}g}{\partial \nu}\Big|_{\Gamma_1} = g$, and

$\Delta(\tilde{N}g) = 0$ in Ω by (3.4). Then (3.6) yields (3.5) □

Step 1 As in [ ], the solution to problem (1.1) can be written abstractly by means of the following "variation of constants" formula. Let $y^0 = y^1 = 0$ in (1.1) and denote the corresponding solutions by $y(t;\, y^0 = y^1 = 0) \equiv y(t)$.

Then

$$(3.7)\qquad \begin{vmatrix} y(T) \\ \\ y_t(T) \end{vmatrix} = \mathcal{L}_T v = \begin{vmatrix} A \int_0^T S(T-t)\tilde{N}v(t)dt \\ \\ A \int_0^T C(T-t)\tilde{N}v(t)dt \end{vmatrix}$$

where $C(\cdot)$ is the strongly continuous cosine operator generated by $-A$ in (2.2) and $S(t) = \int_0^T C(\tau)d\tau$. The operator $\mathcal{L}_T$ in (3.7) with domain

$$(3.8)\qquad \mathcal{D}(\mathcal{L}_T) = \{v \in L^2(\Sigma_1):\ [y(T),\, y_t(T)] \in H^1_{\Gamma_0}(\Omega)xL^2(\Omega)\}$$

is an unbounded, densely defined closed operator. See regularity theory in [ ].

Step 2 By time reversibility, exact controllability of problem (1.1) at time T on the space $H^1_{\Gamma_0}(\Omega)L^2(\Omega)$ by means of controls $v \in L^2(\Sigma_1)$ is equivalent to ontoness of $\mathcal{L}_T$

$$(3.9)\qquad \mathcal{L}_T:\ L^2(\Sigma_1) > \mathcal{D}(\mathcal{L}_T) \overset{\text{onto}}{\rightarrow} H^1_{\Gamma_0}(\Omega)xL^2(\Omega)$$

Let $\mathcal{L}_T^*$ denote the adjoint operator of $\mathcal{L}_T$:

$$(\mathcal{L}_T\, g,\, z)_{\mathcal{D}(A^{1/2})xL^2(\Omega)} = (g,\, \mathcal{L}_T^*\, z)_{L^2(\Sigma_1)}$$

Then, the ontoness property (3.9) for $\mathcal{L}_T$ is equivalent to the property that $\mathcal{L}_T^*$ has a continuous inverse [ ]; i.e.

$$(3.10)\qquad \left\| \mathcal{L}_T^* \begin{vmatrix} z_0 \\ z_1 \end{vmatrix} \right\|_{L^2(\Sigma_1)} > C'_T\, \|\{z_0, z_1\}\|_{\mathcal{D}(A^{1/2})xL^2(\Omega)}$$

for some $C'_T > 0$ and all $\{z_0,\, z_1\} \in \mathcal{D}(\mathcal{L}_T^*)$

Lemma 3.3 Property (3.10) (and hence property (3.9)) is equivalent to inequality (3.1) or inequality (3.3) of Thoerem 3.1.

Proof of Lemma 3.3 By (2.4), if $v \in \mathcal{D}(\mathcal{L}_T)$ and $\{z_o, z_1\} \in \mathcal{D}(\mathcal{L}_T^*)$ we compute from (3.7) by proceeding as in [ ]:

$$(\mathcal{L}_T v, \begin{vmatrix} z_o \\ z_1 \end{vmatrix})_{\mathcal{D}(A^{1/2})xL^2(\Omega)} = (A\int_0^T S(T-t)\tilde{N}v(t)dt, Az_o)_{L^2(\Omega)}$$

$$+ (A\int_0^T C(T-t)\tilde{N}v(t)dt, z_1)_{L^2(\Omega)}$$

$$= (v, \tilde{N}^*AS(T-t)Az_o + \tilde{N}^*AC(T-t)z_1)_{L^2(\Sigma_1)} \tag{3.11}$$

Thus

$$\mathcal{L}_T^* \begin{vmatrix} z_o \\ z_1 \end{vmatrix} = \tilde{N}^*A[S(T-t)Az_o + C(T-t)z_1]$$

$$= \tilde{N}^*A[C(t-T)z_1 - S(t-T)Az_o] \tag{3.12}$$

The solution $\phi(t; \phi^o, \phi^1)$ of problem (3.2) is

$$\phi(t; \phi^o,\phi^1) = C(t-T)\phi^o + S(t-T)\phi^1 \tag{3.13}$$

so that, invoking Lemma 3.2, we have

$$\phi(t; \phi^o, \phi^1)|_{\Gamma_1} = \tilde{N}^*A[C(t-T)\phi^o + S(t-T)\phi^1] \tag{3.14}$$

Comparing (3.12) (3.14) yields

(3.15) a) $$(\mathcal{L}_T^* \begin{vmatrix} z_o \\ z_1 \end{vmatrix})(t) = \phi(t; \phi^o = z_1, \phi^1 = -Az_o)|_{\Gamma_1}$$

b) $$\phi^o = z_1, \phi^1 = -Az_o$$

Thus inequality (3.10) becomes precisely inequality (3.3) since

$$\|\{z_o = -A^{-1}\phi^1, z_1 = \phi^o\}\|_{\mathcal{D}(A^{1/2})xL^2(\Omega)} = \|\{\phi^o, \phi^1\}\|_{L^2(\Omega)x[\mathcal{D}(A^{1/2})]'} \tag{3.16}$$

Instead, if we differentiate (3.13) in t, we obtain

$$\phi_t(t; \phi^o, \phi^1) = C(t-T)\phi^1 - AS(t-T)\phi^o \tag{3.17}$$

and by Lemma 2.2

$$(3.18)\qquad \phi_t(t;\phi^o,\phi^1)\Big|_{\Gamma_1} = \tilde{N}^*A[C(t-T)\phi^1 - S(t-T)A\phi^o]$$

comparing now (3.12) with (3.18) yields

$$(3.19)\qquad \left(\mathcal{L}^*_T\begin{vmatrix} z_o \\ z_1\end{vmatrix}\right)(t) = \phi_t(t;\phi^o = z_o,\ \phi^1 = z_1)\Big|_{\Gamma_1}$$

whereby inequality (3.10) becomes now inequality (3.1) with $C_T = C'^2_T$. Theorem 3.1 is proved □

3.2 The case $\Gamma_0 = \emptyset$

We now briefly indicate the modifications that need to be made on the arguments of subsection 3.1 in order to treat the case $\Gamma_0 = \emptyset$. the main result now is

Theorem 3.4 a) Problem (1.1) is exactly controllable on the space $H^1(\Omega)xL^2(\Omega)$ over the time interval $[0,T]$, $0 < T < \infty$, by means of $L^2(\Sigma) = L^2(0,T;L^2(\Gamma))$ - controls v if and only if the following inequality holds true for the solution ϕ of problem (3.2) with $\Gamma_0 = \emptyset$:

$$(3.20)\qquad \begin{cases} \int_0^T \int_\Gamma |\phi_t - (t-T)\mathcal{M}(\phi^o)|^2 d\Sigma \geq C_T \|\{\phi^o,\phi^1\}\|^2_{H^1(\Omega)xL^2(\Omega)} \\ H^1(\Omega) \text{ - norm equal to } \mathcal{D}((A+I)^{1/2}) \text{ - norm, see (2.5c)} \\ \text{for all } \phi^o,\ \phi^1 \in H^1(\Omega)xL^2(\Omega) \text{ for which the left hand side of the above inequality is finite} \end{cases}$$

where C_T is a positive constant depending on T, but not on ϕ^o, ϕ^1.

b) Inequality (3.20) is, in turn, equivalent to the following inequality for the same problem (3.2):

$$(3.21)\qquad \begin{cases} \int_0^T \int_\Gamma \phi^2 d\Sigma \geq C_T \|\{\phi^o,\phi^1\}\|^2_{L^2(\Omega)x[H^1(\Omega)]'} \\ \text{for all } \phi^o,\ \phi^1 \in L^2(\Omega)x[H^1(\Omega)]' \text{ for which the left hand side is finite} \end{cases}$$

Proof of Theorem 3.4 (Modification of proof of Theorem 3.1)

Step 0 We introduce the operator N_1 (translation by $\lambda = 1$ of corresponding elliptic problem): continuous $L^2(\Gamma) \to H^{3/2}(\Omega)$ defined by

(3.22) $$w = N_1 g \iff \begin{cases} (\Delta-1)w = 0 & \text{in } \Omega \\ \dfrac{\partial w}{\partial \nu} = g & \text{on } \Gamma \end{cases}$$

Let N_1^* denote the adjoint $(N_1 v,u)_{L^2(\Omega)} = (v, N_1^* u)_{L^2(\Gamma)}$

The counterpart of Lemma 3.2 is now (see [ ]):

Lemma 3.5 For $f \in \mathscr{D}(A)$

(3.23) $$N_1^*(A+I)\, f = f|_\Gamma \quad \square$$

Proof As in Lemma 3.2, let $g \in L^2(\Gamma)$ and compute by Green second theorem

$$\begin{aligned} -(N_1^*(A+I)f,g)_\Gamma &= -((A+I)f,\ N_1 g)_\Omega = ((\Delta-1)f,\ N_1 g)_\Omega \\ &= (f,\ (\Delta-1)(N_1 g))_\Omega + (\frac{\partial f}{\partial \nu},\ N_1 g) - (f,\ \frac{\partial(N_1 g)}{\partial \nu})_\Gamma \\ &= -(f,g)_\Gamma \quad \square \end{aligned}$$

Step 1 In place of (3.7) we have now

(3.24) $$\begin{vmatrix} y(T) \\ \\ y_t(T) \end{vmatrix} = \mathscr{L}_T v = \begin{vmatrix} (A+I) \displaystyle\int_0^T S(T-t)N_1 v(t)dt \\ \\ (A+I) \displaystyle\int_0^T C(T-t)N_1 v(t)dt \end{vmatrix}$$

(3.25) $$\mathscr{D}(\mathscr{L}_T) = \{v \in L^2(\Sigma):\ [y(T),\ y_t(T)] \in H^1(\Omega)xL^2(\Omega)\}$$

Step 2 Exact controllability of problem (1.1) at time T on the space $H^1(\Omega)xL^2(\Omega)$ by means of $L^2(\Sigma) = L^2(0,T;L^2(\Gamma))$ - controls v is equivalent to ontoness of $\mathscr{L}_T$

(3.26) $$\mathscr{L}_T:\ L^2(\Sigma) > \mathscr{D}(\mathscr{L}_T) \overset{\text{onto}}{\rightarrow} H^1(\Omega)xL^2(\Omega)$$

in turn equivalent to

$$\left\| \mathscr{L}_T^* \begin{vmatrix} z_0 \\ z_1 \end{vmatrix} \right\|_{L^2(\Sigma)} > C_T' \|\{z_0,\ z_1\}\|_{H^1(\Omega)xL^2(\Omega)}$$

$$(3.27)\qquad C'_T > 0, \text{ all } \{z_o, z_1\} \in \mathscr{D}(\mathscr{L}_T^*)$$

for the adjoint operator: $(\mathscr{L}_T g, z)_{H^1(\Omega)\times L^2(\Omega)} = (g, \mathscr{L}_T^* z)_{L^2(\Sigma)}$.

The counterpart of Lemma 3.3 is now

Lemma 3.6 Property (3.27) (and hence property (3.26)) is equivalent to inequality (3.20) or inequality (3.21) of Theorem 3.4.

Proof of Lemma 3.6 Starting from (3.24) and proceeding as in Lemma 3.3, using this time $H^1(\Omega) = \mathscr{D}((A+I)^{1/2})$ (same norm, see (2.5c)), we find (counterpart of (3.12)):

$$\mathscr{L}_T^* \begin{vmatrix} z_o \\ z_1 \end{vmatrix} = N_1^*(A+I)[C(t-T)z_1 - S(t-T)(A+I)z_o]$$

$$(3.28)\qquad = [C(t-T)z_1 - S(t-T)(A+I)z_o]_\Gamma$$

by Lemma 3.5. Writing according to (2.9)

$$(3.29)\qquad z_o = \overline{z}_o + \overline{k}_o;\ \mathscr{M}(\overline{z}_o) = 0,\ \mathscr{M}(z_o) = \overline{k}_o = \text{const} \in \mathscr{N}(A)$$

and using $A\overline{k}_o = 0$, $S(t)\overline{k}_o \equiv t\overline{k}_o$, we have

$$(\mathscr{L}_T^* \begin{vmatrix} z_o \\ z_1 \end{vmatrix})(t) = N_1^*(A+I)[C(t-T)z_1 - S(t-T)(A+I)\overline{z}_o - (t-T)\overline{k}_o]$$

$$(3.30)\qquad = [C(t-T)z_1 - S(t-T)(A+I)\overline{z}_o]_\Gamma - (t-T)\overline{k}_o$$

On the other hand, the solution ϕ of problem (3.2) is

$$(3.31)\qquad \begin{cases} \phi(t;\phi^o,\phi^1) = C(t-T)\phi^o + S(t-T)\phi^1 \\ \qquad\qquad = C(t-T)\overline{\phi^o} + \mathscr{M}(\phi^o) + S(t-T)\phi^1 \\ \phi^o = \overline{\phi^o} + \overline{c^o},\ \mathscr{M}(\overline{\phi^o}) = 0,\ \mathscr{M}(\phi^o) = \overline{c^o} = \text{const} \in \mathscr{N}(A) \end{cases}$$

Comparing (3.30) with (3.31) yields

$$(3.32) \qquad \left(\mathscr{L}_T^* \begin{vmatrix} z_o \\ z_1 \end{vmatrix}\right)(t) = [\phi(t;\ \phi^o = z_1,\ \phi^1 = -(A+I)z_o)]_\Gamma$$

$$(3.33) \qquad \phi^o = z_1 \qquad \phi^1 = -(A+I)z_o$$

Moreover, by (2.5c); $H^1(\Omega) = \mathscr{D}((A+I)^{1/2})$ (same norms) and thus

$$\|\{z_o,\ z_1\}\|_{H^1(\Omega)\times L^2(\Omega)} = \|\{-(A+I)^{-1}\phi^1,\ \phi^o\}\|_{H^1(\Omega)\times L^2(\Omega)}$$

$$(3.34) \qquad = \|\{\phi^o,\ \phi^1\}\|_{L^2(\Omega)\times[H^1(\Omega)]'}$$

Thus, using (3.32) and (3.34) we see that inequality (3.27) becomes precisely inequality (3.21) of Theorem 3.4. Differentiating (3.31) in t

$$(3.35) \qquad \phi_t(t;\ \phi^o,\ \phi^1) = C(t-T)\phi^1 - S(t-T)A\phi^o$$

Comparing (3.30) with (3.35) gives

$$(3.36) \qquad \left(\mathscr{L}_T^* \begin{vmatrix} z_o \\ z_1 \end{vmatrix}\right)(t) = [\phi_t(t,\ \overline{\phi^o} = A^{-1}(A+I)\bar{z}_o,\ \phi^1 = z_1)]_\Gamma - (t-T)\bar{k}_o$$

$$(3.37) \qquad A\overline{\phi^o} = (A+I)\bar{z}_o \qquad\qquad \phi^1 = z_1$$

By (3.29) and (3.37)

$$(3.38) \qquad z_o = (A+I)^{-1}A\overline{\phi^o} + \bar{k}_o; \qquad \bar{z}_o = (A+I)^{-1}A\overline{\phi^o}$$

Claim. We have

$$(3.39) \qquad \|z_o\|_{H^1(\Omega)} > C\|\overline{\phi^o} + \bar{k}_o\|_{H^1(\Omega)}$$

Proof of Claim. In fact

$$(3.40) \qquad \|z_o\|^2_{H^1(\Omega)} = (\bar{z}_o + \bar{k}_o,\ \bar{z}_o + \bar{k}_o)_{H^1(\Omega)} = \|\bar{z}_o\|^2_{H^1(\Omega)} + \|\bar{k}_o\|^2_{L^2(\Omega)}$$

since via (3.38) and (2.5c): $H^1(\Omega) = \mathscr{D}((A+I)^{1/2})$

$$(3.41) \qquad (\bar{z}_o,\ \bar{k}_o)_{H^1(\Omega)} = (A\overline{\phi^o},\ \bar{k}_o)_{L^2(\Omega)} = 0$$

(orthogonality of $L_0^2(\Omega)$ and $\mathcal{N}(A)$). Then (3.40) becomes via (3.38)

$$\|z_o\|^2_{H^1(\Omega)} = \|(A+I)^{-1}A\overline{\phi^o}\|^2_{\mathcal{D}((A+I)^{1/2})} + \|\bar{k}_o\|^2_{L^2(\Omega)}$$

$$= \|(A+I)^{-1/2}A\overline{\phi^o}\|^2_{L^2(\Omega)} + \|\bar{k}_o\|^2_{L^2(\Omega)}$$

$$\geq c\|A^{1/2}\overline{\phi^o}\|^2_{L^2(\Omega)} + \|\bar{k}_o\|^2_{L^2(\Omega)}$$

(taking $0 < c < 1$, w.l.o.g.)

$$\geq c\{\|A^{1/2}\overline{\phi^o}\|^2_{L^2(\Omega)} + \|\bar{k}_o\|^2_{L^2(\Omega)}\} = c\{\|\overline{\phi^o}\|^2_{H^1(\Omega)} + \|\bar{k}_o\|^2_{H^1(\Omega)}\}$$

$$(3.42) \qquad = c\|\overline{\phi^o} + \bar{k}_o\|^2_{H^1(\Omega)}$$

where in the last step, we have used $(\overline{\phi^o}, \bar{k}_o)_{H^1(\Omega)} = (A^{1/2}\overline{\phi^o}, \bar{k}_o) = 0$. Thus, the claim is proved.

By virtue of (3.39) and (3.37) we obtain for $z_o = \bar{z}_o + \bar{k}_o$:

$$(3.43) \qquad \|\{z_o, z_1\}\|_{H^1(\Omega)\times L^2(\Omega)} \geq C\|\{\overline{\phi^o} + \bar{k}_o, \phi^1\}\|_{H^1(\Omega)\times L^2(\Omega)}$$

Hence, if we set $\phi^o = \overline{\phi^o} + \bar{k}_o$, i.e. $\mathcal{M}(\phi^o) = \overline{c^o} = \bar{k}_o = \mathcal{M}(z^o)$ by (3.29) and (3.31)), then Eqts (3.43) and (3.36) become respectively

$$(3.44) \qquad \|\{z_o, z_1\}\|_{H^1(\Omega)\times L^2(\Omega)} \geq C\|\{\phi^o, \phi^1\}\|_{H^1(\Omega)\times L^2(\Omega)}$$

$$\left(\mathcal{L}_T^*\begin{vmatrix} z_o \\ z_1 \end{vmatrix}\right)(t) = [\phi_t(t; \overline{\phi^o}, \phi^1)]\big|_\Gamma - (t-T)\,\mathcal{M}(\phi^o)$$

$$(3.45) \qquad = [\phi_t(t; \phi^o, \phi^1)]\big|_\Gamma - (t-T)\,\mathcal{M}(\phi^o)$$

Thus, by virtue of (3.44) - (3.45), we see that inequality (3.27) becomes precisely inequality (3.20) with $C_T = C_T'^2$. Theorem 3.4 is proved □

3.3 Completion of the proof of Theorem 1.1 and Theorem 1.2

Let $\Gamma_0 \neq \emptyset$. Then, Theorem 1.1 follows by simply combining Theorem 3.1 and Theorem 2.1. Let now $\Gamma_0 = \emptyset$ and $\mathcal{M}(y^o) = 0, \mathcal{M}(y^1) = 0$. In this case, the dual variables z_o and z_1 in subsection 3.2 satisfy likewise $\mathcal{M}(z^o) = 0 \mathcal{M}(z_1) = 0$. Then, $\mathcal{M}(\varphi^o) = \mathcal{M}(z^o) = 0$, see below (3.43), and $\mathcal{M}(\varphi^1) = \mathcal{M}(z_1) = 0$ see (3.37). Hence, Theorem 3.4 and Theorem 2.1 with $\mathcal{M}(\varphi^o) = \mathcal{M}(\varphi^1) = 0$ combined yield Theorem 1.1.

To prove Theorem 1.2, one replaces Theorem 2.1 with Theorem 2.8 in the above arguments $\square$

4. Exact controllability on $L^2(\Omega)\times[H^1_{\Gamma_0}(\Omega)]'$ with controls $v \in H^{-1}(0,T;L^2(\Gamma_1))$. Equivalence to a-priori inequalities. Ontoness approach of the solution operator.

Again we shall treat the two cases $\Gamma_0 \neq \emptyset$ and $\Gamma_0 = \emptyset$ separately.

4.1 The case $\Gamma_0 \neq \emptyset$

Our goal is to prove the following result.

Theorem 4.1 Problem (1.1) is exactly controllable on the space $L^2(\Omega)\times[H^1_{\Gamma_0}(\Omega)]'$ over the time interval $[0,T]$, $0 < T < \infty$, by means of controls $v \in H^{-1}(0,T;\ L^2(\Gamma_1))$ if and only if the following inequality holds

$$(4.1)\quad \begin{cases} \displaystyle\int_0^T\int_{\Gamma_1} \phi^2 + \phi_t^2\, d\Sigma_1 > C_T \|\{\phi^o,\phi^1\}\|^2_{\mathcal{D}(A^{1/2})\times L^2(\Omega)} = C_T\, E(0) \\ E(0) \text{ equivalent to } \|\{\phi^o,\phi^1\}\|^2_{H^1_{\Gamma_0}(\Omega)\times L^2(\Omega)} \\ \text{for all } \{\phi^o,\phi^1\} \in H^1_{\Gamma_0}(\Omega)\times L^2(\Omega) \text{ for which the left hand side of the above} \\ \text{inequality is finite} \end{cases}$$

where C_T is a positive constant depending on T, but not on ϕ^o,ϕ^1 and where ϕ solves the homogeneous backward problem (3.2) □

Proof of Theorem 4.1 Let $v \in H^{-1}(0,T;\ L^2(\Gamma_1))$ i.e. let

$$(4.2)\qquad v = v_1 + \frac{\partial v_2}{\partial t}, \qquad v_1,\ v_2 \in L^2(0,T;\ L^2(\Gamma_1)) \equiv L^2(\Sigma_1).$$

The exact controllability in question is equivalent to the property that the solution operator $\mathcal{L}_T$ in (3.7), rewritten here as

$$(4.3)\qquad \begin{vmatrix} y(T) \\ y_t(T) \end{vmatrix} = \mathcal{L}_T \begin{vmatrix} v_1 \\ v_2 \end{vmatrix} = \begin{vmatrix} A\int_0^T S(T-t)\tilde{N}[v_1(t) + \frac{\partial v_2}{\partial t}(t)]dt \\ A\int_0^T C(T-t)\tilde{N}[v_1(t) + \frac{\partial v_2}{\partial t}(t)]dt \end{vmatrix},$$

satisfies the ontoness condition

$$(4.4)\qquad \mathcal{L}_T:\ L^2(\Sigma_1)\times L^2(\Sigma_1) > \mathcal{D}(\mathcal{L}_T) \overset{\text{onto}}{\to} L^2(\Omega)\times[H^1_{\Gamma_0}(\Omega)]'$$

$$(4.5)\qquad \mathcal{D}(\mathcal{L}_T) = \{[v_1,\ v_2] \in L^2(\Sigma_1)\times L^2(\Sigma_1):\ [y(T),\ y_t(T)] \in L^2(\Omega)\times[H^1_{\Gamma_0}(\Omega)]'\}$$

or equivalently (cf (2.5a)):

$$\left\| \mathscr{L}_T^* \begin{vmatrix} z_0 \\ z_1 \end{vmatrix} \right\|_{L^2(\Sigma_1)\times L^2(\Sigma_1)} \geq C_T' \|\{z_0, z_1\}\|_{L^2(\Omega)\times[\mathscr{D}(A^{1/2})]'} \tag{4.6}$$

for some $C_T' > 0$ and all $[z_0, z_1] \in \mathscr{D}(\mathscr{L}_T^*)$

By proceeding as in Lemma 3.3, we obtain now

$$\left(\mathscr{L}_T \begin{vmatrix} v_1 \\ v_2 \end{vmatrix}, \begin{vmatrix} z_0 \\ z_1 \end{vmatrix}\right)_{L^2(\Omega)\times[H^1_{\Gamma_0}(\Omega)]'} = \left(A \int_0^T S(T-t)\tilde{N}[v_1(t) + \frac{\partial v_2}{\partial t}(t)]dt, z_o\right)_{L^2(\Omega)}$$

$$+ \left(\int_0^T C(T-t)\tilde{N}[v_1(t) + \frac{\partial v_2}{\partial t}(t)]dt, z_1\right)_{L^2(\Omega)}$$

$$= (v_1, \tilde{N}^*AS(T-t)z_o + \tilde{N}^*C(T-t)z_1)_{L^2(\Sigma_1)}$$

$$+ \left(\frac{\partial v_2}{\partial t}, \tilde{N}^*AS(T-t)z_o + \tilde{N}^*C(T-t)z_1\right)_{L^2(\Sigma_1)} \tag{4.7}$$

(We now restrict to $v_2 \in H^1_0(0,T; L^2(\Gamma_1))$, and thus obtain from (4.7))

$$= (v_1, \tilde{N}^*AS(T-t)z_o + \tilde{N}^*C(T-t)z_1)_{L^2(\Sigma_1)}$$

$$+ \left(v_2, -\frac{d}{dt}[\tilde{N}^*AS(T-t)z_o + \tilde{N}^*C(T-t)z_1\right)_{L^2(\Sigma_1)}$$

$$= \left(\begin{vmatrix} v_1 \\ v_2 \end{vmatrix}, \begin{vmatrix} \tilde{N}^*AS(T-t)z_o + \tilde{N}^*C(T-t)z_1 \\ -\frac{d}{dt}[\tilde{N}^*AS(T-t)z_o + \tilde{N}^*C(T-t)z_1] \end{vmatrix} \right)_{L^2(\Sigma_1)\times L^2(\Sigma_1)} \tag{4.8}$$

Thus

$$\left(\mathscr{L}_T^* \begin{vmatrix} z_o \\ z_1 \end{vmatrix}\right)(t) = \begin{vmatrix} \tilde{N}A[C(t-T)A^{-1}z_1 - S(t-T)z_o] \\ -\frac{d}{dt}\tilde{N}^*A[C(t-T)A^{-1}z_1 - S(t-T)z_o] \end{vmatrix} \tag{4.9}$$

Comparing (4.8) with the solution ϕ of problem (3.2) given by (3.13) - (3.14), we find

$$(4.10)\qquad \left(\mathcal{L}_T^*\begin{vmatrix} z_o \\ z_1\end{vmatrix}\right)(t) = \begin{vmatrix} \phi(t;\phi^o,\phi^1)|_{\Gamma_1} \\ -\phi_t(t;\phi^o,\phi^1)|_{\Gamma_1}\end{vmatrix}$$

where

$$(4.11)\qquad \phi^o = A^{-1}z_1 \qquad\qquad \phi^1 = -z_o$$

Thus, by (4.10)

$$(4.12)\qquad \left\|\mathcal{L}_T^*\begin{vmatrix} z_o \\ z_1\end{vmatrix}\right\|^2_{L^2(\Sigma_1)xL^2(\Sigma_1)} = \int_{\Sigma_1} \phi^2 + \phi_t^2\, d\Sigma_1.$$

Moreover by (4.11)

$$(4.13)\qquad \begin{aligned} \|\{z_o, z_1\}\|_{L^2(\Omega)x[\mathcal{D}(A^{1/2})]'} &= \|\{-\phi^1, A\phi^o\}\|_{L^2(\Omega)x[\mathcal{D}(A^{1/2})]'} \\ &= \|\{\phi^o, -\phi^1\}\|_{\mathcal{D}(A^{1/2})xL^2(\Omega)} \end{aligned}$$

Using (4.12) - (4.13), we see that inequality (4.6) becomes precisely inequality (4.1) with $C_T = C'^2_T$

4.2 The case $\Gamma_0 = \emptyset$

The counterpart of Theorem 4.1 is now

Theorem 4.2 Problem (1.1) is exactly controllable on the space $L^2(\Omega)x[H^1(\Omega)]'$ over the time interval $[0,T]$, $0 < T < \infty$, by means of controls $v \in H^{-1}(0,T; L^2(\Gamma))$ if and only if the following inequality holds for the solution ϕ of problem (3.2) with $\Gamma_0 = \emptyset$:

$$(4.14)\qquad \begin{cases} \int_0^T\int_\Gamma \phi^2 + \phi_t^2\, d\Sigma > C_T\|\{\phi^o, \phi^1\}\|^2_{H^1(\Omega)xL^2(\Omega)} \\ \text{for all } \phi^o, \phi^1 \in H^1(\Omega)xL^2(\Omega) \text{ for which the left hand side of the above inequality is finite} \end{cases}$$

where C_T is a positive constant depending on T but not on ϕ^o, ϕ^1 □

Proof of Theorem 4.2 Let v be as in (4.2). The operator $\mathcal{L}_T$ is now (see (3.24)):

$$(4.15)\qquad \begin{vmatrix} y(T) \\ y_t(T)\end{vmatrix} = \mathcal{L}_T\begin{vmatrix} v_1 \\ v_2\end{vmatrix} = \begin{vmatrix} (A+I)\int_0^T S(T-t)N_1[v_1(t) + \frac{\partial v_2}{\partial t}(t)]\, dt \\ (A+I)\int_0^T C(T-t)N_1[v_1(t) + \frac{\partial v_2}{\partial t}(t)]dt\end{vmatrix},$$

counterpart of (4.3). The exact controllability in question is equivalent to the ontoness property

(4.16) $\mathcal{L}_T : L^2(\Sigma)\times L^2(\Sigma) \supset \mathcal{D}(\mathcal{L}_T) \xrightarrow{\text{onto}} L^2(\Omega)\times[H^1(\Omega)]'$

or equivalently

(4.17) $\left\| \mathcal{L}_T^* \begin{vmatrix} z_o \\ z_1 \end{vmatrix} \right\|_{L^2(\Sigma)\times L^2(\Sigma)} \geq C_T' \| \{z_o, z_1\} \|_{L^2(\Omega)\times[H^1(\Omega)]'}$

for some $C_T' > 0$ and all $[z_o, z_1] \in \mathcal{D}(\mathcal{L}_T^*)$

Using this time $\mathcal{D}((A+I)^{1/2}) = H^1(\Omega)$ (cf 2.5c), same norms, and proceeding as in Theorem 4.1, we likewise obtain

(4.18) $\left(\mathcal{L}_T^* \begin{vmatrix} z_o \\ z_1 \end{vmatrix}\right)(t) = \begin{vmatrix} \phi(t;\, \phi^o, \phi^1)|_\Gamma \\ \\ -\phi_t(t;\, \phi^o, \phi^1)|_\Gamma \end{vmatrix}$

(4.19) $\phi^o = (A+I)^{-1} z_1 \qquad\qquad \phi^1 = -z_o,$

counterparts of (4.10) - (4.11). Thus, by (4.18)

(4.20) $\left\| \mathcal{L}_T^* \begin{vmatrix} z_o \\ z_1 \end{vmatrix} \right\|^2_{L^2(\Sigma)\times L^2(\Sigma)} = \int_\Sigma \phi^2 + \phi_t^2 \, d\Sigma$

counterpart of (4.1). Moreover, by (4.19) and (2.5c)

$$\|\{z_o, z_1\}\|_{L^2(\Omega)\times[H^1(\Omega)]'} = \|\{-\phi^1, (A+I)\phi^o\}\|_{L^2(\Omega)\times[\mathcal{D}((A+I)^{1/2})]'}$$

$$= \|\{\phi^o, \phi^1\}\|_{H^1(\Omega)\times L^2(\Omega)},$$

(4.21)

counterpart of (4.13). Then, by virtue of (4.20) - (4.21), we see that inequality (4.17) becomes precisely inequality (4.14) with $C_T = C_T'^2$. Theorem 4.2 is proved □

4.3 Completion of the proof of Theorem 1.3

Let $\Gamma_0 \neq \emptyset$. Then, Theorem 1.3 follows by simply combining Theorem 4.1 with Theorem 2.5. If, instead, $\Gamma_0 = \emptyset$, then Theorem 4.2 and Theorem 2.6 combined produce Theorem 1.3 □

5. Exact controllability on $L^2(\Omega)x[H^1(\Omega)]'$ in the Neumann case ($\Gamma_0 = \emptyset$) in the absence of geometrical conditions on Ω.

5.1 Equivalence to an a-priori inequality. Ontoness approach.

We now establish exact controllability of problem (1.1) in the Neumann case ($\Gamma_0 = \emptyset$) on the space $L^2(\Omega)x[H^1(\Omega)]'$ without requiring geometrical conditions on Ω. We shall see that in order to achieve this goal, a larger class of controls v in (1.1d) (with $\Gamma_0 = \emptyset$) is needed, which moreover possesses a special structure. This class is introduced as follows. Given Ω, we fix a point x_o in R^n and set

$$(5.1)\qquad \begin{aligned} &a) \quad \Gamma(x^o) = \{x \in \Gamma: \ (x - x^o)\cdot\nu(x) > 0\} \\ &b) \quad \Gamma_*(x^o) = \Gamma\backslash\Gamma(x^o) \end{aligned}$$

as in (1.14). Then, the class of controls v in (1.1d) with $\Gamma_0 = \emptyset$ is defined by

$$(5.2)\qquad v = \begin{cases} v_1 + \dfrac{\partial v_2}{\partial t}, & v_1, v_2 \in L^2(0,T;\ \Gamma(x^o)) \\ v_3, & v_3 \in L^2(0,T;\ H^{-1}(\Gamma_*(x^o))) \end{cases}$$

We shall henceforth set in this section the following notation

$$(5.3)\qquad L^2(\Sigma(x^o)) \equiv L^2(0,T;\ L^2(\Gamma(x^o));$$

In the main statement below, we shall need the concept of tangential gradient $\nabla_\sigma\phi$ for a function $\phi \in C^1(\overline{\Omega})$ on Γ (or part thereof; in our case below: $\Gamma_*(x^o)$).

At each point of Γ, (sufficiently smooth) consider the unit outward normal ν and a, say, orthogonal system of unit vectors $\tau_1, \dots \tau_{n-1}$ on the tangent plane. We have

$$(5.4)\qquad \nabla\phi = (\nabla\phi\cdot\nu)\nu + \sum_{i=1}^{n-1} (\nabla\phi\cdot\tau_i)\tau_i = \frac{\partial\phi}{\partial\nu}\nu + \sum_{i=1}^{n-1} \frac{\partial\phi}{\partial\tau_i} z_i.$$

Let, in addition $\frac{\partial\phi}{\partial\nu} = 0$ on Γ (or part thereof), so that

$$(5.4)\qquad \begin{cases} \phi_i = \sigma_i\phi \\ \sigma_i = \text{first order tangential operator on } \Gamma \end{cases}$$

Then we define

$$(5.5)\qquad \|\nabla\phi\|^2_{L^2(\Gamma)} = \sum_{i=1}^{n-1} |\sigma_i\phi|^2 \equiv \|\nabla_\sigma\phi\|^2_{L^2(\Gamma)}$$

The main result of this section is

<u>Theorem</u> 5.1 Problem (1.1) with $\Gamma_0 = \emptyset$ is exactly controllable on the space $L^2(\Omega)\times[H^1(\Omega)]'$ over the time interval $[0,T]$, $0 < T < \infty$, by means of controls v given by (5.2) if and only if the following inequality holds for the solution ϕ of problem (3.2) with $\Gamma_0 = \emptyset$:

$$(5.6)\quad \begin{cases} \int_0^T \int_{\Gamma(x^o)} \phi_t^2 \, d\Gamma(x^o)dt + \int_0^T \int_{\Gamma_*(x^o)} |\nabla_\sigma \phi|^2 d\Gamma_*(x^o)dt + \int_0^T \int_\Gamma \phi^2 d\Sigma \\ \qquad \geq C_T \, \|\{\phi^o, \phi^1\}\|^2_{H^1(\Omega)\times L^2(\Omega)} \\ \text{for all } \{\phi^o, \phi^1\} \in H^1(\Omega)\times L^2(\Omega) \text{ for which the left hand side is finite} \end{cases}$$

where C_T is a positive constant depending on T but not on ϕ^o, ϕ^1.

<u>Proof of Theorem</u> 5.1 (Modification of proof of Theorem 1.3)
<u>Step</u> 0 We introduce the operator $\tilde{N}_1 : L^2(\Gamma(x^o))\times H^{-1}(\Gamma_*(x^o)) \to H^{1/2}(\Omega)$ defined by

$$(5.7)\quad w = \tilde{N}_1 \begin{vmatrix} g_1 \\ g_2 \end{vmatrix} \iff \begin{cases} (\Delta - 1)w = 0 & \text{on } \Omega \\ \frac{\partial w}{\partial \nu} = g_1 & \text{on } \Gamma(x^o) \\ \frac{\partial w}{\partial \nu} = g_2 & \text{on } \Gamma_*(x^o) \end{cases}$$

Let $\tilde{N}_1^*$ be the adjoint of $\tilde{N}_1$:

$$(5.8)\quad \left(\tilde{N}_1 \begin{vmatrix} g_1 \\ g_2 \end{vmatrix}, v\right)_{L^2(\Omega)} = \left(\begin{vmatrix} g_1 \\ g_2 \end{vmatrix}, \tilde{N}_1^* v\right)_{L^2(\Gamma(x^o))\times H^{-1}(\Gamma_*(x^o))}$$

<u>Lemma</u> 5.2 For $f \in \mathcal{D}(A)$ (cf(2.1) with $\Gamma_0 = \emptyset$), we have

$$\tilde{N}_1^*(A+I)\,f = ([\tilde{N}_1^*(A+I)f]_1, [\tilde{N}_1^*(A+I)f]_2) \in L^2(\Gamma(x^o))\times H^{-1}(\Gamma_*(x^o))$$

and

$$(5.9a)\qquad [\tilde{N}_1^*(A+I)f]_1 = f|_{\Gamma(x^o)} \in L^2(\Gamma(x^o))$$

$$(5.9b)\qquad ([\tilde{N}_1^*(A+I)f]_2,\ g_2)_{H^{-1}(\Gamma_*(x^o))} = (f,\ g_2)_{L^2(\Gamma_*(x^o))}$$

$$\forall\ g_2 \in H^{-1}(\Gamma_*(x^o))$$

$$(5.9c)\qquad [\tilde{N}_1^*(A+I)f]_2 = \Lambda\Lambda^* f\Big|_{\Gamma_*(x^o)} \qquad \in H^{-1}(\Gamma_*(x^o))$$

where Λ is a first order tangential operator on $\Gamma_*(x^o)$ (with smooth coefficients) which defines an isomorphism Λ: $H^s(\Gamma_*(x^o)) \to H^{s-1}(\Gamma_*(x^o))$, with (bounded) inverse Λ^{-1}, and Λ^* is the L^2-adjoint

$$(5.10)\qquad (\Lambda u_1,\ u_2)_{L^2(\Gamma_*(x^o))} = (u_2,\ \Lambda^* u_2)_{L^2(\Gamma_*(x^o))},\ u_i \in L^2(\Gamma_*(x^o))$$

Proof of Lemma 5.2 With $[g_1,\ g_2] \in L^2(\Gamma(x^o)) \times H^{-1}(\Gamma_*(x^o))$, we proceed as in the proof of Lemma 3.5 and find by Green second theorem and (5.7)

$$-\left(\tilde{N}_1^*(A+I)f,\ \begin{vmatrix} g_1 \\ g_2 \end{vmatrix}\right)_{L^2(\Gamma(x^o))\times H^{-1}(\Gamma_*(x^o))} = -\left((A+I)f,\ \tilde{N}_1 \begin{vmatrix} g_1 \\ g_2 \end{vmatrix}\right)_{L^2(\Omega))}$$

$$= \left((\Delta-1)f,\ \tilde{N}_1 \begin{vmatrix} g_1 \\ g_2 \end{vmatrix}\right)_{L^2(\Omega)}$$

$$= -(f,g_1)_{L^2(\Gamma(x^o)} - (f,g_2)_{L^2(\Gamma_*(x^o))}$$

from which (5.9a-b) follow at once. Using

$$(5.11)\qquad (u_1,\ u_2)_{H^{-1}(\Gamma_*(x^o))} = (\Lambda^{-1}u_1,\ \Lambda^{-1}u_2)_{L^2(\Gamma_*(x^o))},\ u_i \in H^{-1}(\Gamma_*(x^o))$$

in (5.9b) yields (5.9c) by (5.10)

Step 1 In place of (4.15) we now have

$$\begin{vmatrix} y(T) \\ y_t(T) \end{vmatrix} = \mathcal{L}_T \begin{vmatrix} v_1 \\ v_2 \\ v_3 \end{vmatrix} = \begin{vmatrix} (A+I) \int_0^T S(T-t)\tilde{N}_1 \begin{vmatrix} v_1(t) + \frac{\partial v_2(t)}{\partial t} \\ v_3(t) \end{vmatrix} dt \\ (A+I) \int_0^T C(T-t)\tilde{N}_1 \begin{vmatrix} v_1(t) + \frac{\partial v_2}{\partial t}(t) \\ v_3(t) \end{vmatrix} dt \end{vmatrix}$$

$$\mathcal{D}(\mathcal{L}_T) = \{[v_1, v_2, v_3] \in [L^2(\Sigma(x^o))]^2 x L^2(0,T; H^{-1}(\Gamma_*(x^o))): \\ [y(T), y_t(T)] \in L^2(\Omega) x [H^1(\Omega)]'\}$$

Exact controllability in question is equivalent to

$$(5.13) \qquad \mathcal{L}_T : \mathcal{D}(\mathcal{L}_T) \xrightarrow{\text{ONTO}} L^2(\Omega) x [H^1(\Omega)]',$$

which in turn is equivalent to

$$(5.14) \qquad \left\| \mathcal{L}_T^* \begin{vmatrix} z_0 \\ z_1 \end{vmatrix} \right\|_{[L^2(\Sigma(x^o))]^2 x L^2(0,T; H^{-1}(\Gamma_*(x^o)))} \geq C_T' \| \{z_o, z_1\} \|_{L^2(\Omega) x [H^1(\Omega)]}$$

for some $C_T' > 0$ and all $[z_o, z_1] \in \mathcal{D}(\mathcal{L}_T^*)$

We now proceed as below (4.6) or (4.17). Setting

$$(5.15) \qquad \phi(t, \phi^o, \phi^1) = C(t-T)\phi^o + S(t-T)\phi^1$$

for the solution of problem (3.2) with

$$(5.16) \qquad \phi^o = (A+I)^{-1} z_1, \ \phi^1 = -z_o$$

as in (4.19), we find recalling (5.9a-c)

$$\left(\mathscr{L}_T\begin{vmatrix}v_1\\v_2\\v_3\end{vmatrix}, \begin{vmatrix}z_o\\z_1\end{vmatrix}\right)_{L^2(\Omega)\times[H^1(\Omega)]'} = (v_1, \phi|_{\Gamma(x^o)})_{L^2(\Sigma(x^o))}$$

$$+ (v_2, -\phi_t|_{\Gamma(x^o)})_{L^2(\Sigma(x^o))}$$

(5.16)

$$+ (v_3, \Lambda\Lambda^*\phi|_{\Gamma_*(x^o)})_{L^2(0,T;\ H^{-1}(\Gamma_*(x^o)))}$$

so that

(5.17) $$(\mathscr{L}_T^*\begin{vmatrix}z_0\\z_1\end{vmatrix})(t) = \begin{vmatrix}\phi|_{\Gamma(x^o)}\\-\phi_t|_{\Gamma(x^o)}\\\Lambda\Lambda^*\phi|_{\Gamma_*(x^o)}\end{vmatrix}$$

Recalling (5.10) for $u_1 = u_2$, we finally obtain by (5.17)

$$\left\|\mathscr{L}_T^*\begin{vmatrix}z_0\\z_1\end{vmatrix}\right\|^2_{[L^2(\Sigma(x^o))]^2\times L^2(0,T;\ H^{-1}(\Gamma_*(x^o)))} =$$

$$= \int_0^T\int_{\Gamma(x^o)} \phi^2 + \phi_t^2\, d\Gamma dt + \int_0^T\int_{\Gamma_*(x^o)} |\Lambda^*\phi|^2 d\Gamma dt$$

Moreover, as in (4.21), we have by (5.16)

(5.19) $$\|\{z_o, z_1\}\|_{L^2(\Omega)\times[H^1(\Omega)]'} = \|\{\phi^o, \phi^1\}\|_{H^1(\Omega)\times L^2(\Omega)}$$

since Λ^* is a first order tangential operator, then (5.14), (5.18), (5.19) yield (5.6) as desired □

5.2 Completion of the proof of Theorem 1.4

It suffices to combine Theorem 5.1 with Theorem 2.9 □

REFERENCES

[C.1] G. Chen, Energy decay estimates and exact boundary value controllability for the wave equation in a bounded domain, J. Math. Pures et Appliques (9) 58 (1979), 249-274.

[C.2] G. Chen, A note on the boundary stabilization of the wave equation, SIAM J. Control & Optimiz. 19 (1981), 106-113.

[G-R.1] K. Graham and D. L. Russell, Boundary value control of the wave equation, in a spherical region, SIAM J. Control 13 (1975), 174-196.

[H.1] L. F. Ho, Obserbilite frontiere de l'equation des CRAS, 302, Paris, 1986.

[J.1] F. John, On linear partial differential equations with analytic coefficients - Unique continuation of data. Comm. Pure Appl. Math 2 (1949) 209-253.

[L-1] J. Lagnese, Decay of solutions of wave equations in a bounded region with boundary dissipation J. Diff. Equats. 50 (1983), 163-182.

[L.2] J. L. Lions, Controle des systemes distribués singuliers, Gauthier Villars, 1983.

[L.3] J. L. Lions, Controlabilité exacte de systemes distribues C.R.A.S. 302, Paris (1986), 471-475.

[L.4] J. L. Lions, Controlabilite exacte de systemes distribues: remarques sur le theorie generale et les applications, Proceedings of 7th International Conference on Analysis & Optimization of Systems, Antibes, France, June 25-27, 1986; Lecture Notes in CIS, 1-13.

[L.5] J. L. Lions, Exact controllability of distributed systems. An introduction Proceedings of 25th Conference on Decision and Control, Athens, Greece, December 1986.

[L.6] J. L. Lions, Exact controllability, stabilization and perturbations, J. Von Neumann Lecture, SIAM July 1986.

[L.7] P. L. Lions, private communication.

[L-7] W. Littman, Boundary control theory for hyperbolic and parabolic partial differential equations with constant coefficients, Anneli Scuole Normale Superiore de Pisa, Classe Suarze, Serie IV, Vol. 3 (1978), 567-580.

[L-8] W. Littman, Near optimal time boundary controllability for a class of hyperbolic equations Proceedings of the IFIP WG7.2 Working conference on 'Control Systems governed by partial differential equations' held at the University of Florida, Gainesville, February 3-6, 1986, to appear in Springer Verlag series.

[L-9] W. Littman, private communication.

[L-L-T.1] I. Lasiecka, J. L. Lions, and R. Triggiani, Non homogeneous boundary value problems for second order hyperbolic operators, Journ. de Mathematiques Pures et Appliques, 65, (1986), 149-192.

[L-T.1] I. Lasiecka, and R. Triggiani, A cosine opoerator approach to modeling $L_2(0,T; L_2(\Gamma))$ - boundary input hyperbolic equations, Applied Math & Optimization 7(1981), 35-83.

[L-T.2] I. Lasiecka, and R. Triggiani, Uniform exponential energy decay of the wave equation in a bounded region with $L_2(0,\infty; L_2(\Gamma))$ - feedback control in the Dirichlet boundary conditions, J. Diff. Eqts. 66 (1987), 340-390.

[L-T.3] I. Lasiecka, and R. Triggiani, Sharp regularity results for second order hyperbolic equations of Neumann type, preprint 1986, to appear.

[R.1] D. L. Russell, Controllability and stabilizability theory for linear partial differential equations. Recent progress and open questions.

SIAM Review 20 (1978), 639-739.

[T.1] R. Triggiani, Exact boundary controllability on $L_2(\Omega)xH^{-1}(\Omega)$ for the wave equation with Dirichlet control acting on a portion of the boundary, and related problems, preprint 1986, submitted.

[T.2] R. Triggiani, Wave equation on a bounded domain with boundary dissipation: an operator approach, to appear in "Operator Methods for Optimal Control Problems" Proceedings of special session at Annual Meeting of the A.M.S. held at New Orleans, January 1986, Marcel Dekker.

Shape stabilization of wave equation

J.P. ZOLESIO

CNRS

U.S.T.L.

Place E.Bataillon

Montpellier 34060

France

C.TRUCHI

Centre de Recherche Mathematique

Ecole Nationale Superieure des Mines

de Paris, Sophia Antipolis .

Valbonne 06565

France

Introduction

Ω_t being a smooth bounded domain which moves in $\mathbb{R}^n$ with the time t, we select this motion on a time interval $[0,T]$, such that the following hyperbolic problem is well posed :

$$
\begin{aligned}
&-\Delta y + \partial^2_{tt} y = 0 \quad \text{in } Q_T \text{ the evolutionary domain in } [0,T] \times \mathbb{R}^n \\
&y(0) = y_0 \quad \text{on } \Omega_0 \qquad\qquad (0.1)\\
&\partial_t y(0) = y_1 \quad \text{on } \Omega_0 \\
&y = 0 \quad \text{on a part of the lateral boundary } \Sigma_T \text{ of } Q_T
\end{aligned}
$$

We show that the energy term $W(t)$ has its time derivative negative by the same technic as developed in J.P. Quinn and D.L. Russel [1], but including the shape derivative contribution leading to the cubic derivative introduced in J.P. Zolésio [2]. The decay of energy is obtain without any kind of absorbing condition on the boundary of Ω_t. In the first section, we formulate the problem and obtain $W'(t)$ negative by a periodical motion of Ω_t. In the second section, we get existence and uniqueness results for the problem (0.1) with smooth initial datas. Assuming additional regularity on the solution y, we prove the exponential decay of $W(t)$. The third section is devoted to numerical results.

1. Wave equation in periodical time varying domain

We shall be concerned with the classical wave equation formulated in a non cylindrical domain Q_T in $\mathbb{R}^3$. The current point being (t,x_1,x_2), the time t ranging in $[0,T]$ and at each time t, $x = (x_1,x_2)$ being in an open set Ω_t of $\mathbb{R}^2$

, we define :

$$Q_T = \bigcup_{0 < t < T} \{t\} \times \Omega_t \tag{1.1}$$

If Γ_t is the boundary of Ω_t (which is assumed to be smooth enough, say piecewise C^2, lying on one side of its boundary and bounded), the lateral boundary of Q_T will be denoted by :

$$\Sigma_T = \bigcup_{0 < t < T} \{t\} \times \Gamma_t \tag{1.2}$$

The boundary of Q_T is then composed of five parts :

$$\delta Q_T = \Sigma_T \cup \Omega_0 \cup \Omega_T \cup \Gamma_0 \cup \Gamma_T \tag{1.3}$$

At each time, n_t is the unitary normal field to Γ_t and ν is the unitary normal field to Σ_T at a point (t,x) of Σ_T . We suppose that $(t,x) \to n_t(x)$, $\Sigma_T \to \mathbb{R}^2$ is continuous, while $(t,x) \to \nu(t,x)$, $\Sigma_T \to \mathbb{R}^3$ can be discontinuous. ν can be expressed as :

$$\nu(t,x) = \frac{1}{(1+v^2(t,x))^{1/2}} \begin{bmatrix} v(t,x) \\ n_t(x) \end{bmatrix} \tag{1.4}$$

where $v(t,x)$ is the normal component of the (boundary) velocity on Γ_t , as it is usually considered in shape variation problems. We suppose that a velocity field $V = V(t,x)$ is given :

$$V \in C^0([0,T] \text{ , } C^{k,\infty}) \text{ with } C^{k,\infty} = L^\infty(\mathbb{R}^2,\mathbb{R}^2) \cap C^k(\mathbb{R}^2,\mathbb{R}^2) \tag{1.5}$$

or more generally $V(t) = V(t,.)$ can be piecewise continuous in time :

$$0 < t_1 < t_2 < \ldots < t_n = T \quad \text{and} \quad V_{|[t_{i-1}, t_i]} \in C^0([t_{i-1}, t_i] \text{ , } C^{k,\infty}) \tag{1.6}$$

V can be discontinuous at time t_i , $1 \leq i \leq n$, but do have limits on both sides.

We associate to V the transformation $T_t = T_t(V)$ defined as :

$$T_t(V) \ : \ X \to x(t,X) \text{ solution of } \frac{d}{dt}x(t,X) = V(t \text{ , } x(t,X)) \text{ , } x(0,X) = X$$

X can be looked as the Lagrangian coordinate, while x is the materiel coordinate associated to the flow V. From J.P. Zolésio [3], we know that with (1.5) or (1.6), T_t is defined for all $t \in [0,T]$.

Then we handle all the needed properties for computations : considering the perturbed domain $\Omega_t = T_t(V)(\Omega_0)$ and its boundary $\Gamma_t = T_t(V)(\Gamma_0)$, where Ω_0 is the initial domain, we have :

$$\frac{\partial}{\partial t} T_t(V)(X) = V(t,x) = V(t\,,\, T_t(V)(X))$$

$$n_t(x) = \alpha \; {}^*(DT_t)^{-1} \, . \, n_0(X) \tag{1.7}$$

where n_0 is the normal field on Γ_0 , DT_t the Jacobian matrix of T_t and α the normalisation factor.

$v = v(t,x) = V(t,x) \, . \, n_t(x)$ is the normal component of the speed on the boundary.

$\nu(t,x)$ is the normal field on the lateral boundary Σ_T on which the superficial measure can be expressed as :

$$d\Sigma = (1 + v^2(t,x))^{1/2} \, d\Gamma_t \, dt \tag{1.8}$$

where $d\Gamma_t$ is the superficial measure on Γ_t .

Periodical domain in time can be obtained by taking $v = V(t) \, . \, n_t$ periodical in time. One says that Σ_T is timelike if $|v(t,x)| < 1$ and spacelike if $|v(t,x)| \geq 1$. The domain Ω_t is in expansion if (and only if) $v(t,.) \geq 0$ on Γ_t and is in contraction if (and only if) $v(t,.) \leq 0$ on Γ_t.

Considering λ as the time period of the domain, we shall consider two parts in each period : an expansion for :

$$k\lambda \leq t < \mu + k\lambda \quad , \text{with } 0 < \mu < \lambda$$

$$v(t,x) = \gamma > 0 \quad , \forall \, x \in \Gamma_t \quad , \; 0 < \gamma < 1 \tag{1.9}$$

this expansion will be timelike, then a contraction which will be spacelike on the remaining of the time period :

$$v(t,x) = -\;(1+\alpha) < 0 \quad , \mu + k\lambda \leq t < (k+1)\lambda \;\;, \;\; \forall \, x \in \Gamma_t \;\;, \;\; \alpha > 0 \tag{1.10}$$

Obviously, to such a function v defined on Σ_T , corresponds several velocity fields V such as (1.6), the discontinuity points being the $k\lambda$ and $k\lambda + \mu$, $0 \leq k \leq n$ ($T = n\lambda$).

To get a periodical domain Ω_t, we have to impose some extra conditions on the four constants α, γ, λ and μ, see Ch. Truchi [4].

Differentiation of an integral :

Let $\Phi(t,x)$ be a "smooth enough" function defined on a neighbourhood of $\bar{Q}_T$. We classically have, for any $\varepsilon > 0$:

$$\int_{\Omega_{t+\varepsilon}} \Phi(t+\varepsilon\ , x)\ dx = \int_{\Omega_t} \Phi(t+\varepsilon\ , F_t^\varepsilon(x))\ J_t^\varepsilon(x)\ dx$$

where $F_t^\varepsilon = T_{t+\varepsilon} \circ (T_t)^{-1}$ and $J_t^\varepsilon = \det(DF_t^\varepsilon)$, $F_t^0 = Id$. But this mapping, which maps Ω_t on to $\Omega_{t+\varepsilon}$, can be expressed as :

$$F_t^\varepsilon(V) = T_\varepsilon(V_t) \quad \text{with} \quad V_t(s) = V(t+s)$$

then :

$$\partial_\varepsilon F_t^\varepsilon(V)_{|\varepsilon=0} = \partial_\varepsilon T_\varepsilon(V_t)_{|\varepsilon=0} = V_t(0) = V(t)$$

and :

$$(\partial_\varepsilon J_t^\varepsilon)_{|\varepsilon=0} = div V_t(0) = div V(t)$$

finally we get :

$$\frac{d}{dt} \int_{\Omega_t} \Phi(t,x)\ dx = \int_{\Omega_t} \partial_t \Phi(t,x)\ dx + \int_{\Omega_t} [\ \nabla\Phi(t,x).V(t,x)$$

$$+ \Phi\ div V(t,x)]\ dx$$

$$= \int_{\Omega_t} \partial_t \Phi\ dx + \int_{\Gamma_t} \Phi\ v\ d\Gamma_t \tag{1.11}$$

The "Cubic derivative" of the energy :

For any smooth function, say $\Phi \in C^\infty(\bar{Q}_T)$, we consider the energy term :

$$W(t) = \frac{1}{2} \int_{\Omega_t} (\ |\nabla_x \Phi|^2 + |\partial_t \Phi|^2\ )\ dx \tag{1.12}$$

then from (1.11) we have :

$$W'(t) = \text{Re} \int_{\Omega_t} [\ \nabla\Phi\ .\ \nabla(\partial_t \bar{\Phi}) + \partial_t \bar{\Phi}\ .\ \partial_{tt}^2 \Phi\ ]\ dx$$

$$+ \frac{1}{2} \int_{\Gamma_t} [\ |\nabla\Phi|^2 + |\partial_t \Phi|^2\ ]\ v(t)\ d\Gamma_t$$

We introduce the hyperbolic operator :

$$H\Phi = - \Delta\Phi + \partial_{tt}^2\Phi \tag{1.13}$$

which is the wave operator; then using the Stokes formula in the expression of $W'(t)$, we get :

$$W'(t) = \mathrm{Re} \int_{\Omega_t} H\Phi \,.\, \partial_t\overline{\Phi}\, dx + \int_{\Gamma_t} \left[\mathrm{Re} \frac{\partial\Phi}{\partial n_t} \,.\, \partial_t\overline{\Phi} + \frac{1}{2} \left(|\nabla\Phi|^2 + |\partial_t\Phi|^2 \right) v \right] d\Gamma_t \tag{1.14}$$

Suppose now that on a piece of the lateral boundary Σ_T , Φ verifies the homogeneous Dirichlet condition, that is :

$$\text{for } t \text{ such that} ,\ a < t < b \ ,\ \Phi(t,x) = 0 \quad \forall\, x \in \Gamma_t \tag{1.15}$$

Then, for any t , $a < t < b$, and ε such that $a < t+\varepsilon < b$ we have :

$$\Phi(t+\varepsilon\ ,\ x(t+\varepsilon\ ,\ X)) = 0 \qquad \text{on } \Gamma_t$$

By differentiation on ε , we get, at $\varepsilon = 0$:

$$\partial_t\Phi(t,x) + \nabla\Phi(t,x) \,.\, V(t,x) = 0 \qquad \text{on } \Gamma_t$$

But :

$$\nabla\Phi = \frac{\partial\Phi}{\partial n_t} \,.\, n_t$$

then, we obtain :

$$\partial_t\Phi(t,x) = - \frac{\partial\Phi(t,x)}{\partial n_t}\, v(t,x) \tag{1.16}$$

$$\forall\, t \ ,\ a < t < b \ ,\ \forall\, x \in \Gamma_t$$

In particular with (1.16) and (1.4) we get :

$$\frac{\partial}{\partial\nu}\Phi(t,x) = \frac{1}{(1+v^2)^{1/2}} \left(v\, \partial_t\Phi + \frac{\partial\Phi}{\partial n_t} \right)$$

$$= \frac{1}{(1+v^2)^{1/2}} (1-v^2) \frac{\partial\Phi}{\partial n_t} \tag{1.17}$$

$\forall\, t\,;\ a < t < b \ ,\ \forall\, x \in \Gamma_t$

Going back to the expression of $W'(t)$ we have, for any t , $a < t < b$, such that Φ verifies (1.15) :

$$\text{Re}\ \frac{\partial \Phi}{\partial n_t}\ .\ \partial_t \overline{\Phi} + \frac{1}{2}\ (\,|\nabla \Phi\,|^2 + |\partial_t \Phi\,|^2)\, v$$

$$= - \left| \frac{\partial \Phi}{\partial n_t} \right|^2 v + \frac{1}{2} \left[\left| \frac{\partial \Phi}{\partial n_t} \right|^2 + \left| \frac{\partial \Phi}{\partial n_t} \right|^2 v^2 \right] v$$

Finally, we get the cubic expression for the derivative introduced in J.P. Zolesio [5] :

$$W'(t) = \int_{\Gamma_t} (v^3(t) - v(t)) \left| \frac{\partial \Phi}{\partial n_t} \right|^2 d\Gamma_t + \text{Re} \int_{\Omega_t} H\Phi\ .\ \overline{\partial_t \Phi}\ dx \quad (1.18)$$

The idea is now to choose $v(t)$ such that the domain is periodical in time and such that the cubic is negative. In extension, it is sufficient to have $0 < v(t) < 1$ and in contraction, it turns out that no boundary condition is needed to obtain $W'(t)$ negative, this point will be developed at the next section.

The choice of Q_T such that the normal component of the speed verifies (1.9) and (1.10) with $0 < \gamma < 1$ and $\alpha > 0$ leads to an a priori estimate from which we obtain, at theorems 1 and 2, existence and uniqueness results for the non cylindrical problem (0.1), introduced at the beginnning. We just now give the idea of the technic developed at the next section. We note Σ_T^+ the part of the lateral boundary Σ_T which corresponds to the expansion of the domain and Σ_T^- the part corresponding to the contraction of the domain. So, we have $\Sigma_T = \Sigma_T^- \cup \Sigma_T^+$ (on Σ_T^+ we have the homogeneous Dirichlet condition, and on Σ_T^- we have no boundary condition) . On Σ_T^-, using directely (1.14) and (1.10), we get for $\alpha > 0$:

$$W'(t) = -\frac{1}{2} \int_{\Gamma_t} \left| \frac{\partial \Phi}{\partial n_t} - \partial_t \Phi \right|^2 d\Gamma_t - \frac{1+\alpha}{2} \int_{\Gamma_t} |\nabla_{\Gamma_t} \Phi\,|^2\, d\Gamma_t$$

$$- \frac{\alpha}{2} \int_{\Gamma_t} \left| \frac{\partial \Phi}{\partial n_t} \right|^2 d\Gamma_t + \text{Re} \int_{\Omega_t} H\Phi\ .\ \overline{\partial_t \Phi}\ dx \qquad (1.19)$$

Now, if Φ is solution of the problem (0.1), then $H\Phi = 0$ and by (1.18) and (1.19) we see that for $t \neq t_i$, we have $W'(t) < 0$ which is the required property to obtain the strong stability, that is, $W(t) \to 0$ as $t \to 0$.
This decay will be obvious on numerical experiment at section 3

2. Existence and uniqueness results

Let $T = n\lambda$ and for any interger m, u_m is the solution of an ordinary second order differential system on each period. Let $\alpha \in \mathbb{C}$ and :

$$u_m^1(t,x) = \sum_{i=1}^{m} \varphi_{i,m}^1(t)\, w_i(t,x) + \alpha \frac{t^2}{2}\, w_0(x) \tag{2.1}$$

$$\text{for } 0 < t < \lambda \ , \ x \in \Omega_t$$

and for $2 \leq k \leq n$:

$$u_m^k(t,x) = \sum_{i=1}^{m} \varphi_{i,m}^k(t)\, w_i(t,x) + \alpha \frac{t^2}{2}\, w_0(x)$$

$$+ \sum_{i=1}^{m} \varphi_{i,m}^{k-1}((k-1)\lambda)\, e_i(t-(k-1)\lambda \ , \ x) \tag{2.2}$$

$$\text{for } (k-1)\lambda < t < k\lambda \ , \ x \in \Omega_t$$

where w_0 is given in $D(\Omega_0)$ so that :

$$w_0 \in \bigcap_{0<t<T} D(\Omega_t) \tag{2.3}$$

The initial values for (2.1) are such that :

$$\varphi_{i,m}^1(0) = 0 \tag{2.4}$$

$$\dot{\varphi}_{i,m}^1(0) = 0 \tag{2.5}$$

We suppose also that the family :

$$w_i(0) \ , \ i \in \mathbb{N} \text{ is dense in } H_0^1(\Omega_0) \text{ , then in } L^2(\Omega_0) \tag{2.6}$$

The initial values for the others periods are :

$$\varphi_{i,m}^{k}((k-1)\lambda) = \varphi_{i,m}^{k-1}((k-1)\lambda) \tag{2.7}$$

$$\dot{\varphi}_{i,m}^{k}((k-1)\lambda) = \dot{\varphi}_{i,m}^{k-1}((k-1)\lambda) \tag{2.8}$$

The functions $\varphi_{m}^{k}(t) = \{\varphi_{i,m}^{k}(t)\}_{1\leq i\leq m}$ are solutions of the linear systems : $1 \leq k \leq n$

$$\int_{\Omega_t} Hu_m^k \cdot \bar{w}_j(t,x)\, dx = \int_{\Omega_t} f \cdot \bar{w}_j(t,x)\, dx \tag{2.9}$$

$$f \in L^2(Q_T) \quad , \quad 1 \leq j \leq m \quad , \quad (k-1)\lambda < t < k\lambda$$

which are equivalent to :

$$M\, \ddot{\varphi}_m^k(t) + B\, \dot{\varphi}_m^k(t) + C\, \varphi_m^k(t) = F_k(t) \quad , (k-1)\lambda < t < k\lambda \tag{2.10}$$

where the three matrix $M(t)$, $B(t)$ and $C(t)$ are λ - periodical in time and are in $C^0([0,\lambda])$ under the following assumptions :

$$w_j(t,x) \text{ is time periodical} \quad , \quad w_{j\,|Q_\lambda} \in C^3(\bar{Q}_\lambda) \tag{2.11}$$

where $Q_\lambda = \bigcup\limits_{0<t<\lambda} \{t\} \times \Omega_t$.

The right hand sides in (2.10) are given by :

$$F_1(t) = \int_{\Omega_t} [\, f - \alpha\, w_0 + \alpha \frac{t^2}{2} \Delta w_0 \,] \cdot \bar{w}_j\, dx$$

and for $k \geq 2$:

$$F_k(t) = \int_{\Omega_t} [\, f - \alpha w_0 + \alpha \frac{t^2}{2} \Delta w_0 - \sum_{i=1}^{m} He_i\, \varphi_{i,m}^{k-1}((k-1)\lambda)\,] \cdot \bar{w}_j\, dx \tag{2.12}$$

From (2.12), (2.7) and (2.8) it is obvious that the mapping $\alpha \rightarrow (\varphi_m^1, \varphi_m^2, \ldots\ldots, \varphi_m^n)$ from $\mathbb{C}$ in $\prod\limits_{k=1}^{n} C_{\mathbb{C}}^2([(k-1)\lambda\, ,\, k\lambda])^m$ is polynomial.

We now introduce the space $R(Q_T)$, where u_m will be built :

$R(Q_T)$ is the set of functions φ characterized by the five following properties :

(P1) $\varphi \in C^0(\bar{Q}_T)$, $\varphi_{|\Sigma_T^+} = 0$ and $\varphi(0) = 0$ on Ω_0 .

(P2) $\nabla \varphi$ and $\dot{\varphi}$ exist and are bounded on $\bar{Q}_T$, are continuous on each $\bar{Q}_k$ - $(k-1)\lambda \times \Gamma_0$ and $\dot{\varphi}(0) = 0$ on Ω_0 .

(P3) $\ddot{\varphi}$ exists on $\bar{Q}_T$ and is continuous on each $\bar{Q}_k$ - $(k-1)\lambda \times \Gamma_0$.

(P4) for any t , $0 < t < T$, $\Delta\varphi(t)$ is in $L^2(\Omega_t)$ and for any function h in $C^0(\bar{Q}_T)$, $t \rightarrow \int_{\Omega_t} H\varphi \cdot \bar{h}\, dx$ and $t \rightarrow \int_{\Omega_t} H\varphi \cdot \dot{\bar{\varphi}}\, dx$ are continuous on each interval $[(k-1)\lambda\ ,\ k\lambda]$ and then do have right and left side limit at each point on $[0,T]$.

(P5) $\left[\dfrac{\partial\varphi}{\partial n_t}\right]_{|\bar{\Sigma}_k}$ belongs to $C^0(\bar{\Sigma}_k)$

where we wrote :

$$Q_k = \bigcup_{(k-1)\lambda < t < k\lambda} \{t\} \times \Omega_t$$

Furthermore, we suppose that :

$$w_i(t,x)_{|\Sigma_T^+} = 0 \tag{2.13}$$

to verify that u_m lies in $R(Q_T)$.

By selecting specific functions e_i in (2.2) such that :

$$e_i \text{ belongs to } R(Q_T)\ ,\ \text{ support of } e_i \subset [0,\mu[\tag{2.14}$$

with the following property :

$$\dot{e}_i(0,x) = \dot{w}_i(\lambda,x) - \dot{w}_i(0,x) \qquad x \in \Omega_0$$

such functions e_i can be built on $[0,\lambda] \times \Omega_0$ by $e_i(t,x) = [\ \dot{w}_i(\lambda,x) - \dot{w}_i(0,x)\ ]\ t$ and then extended as elements of $R(Q_T)$ verifying (2.14).

It is important to remark that such a construction can only be done at the points $t = \lambda, 2\lambda, \ldots, k\lambda$, for which the space domain is $\Omega_0 \subset \Omega_t$, $\forall t$, so that this construction can be handle by a suitable extension to Q_1 .

Introducing $I_k =](k-1)\lambda \,,\, k\lambda[$, we consider, for any function Φ in $R(Q_T)$, the restriction $W_k(t)$ to $\bar{I}_k$ of the energy function $W(t) = \frac{1}{2} \int_{\Omega_t} (|\nabla \Phi|^2 + |\dot{\Phi}|^2)\, dx$. We shall use the :

Lemma 1 : $\forall\, \Phi \in R(Q_T)$, $W : [0,T] \to \mathbb{R}^+$ is continuous

Proof :

It is enough to verify this continuity at $t = \lambda$:

$$W(\lambda+t) = \frac{1}{2} \int_{\Omega_{\lambda+t}} (|\nabla \Phi|^2 + |\dot{\Phi}|^2)(\lambda+t \,,\, x)\, dx$$

By a change of variable $T_t : \Omega_\lambda \to \Omega_{\lambda+t}$ we get :

$$W(\lambda+t) = \int_{\Omega_{\lambda+t}} f(t,x)\, dx$$

where :

$$f(t,x) = \frac{1}{2} (|\nabla \Phi|^2 + |\dot{\Phi}|^2)(\lambda+t \,,\, T_t(x))\, J_t(x)$$

$\forall\, x \in (\Omega_\lambda - \Gamma_\lambda)$, $f(t,x) \to \frac{1}{2} (|\nabla \Phi|^2 + |\dot{\Phi}|^2)(\lambda,x)$ when $t \to 0$ and $\exists$ $M > 0$, such that $|f(t,x)| \leq M$ for $(t,x) \in Q_T$.

Then, the conclusion derives from the Lebesgue convergence theorem.

Obviously, $\Phi \in R(Q_T)$, $W_k \in C^1(\bar{I}_k)$ and the direct computation (1.14) leads to :

$$W'_k(t) = \mathrm{Re} \int_{\Omega_t} H\Phi \,.\, \dot{\bar{\Phi}}\, dx$$

$$+ \mathrm{Re} \int_{\Gamma_t} \left[\frac{\partial \Phi}{\partial n_t} \,.\, \dot{\bar{\Phi}} + \frac{1}{2} (|\nabla \Phi|^2 + |\dot{\Phi}|^2)\, v \right] d\Gamma_t \tag{2.15}$$

Now, each time period $\bar{I}_k = \bar{A}_k \cup \bar{B}_k$ with $\bar{A}_k = [(k-1)\lambda \,,\, (k-1)\lambda+\mu]$ and $\bar{B}_k = [(k-1)\lambda+\mu \,,\, k\lambda]$.

Now using (1.18) and (1.19) and some obvious considerations, we get :

for $t \in A_k$:

$$W'_k(t) + \frac{1}{2} \int_{\Gamma_t} (v - v^3) \left| \frac{\partial \Phi}{\partial n_t} \right|^2 d\Gamma_t = \mathrm{Re} \int_{\Omega_t} H\Phi \, . \, \dot{\bar{\Phi}} \, dx \qquad (2.16)$$

and for $t \in B_k$:

$$W'_k(t) + \frac{\alpha}{2} \int_{\Gamma_t} \left(\left| \frac{\partial \Phi}{\partial n_t} \right|^2 + |\dot{\Phi}|^2 \right) d\Gamma_t + \frac{1+\alpha}{2} \int_{\Gamma_t} |\nabla_{\Gamma_t} \Phi|^2 \, d\Gamma_t$$

$$\leq \mathrm{Re} \int_{\Omega_t} H\Phi \, . \, \dot{\bar{\Phi}} \, dx \qquad (2.17)$$

$W'(t)$ is discontinuous over $[0, T]$, continuous on each I_k, and we get the :

Lemma 2 : For each Φ in $R(Q_T)$, we have :

$$W(t) = W(0) + \int_0^t W'(s) \, ds \quad \text{for } t \in [0, T] \qquad (2.18)$$

Proof :

We consider the two parts A_k and B_k of the period I_k . For $t \in \bar{B}_k$, Φ is in $C^1(\bar{Q}_{B_k})$ then if $t \in \bar{B}_k$, obviously, we have :

$$W(t) = W((k-1)\lambda + \mu) + \int_{(k-1)\lambda + \mu}^t W'(s) \, ds \qquad (k-1)\lambda \leq t \leq k\lambda$$

Now, for t , $(k-1)\lambda < t \leq (k-1)\lambda + \mu$, we first introduce an arbitrary positive ε , and :

$$W(t) = W((k-1)\lambda + \varepsilon) + \int_{(k-1)\lambda + \varepsilon}^t W'(s) \, ds$$

In this situation $W'(s)$ is given by (2.16) and by Lemma 1, and P4, P5 we can get the limit when ε goes to zero.

To obtain a suitable a priori estimate from (2.16) and (2.17) we suppose that for $t \in \bar{A}_k$, $v \leq -(1+\alpha)$ on Σ_T^- , for some number $\alpha > 0$ (which means a fast contraction) and on $\bar{B}_k$, $v(t) \in [a, b]$ for some $0 < a < b < 1$ (which means a slow expansion) so that $v - v^3 > c > 0$. Now introducing

$\beta = Min(\frac{c}{2}, \frac{\alpha}{2})$, from (2.16) and (2.17) we get :

$$W'(t) + \beta \int_{\Gamma_t} \left| \frac{\partial \Phi}{\partial n_t} \right|^2 d\Gamma_t \leq \text{Re} \int_{\Omega_t} H\Phi \, . \, \dot{\bar{\Phi}} \, dx \qquad t \in [0,T] \quad (2.19)$$

In particular $W'(t) \leq \text{Re} \int_{\Omega_t} H\Phi \, . \, \dot{\bar{\Phi}} \, dx$, integrating twice in time, we get the :

Lemma 3 : $\forall \, \Phi \in R(Q_T)$:

$$\int_0^T W(t) \, dt \leq TW(0) + \text{Re} \int_0^T (T-t) \int_{\Omega_t} H\Phi \, . \, \dot{\bar{\Phi}} \, dx \, dt \qquad (2.20)$$

By using the Cauchy Schwartz inequality in (2.20), we obtain the following estimate, which will be useful, for uniqueness purpose :

Lemma 4 : $\forall \, \Phi \in R(Q_T)$, we have :

$$\int_0^T W(t) \, dt \leq T^2 \, \|H\Phi\|^2_{L^2(Q_T)} \qquad (2.21)$$

Integrating once (2.19) in time, we get the :

Lemma 5 : $\forall \, \Phi \in R(Q_T)$:

$$\int_{\Sigma_T} \left| \frac{\partial \Phi}{\partial n_t} \right|^2 d\Gamma_t \, dt \leq \beta^{-1} \, \text{Re} \int_{Q_T} H\Phi \, . \, \dot{\bar{\Phi}} \, dx \, dt \qquad (2.22)$$

From (2.17) multiplied by the function $\chi_-(t)$ (equal to 1 if v is negative, zero if v is positive) and integrated in time, we get :

$$\frac{\alpha}{2} \int_{\Sigma_T} \chi_-(t) \, [\, |\dot{\Phi}|^2 + |\nabla_{\Gamma_t} \Phi|^2] \, d\Gamma_t \, dt \leq \text{Re} \int_{Q_T} \chi_-(t) \, H\Phi \, . \, \dot{\bar{\Phi}} \, dx \, dt$$

$$+ \sum_{i=1}^{n} [\; W(k\lambda) - W((k-1)\lambda+\mu) \;] \qquad (2.23)$$

By (2.19), we get :

$$W(k\lambda)- W((k-1)\lambda+\mu) = \int_{(k-1)\lambda+\mu}^{k\lambda} W'(t)\, dt \le \mathrm{Re}\int_{(k-1)\lambda+\mu}^{k\lambda} \int_{\Omega_t} H\Phi \cdot \dot{\bar{\Phi}}\, dxdt$$

and together with (2.23), we obtain the :

Lemma 6 : $\forall\, \Phi \in R(Q_T)$, we have :

$$\int_{\Sigma_T} \chi_-(t)\,[\,|\dot\Phi|^2 + |\nabla_{\Gamma_t}\Phi|^2]\, d\Gamma_t\, dt \le \frac{4}{\alpha}\, \mathrm{Re} \int_{Q_T} \chi_-(t)\, H\Phi \cdot \dot{\bar{\Phi}}\, dx\, dt \qquad (2.24)$$

We turn back to the element u_m of $R(Q_T)$, defined by (2.1) to (2.10), for which we get the following boundness result :

Proposition 1 : $\forall\, m$, $\exists\, \alpha_m \in \mathbb{C}$ and $M > 0$ such that :

$$\|u_m\|^2_{H^1(Q_T)} + \|\frac{\partial u_m}{\partial n_t}\|^2_{L^2(\Sigma_T)} + \|\chi_-\, \dot u_m\|^2_{L^2(\Sigma_T)} + \|\chi_-\, \nabla_{\Gamma_t} u_m\|^2_{L^2(\Sigma_T)} \le M \qquad (2.25)$$

Proof :

By adding (2.20), (2.22), (2.24), the left hand side of (2.25) is dominated by $\mathrm{Re}\int_{Q_T} g(t)\, Hu_m \cdot \dot{\bar u}_m\, dx\, dt$, where $g = c_1 + c_2\, \chi_-$. From (2.9), we have :

$$\int_{Q_T} g(t)\, Hu_m \cdot \dot{\bar u}_m\, dx\, dt = \int_{Q_T} g(t)\, f \cdot \dot{\bar u}_m\, dx\, dt + R$$

where the correcting term R is polynomial in α, α_m is chosen such that $R(\alpha_m) = 0$, then, for this choice, the left hand side of (2.25) is dominated by $\mathrm{Re}\int_{Q_T} g(t)\, f \cdot \dot{\bar u}_m\, dx\, dt$ and in particular the first term :

$$\|u_m\|^2_{H^1(Q_T)} \le \mathrm{Re}\int_{Q_T} g\, f \cdot \dot{\bar u}_m\, dx\, dt \le c\, \|f\|_{L^2(Q_T)}\, \|u_m\|_{H^1(Q_T)}$$

Then $\|u_m\|_{H^1(Q_T)}$ is bounded, then $\mathrm{Re}\int_{Q_T} g\, f \cdot \dot{\bar u}_m\, dx\, dt \le M$

We turn now to the existence result by looking to the limiting point u of the sequence u_m . For this purpose we formulate the following assumption on the family w_i ; introducing first :

$$U = \{\, \Phi \in H^1(Q_T)\ ;\ \Phi v^+ = 0 \text{ on } \Sigma_T\ ;$$
$$\Phi(0) = \Phi(T) = 0 \text{ in } H^{1/2}(\Omega_0)\, \} \qquad (2.26)$$

Let :

$$S = \{ \varphi = \sum_{i=1}^{N} \varphi_i(t)\ w_i(t,x)\ ;\ N \in \mathbb{N}\ ;$$

$$\varphi_i \in C^{\infty}([0,T])\ ;\ \varphi_i(0) = \dot{\varphi}_i(0) = 0\ \}$$

We suppose that S is dense in U (2.27)

This assumption will be discussed below at remark 2.

Using the Green formula and (2.9), we obtain at the limit u of a converging subsequence : $u \in H^1(Q_T)$ such that :

$$\int_{Q_T} [\nabla u\ .\ \nabla \bar{\psi} - \dot{u}\ .\ \dot{\bar{\psi}}]\ dxdt = \int_{Q_T} f\ .\ \bar{\psi}\ dxdt + \int_{\Sigma_T} [l+vz]\ \chi_-\ \psi\ d\Gamma_t\ dt \quad (2.28)$$

where we have supposed that $u_{m,k} \longrightarrow u$ weakly in $H^1(Q_T)$ and :

$$\frac{\partial u_{m,k}}{\partial n_t} \longrightarrow l \text{ weakly in } L^2(\Sigma_T) \text{ and } \dot{u}_{m,k} \longrightarrow z \text{ weakly in } L^2(\Sigma_T^-) \quad (2.29)$$

Now, by the density (2.27), the equality (2.28) is true for any ψ in U. As a first result we get $Hu = f$ in $D'(Q_T)$, then $u \in \mathbf{H}(Q_T)$ with :

$$\mathbf{H}(Q_T) = \{ \Phi \in L^2(Q_T)\ ;\ H\Phi \in L^2(Q_T)\ \}$$

Now, on Σ_T , we have $0 = u_{m,k} v^+ \longrightarrow u\ v^+$ weakly in $H^{1/2}(\Sigma_T)$, then $u\ v^+ = 0$ on Σ_T . Now, for any element u of $H^1(Q_T) \cap \mathbf{H}(Q_T)$, it is immediat to define, by the Green formula, the weak trace $\dot{u}(0)$ in $H_{00}^{1/2}(\Omega_0)'$. Then by (2.4) and (2.5) we obtain $u(0) = \dot{u}(0) = 0$. Also by the Green formula we get the existence of $\frac{\partial u}{\partial n_t}$ in $H_{00}^{1/2}(\Sigma_T)'$, but from (2.29) it is readily verified that this term is in $L^2(\Sigma_T)$. Finally we summerize by the :

Proposition 2 : For any f in $L^2(Q_T)$, there exists $u \in \Lambda(Q_T)$ with :

$$\Lambda(Q_T) = \{ \Phi \in H^1(Q_T) \cap \mathbf{H}(Q_T)\ ;\ \frac{\partial \Phi}{\partial n_t} \in L^2(\Sigma_T)\ \}$$

and u is solution of the problem :

$$Hu = f \text{ in } Q_T\ ,\ u_{|\Sigma_T^+} = 0\ ,\ u(0) = \dot{u}(0) = 0 \quad (2.30)$$

We turn now to uniqueness :

The tool is that, if u_1 and u_2 are two solution of (2.30), then $u = u_1 - u_2$ is a solution for the problem (2.30) with $f = 0$. Now, by regularization we can show the :

Proposition 3 : The two following subspaces of Λ are equal :

$\tilde{L}_0(Q_T) = \{ \Phi \in \Lambda(Q_T)$ such that $\Phi(0) = \dot{\Phi}(0) = 0$ and $\Phi v^+ = 0$ on $\Sigma_T \}$

$L_0(Q_T) =$ the closure of $R(Q_T)$ in $\Lambda(Q_T)$

It can also be verified directely that the estimate (2.21) established at Lemma 4, true for any Φ in $R(Q_T)$, holds for any Φ in $L_0(Q_T)$, then by the proposition 3, for any Φ in $\tilde{L}_0(Q_T)$, in particular for u . Finally, we get the :

Theorem 1 : Under assumptions (2.11), (2.13), (2.14), (2.27), given f in $L^2(Q_T)$ there exists a unique u in $\Lambda(Q_T)$ solution of the problem (2.30).

Remark 1 : The construction of the family verifying (2.14) is done in Ch. Truchi [4] , in the case $n = 1$ and we obtain :

$$e_i(t,x) = \left| \begin{array}{ll} \alpha(t)\ t\ [\dot{w}_i(x)]_0 & \text{in } [0,\lambda] \times \Omega_0 \\ \alpha(t)\ P_i(t,x) & \text{if } x/t \leq a' \end{array} \right.$$

with :

$$* \ [\dot{w}_i(x)]_0 = \dot{w}_i(\lambda,x) - \dot{w}_i(0,x)$$

$$* \ \alpha(t) \in C^\infty([0,\lambda]\ ,\ \mathbb{R}) \ ; \ \alpha(t) = 1 \text{ for } 0 \leq t \leq \varepsilon$$

and support of $\alpha \subset [0,\mu[$

$$* \ P_i(t,x) = -\left[\frac{1}{a'} \partial_x [\dot{w}_i(0)]_0 + \frac{1}{a'^2 t} [\dot{w}_i(0)]_0 \right] x^2$$

$$+ x\ t\ \partial_x [\dot{w}_i(0)]_0 + t\ [\dot{w}_i(0)]_0$$

* $a(t)$ is the length of the structure and is piecewise linear

Remark 2 : We consider $\hat{Q}_k$ of class C^2 , with :

$$\hat{Q}_k = \bigcup_{(k-1)\lambda < t < k\lambda} \{t\} \times \hat{\Omega}_t$$

$$\hat{\Sigma}_k = \bigcup_{(k-1)\lambda < t < k\lambda} \{t\} \times \hat{\Gamma}_t$$

such that :

$$Q_k \subset \hat{Q}_k \quad \text{and} \quad \Sigma_k^+ = \Sigma_k \cap \hat{\Sigma}_k$$

We define a family $(\hat{w}_i(t,x))_{i\geq 1}$ in $H^1(\hat{Q}_T)$, with $\hat{Q}_T = \bigcup_{0<t<T} \{t\} \times \hat{\Omega}_t$, λ - periodical in time, such that :

$$\hat{w}_i \in C^3(\bar{\hat{Q}}_k) \quad \text{and} \quad \hat{w}_{i\,|\hat{\Sigma}_r} = 0$$

Let :

$$w_i = \hat{w}_{i\,|Q_r} \quad \text{and} \quad S = \hat{S}_{|Q_r}$$

with :

$$\hat{S} = \{ \hat{\psi} = \sum_{i=1}^{N} \xi_i(t)\ \hat{w}_i(t,x)\ ;\ N \in \mathbb{N}\ ;$$
$$\xi_i \in C^{\infty}([0,T]\ , \mathbb{C})\ ;\ \xi_i(0) = \xi_i(T) = 0 \}$$

so, we have the :

Lemma 7 : $\hat{S}$ is dense in $H_0^1(\hat{Q}_T)$

Proof :

1/ The cylindrical case :

Let $\tilde{Q}_T = [0,T] \times \Omega_0$, a cylindrical domain, and $(\tilde{w}_i(x))_{i\geq 1}$, a family in $H_0^1(\tilde{\Omega}_0)$, we have now the classical result :

$$\tilde{S} \text{ is dense in } H_0^1(\tilde{Q}_T)$$

with :

$$\tilde{S} = \{ \tilde{\psi} = \sum_{i=1}^{N} \xi_i(t)\, \tilde{w}_i(x) \ ; \ N \in \mathbb{N} \ ;$$
$$\xi_i(t) \in C^\infty([0,T]\ , \mathbb{C}) \ ; \ \xi_i(0) = \xi_i(T) = 0 \ \}$$

2/ The non cylindrical case :

We consider $\hat{T}_t : \mathbb{R}^n \to \mathbb{R}^n$ and $\hat{\Omega}_0 \to \hat{\Omega}_t$ a one to one mapping and $\hat{w}_i(t,x) = \tilde{w}_i \circ \hat{T}_t^{-1}(x)$. For any $\hat{\psi}$ in $H_0^1(\hat{Q}_T)$, $\tilde{\psi} = \hat{\psi} \circ T_t$ belongs to $H_0^1(\tilde{Q}_T)$. Then, from 1/ , there exists a sequence $\tilde{\psi}_n \in \tilde{S}$ and converging to $\tilde{\psi}$ in $H_0^1(\tilde{Q}_T)$. It is easily verified that $\hat{\psi}_n = \tilde{\psi}_n \circ \hat{T}_t^{-1}$ belongs to $H_0^1(\hat{Q}_T)$ and $\hat{\psi}_n$ converges to $\hat{\psi}$ in $H_0^1(\hat{Q}_T)$.

Now, any element ψ of $H^1(Q_T)$ can be expressed as the restriction to Q_T of an element $\hat{\psi}$ of $H_0^1(\hat{Q}_T)$, then $\psi_n = \hat{\psi}_{n|Q_T}$ belongs to S and converges to ψ in $H^1_{\Sigma_T^+}(Q_T)$.

To consider the stabilization problem (0.1), we have now to give a similar result when f is zero but with non zero initial datas. For smooth initial datas, introducing $y = u + (y_0 \circ T_t^{-1})\ \alpha(t)$, (where α is a smooth function with $\alpha(t) = 1$ for $0 \leq t < \varepsilon$ and support of α is including in $[0,\mu[$) we easily obtain the :

Theorem 2 : Under assumptions (2.11), (2.13), (2.14), (2.27), given y_1 in $H^1(\Omega_0)$ and y_0 in $H^2(\Omega_0) \cap H_0^1(\Omega_0)$, verifying the compatibility condition :

$$v^+ \left(y_1 + \frac{\partial y_0}{\partial n_t} v\right) = 0 \quad \text{on } \Sigma_T \tag{2.31}$$

there exists a unique solution y in $\Lambda(Q_T)$ for the problem :

$$Hy = 0 \ \text{in } Q_T \ , \ y\, v^+ = 0 \ \text{on } \Sigma_T \ , \ y(0) = y_0 \ , \ \dot{y}(0) = y_1 \ \text{on } \Omega_0 \tag{2.32}$$

Remark on the exponential stability :

Assuming that for $(y_0, y_1) \in H_0^1(\Omega_0) \times L^2(\Omega_0)$ (with (2.31)), the problem (2.32) is well posed, with $y \in H^{3/2}(Q_T)$, then, at each $k\lambda$ the traces $y(k\lambda)$ and $\dot{y}(k\lambda)$ would be classically defined and introducing the linear mapping $\Theta : (y_0, y_1) \to (y(\lambda), \dot{y}(\lambda))$, Θ would be a contraction of $H_0^1(\Omega_0) \times L^2(\Omega_0)$ and :

$$W(k\lambda) \leq \|\Theta\|^k \, W(0)$$

which is an exponential decay for $\|\Theta\| < 1$.

In the next section, we investigate numerically this decay.

3. Discretized problem and numerical results

In this section, the domain Ω_t is supposed to be a rectangular membrane. The length, $a(t)$, of the membrane is varying periodically in time.

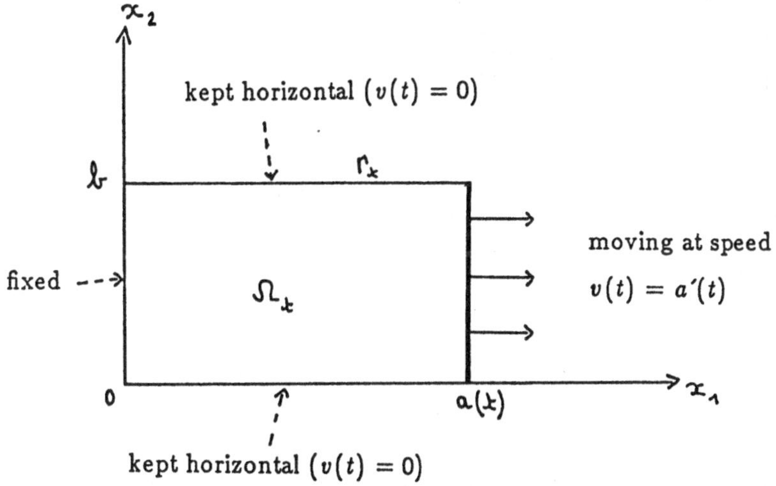

y = the deflection of the membrane.

Using the bijective mapping $T_t : \tilde{\Omega} = \Omega_0 \to \Omega_t$, $(X_1, x_2) \to (X_1 \dfrac{a(t)}{A} = x_1, x_2)$ with $A = a(0)$ and $y(t,x) = Y(t,X)$, in the problem (0.1), we have now to solve the problem in a fixed domain :

$$\left[X_1^2 \frac{a'(t)^2}{a(t)^2} - \frac{A^2}{a(t)^2} \right] \frac{\partial^2 Y}{\partial X_1^2} - \frac{\partial^2 Y}{\partial x_2^2} + \frac{\partial^2 Y}{\partial t^2} - 2 X_1 \frac{a'(t)}{a(t)} \frac{\partial^2 Y}{\partial X_1 \partial t}$$

$$+ X_1 \frac{2\,a'(t)^2 - a''(t)\,a(t)}{a(t)^2} \frac{\partial Y}{\partial X_1} = 0 \quad \text{in } \tilde{Q}_T = \bigcup_{0<t<T} \{t\} \times \tilde{\Omega}$$

$$Y(t,X) = 0 \quad \text{on } \tilde{\Sigma}_T^+$$

$$Y(0,X) = Y_0(X) \quad \text{on } \Omega_0$$

$$\dot{Y}(0,X) = Y_1(X) \quad \text{on } \Omega_0$$

The space discretization is achieved by a finite element method (Q_1 Lagrange) and the time discretization by the Euler's implicit method (to anticipate the jumps of the boundary speed $v(t)$). We can find all the details in Ch. Truchi [4].

As the function $a'(t) = v(t)$ has discontinuities in μ and λ , we appoximate the function $a''(t)$ by :

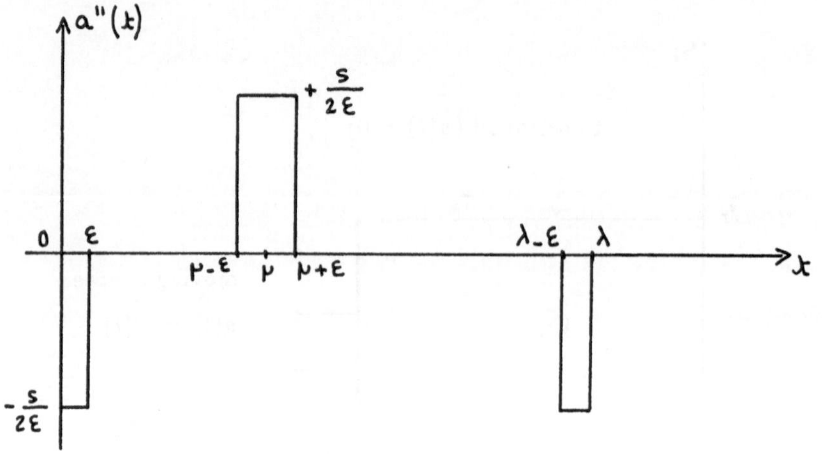

By integrating twice in time, we can compute the function $a(t)$. So, the numerical results are obtained with the following length $a(t)$ of the membrane, which makes the energy $W(t)$ decrease :

$$0 \leq t \leq \varepsilon \qquad a(t) = -\frac{s}{4\varepsilon} t^2 + \frac{s}{2} t - (1+c)t + A$$

$$\varepsilon \leq t \leq \mu - \varepsilon \qquad a(t) = -(1+c)t + \frac{s\varepsilon}{4} + A$$

$$\mu-\varepsilon \le t \le \mu+\varepsilon \qquad a(t) = \frac{s}{4\varepsilon}t^2 - (1+c)t - \frac{s}{2\varepsilon}(\mu-\varepsilon)t + \frac{s\varepsilon}{4} + \frac{s}{4\varepsilon}(\mu-\varepsilon)^2 + A$$

$$\mu+\varepsilon \le t \le \lambda-\varepsilon \qquad a(t) = (1+c)\frac{\mu}{\lambda-\mu}t + s(\frac{\varepsilon}{4}-\mu) + A$$

$$\lambda-\varepsilon \le t \le \lambda \qquad a(t) = -\frac{s}{4\varepsilon}t^2 + (1+c)\frac{\mu}{\lambda-\mu}t + \frac{s}{2\varepsilon}(\lambda-\varepsilon)t + s(\frac{\varepsilon}{4}-\mu) - \frac{s}{4\varepsilon}(\lambda-\varepsilon)^2 + A$$

with :

$$0 < \mu < \frac{\lambda}{2} \quad ; \quad (1+c)\frac{\mu}{\lambda-\mu} < 1 \quad ; \quad c > 0$$

$$s = (1+c)(1+\frac{\mu}{\lambda-\mu}) = \text{the jump of the boundary speed}$$

$$A = a(0) \quad ; \quad \varepsilon > 0 \; ; \; \varepsilon << 1$$

Numerically, we observe that $W(t)$ and $\|y\|_{L^2(\Omega_t)}$ decrease but with jumps in μ and λ . These jumps are alternatively positive and negative and they neutralize each other. The following results represent the local maximums of $W(t)$ and $\|y\|_{L^2(\Omega_t)}$. These results are obtained with :

$*$Dimensions of the membrane : length : $a(0) = 5{,}5$; width : 3

At the end of the contraction : length $= 5.294$

$*$Initial datas : $\dot{y}(0,x) = 0$; $y(0,x)$ solution of the problem :

$$-\Delta y(0,x) = f(x) \qquad \text{in } \Omega_0$$

$$y(0,x) = 0 \qquad \text{on } \Gamma_0$$

$\lambda = 0.5 \qquad \mu = 0.2 \qquad dx_1 = dx_2 = 0.5 \qquad \Delta t = 0.05$

We have an initial uniform force = 1.

Thus we observe a fairly exponential decay of the elastic energy $W(t)$

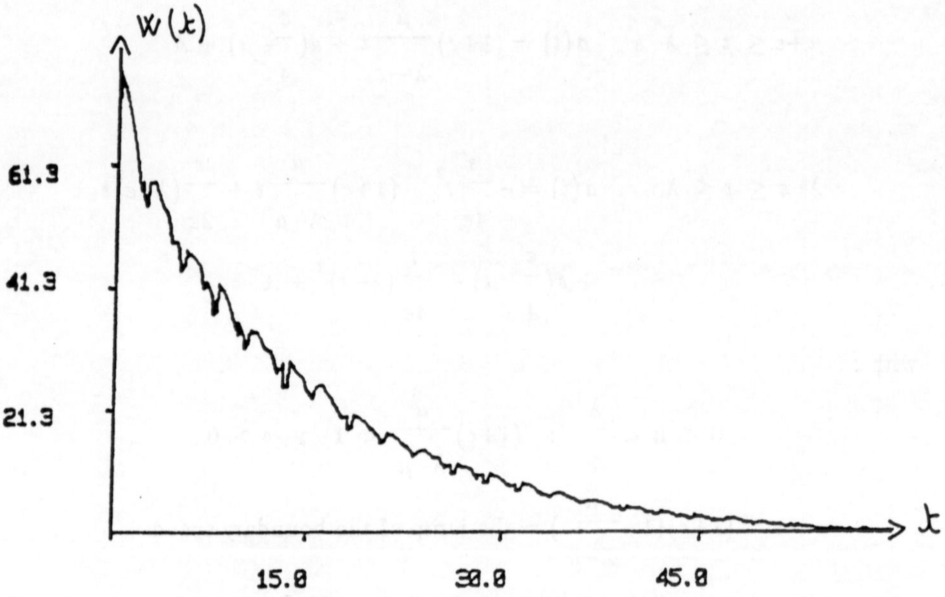

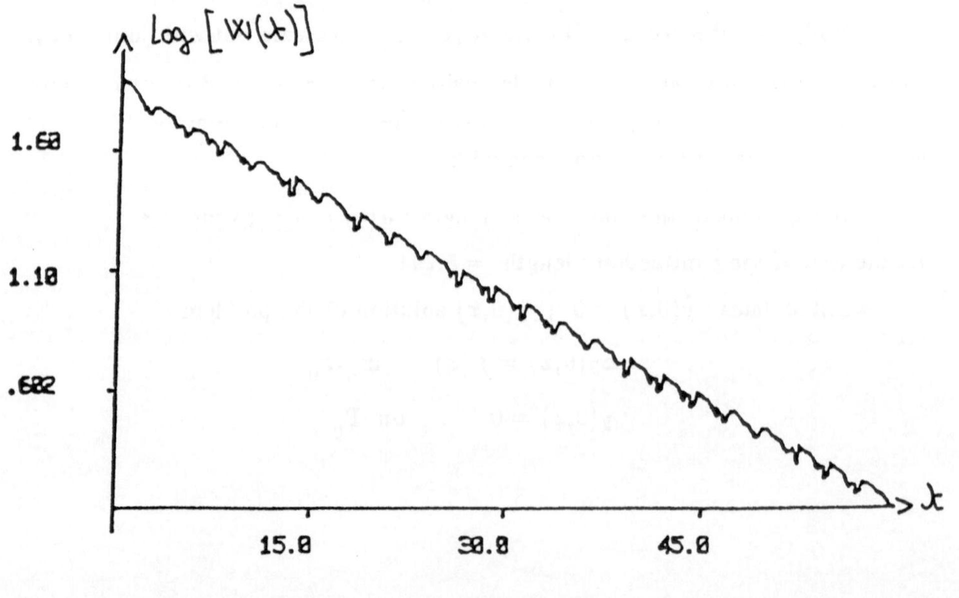

$$\lambda = 0.5 \qquad \mu = 0.2 \qquad dx_1 = dx_2 = 0.5 \qquad \Delta t = 0.05$$

We have an initial uniform force = 1.

Thus we observe a fairly exponential decay of $\|y\|_{L^2(\Omega_t)}$, $y =$ deflection of the membrane.

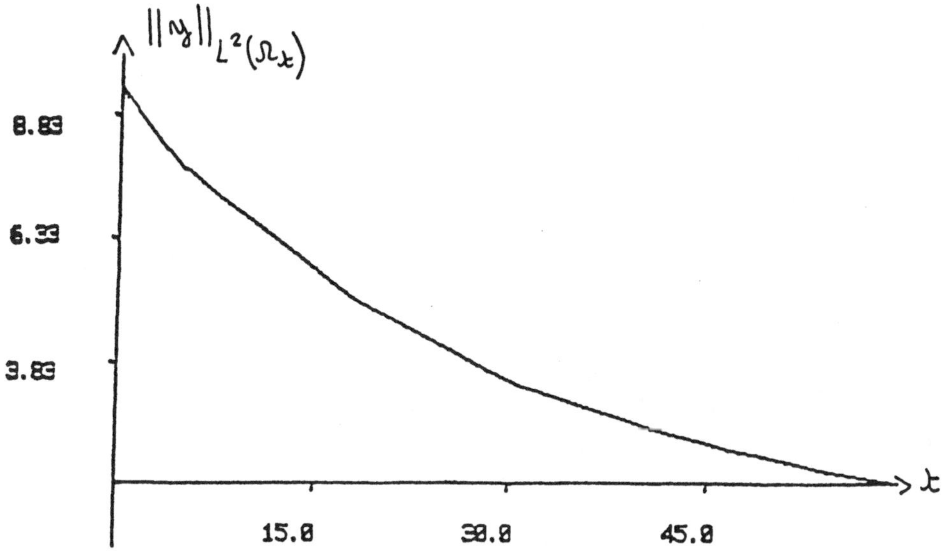

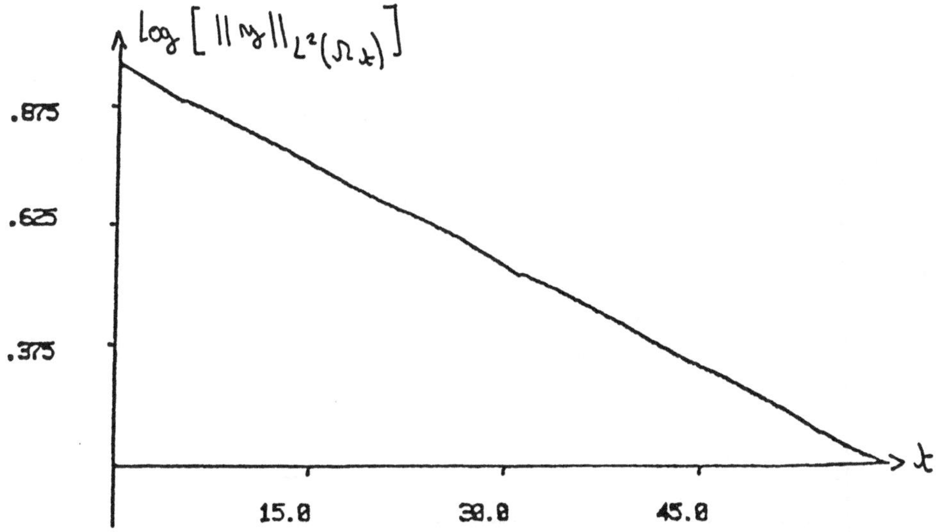

$\lambda = 1. \quad \mu = 0.2 \quad dx_1 = dx_2 = 0.5 \quad \Delta t = 0.05$

We have an initial uniform force = 1.

Thus we observe a fairly exponential decay of $W(t)$ and $\|y\|_{L^2(\Omega_t)}$

$y =$ deflection of the membrane.

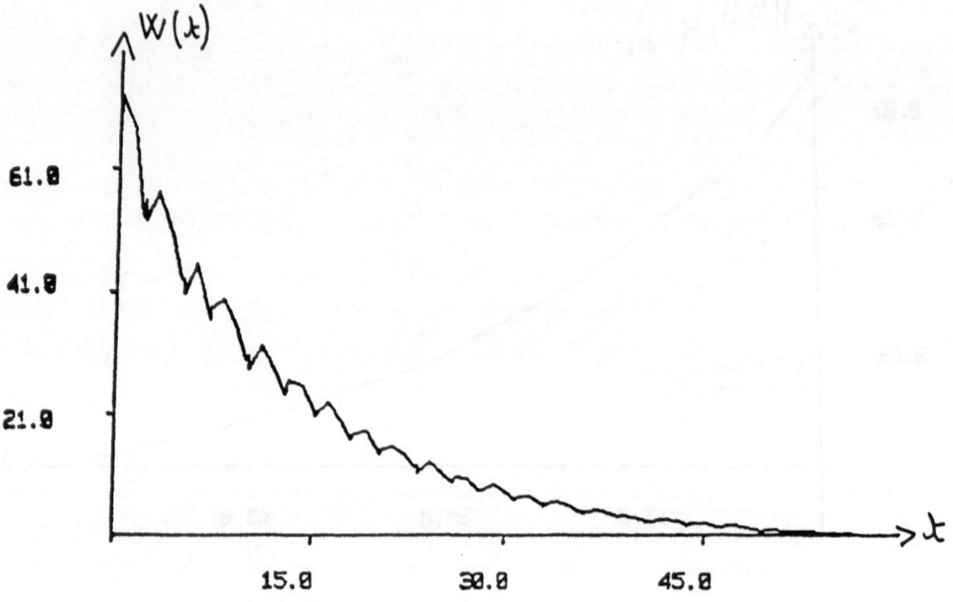

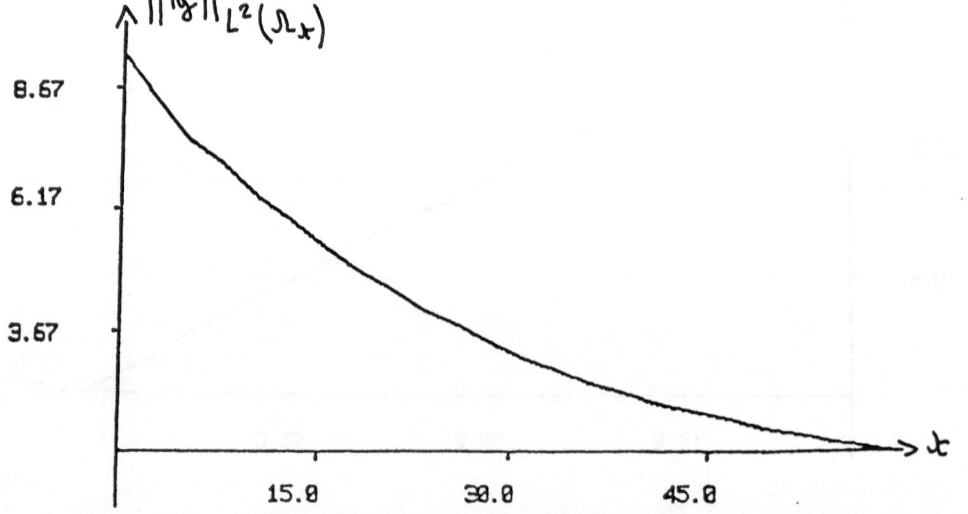

$\lambda = 0.5 \qquad \mu = 0.2 \qquad dx_1 = dx_2 = 0.5 \qquad \Delta t = 0.05$

Initial uniform force $f(x_1, x_2) = \cos(x_2)$

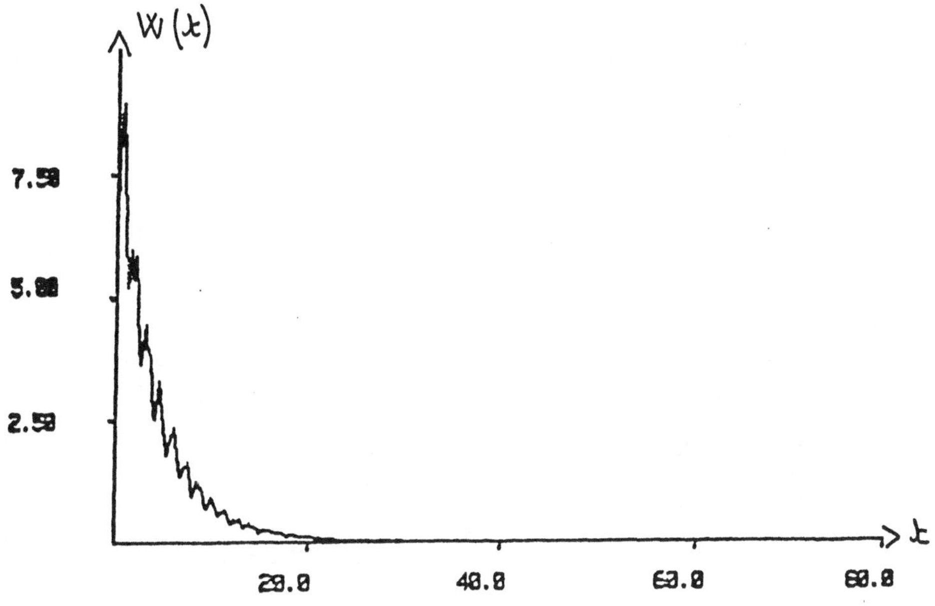

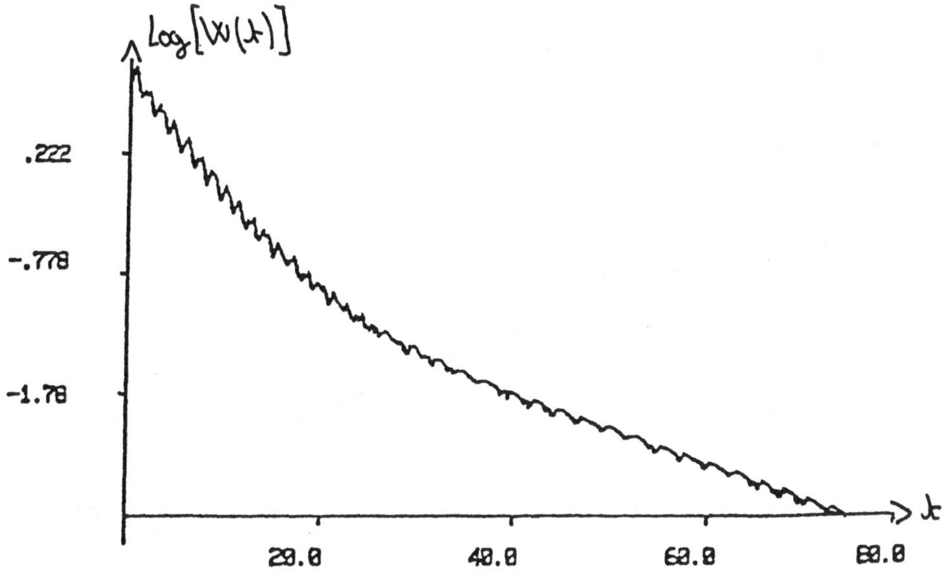

$\lambda = 0.5 \quad \mu = 0.2 \quad dx_1 = dx_2 = 0.5 \quad \Delta t = 0.05$

Initial uniform force $f(x_1, x_2) = \cos(x_2)$

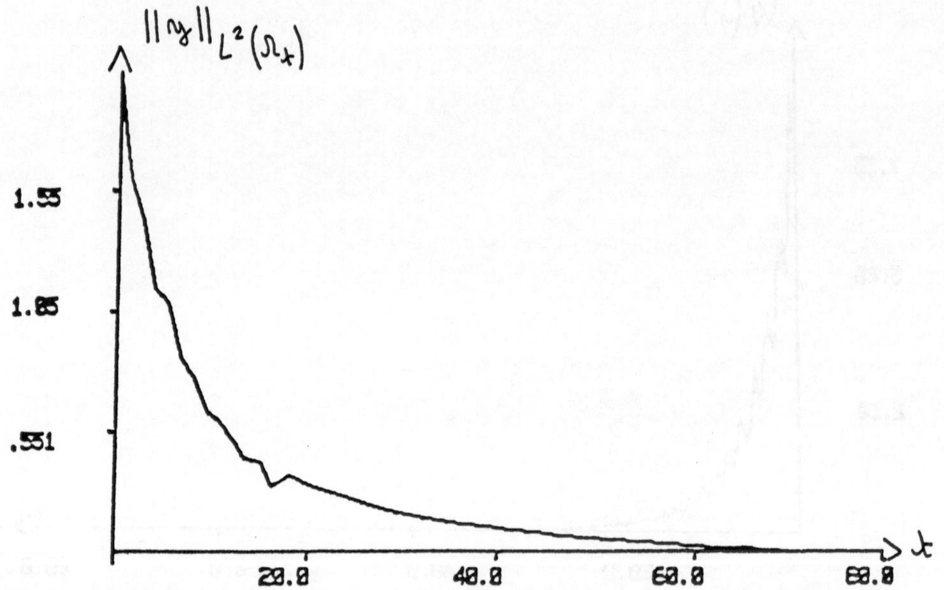

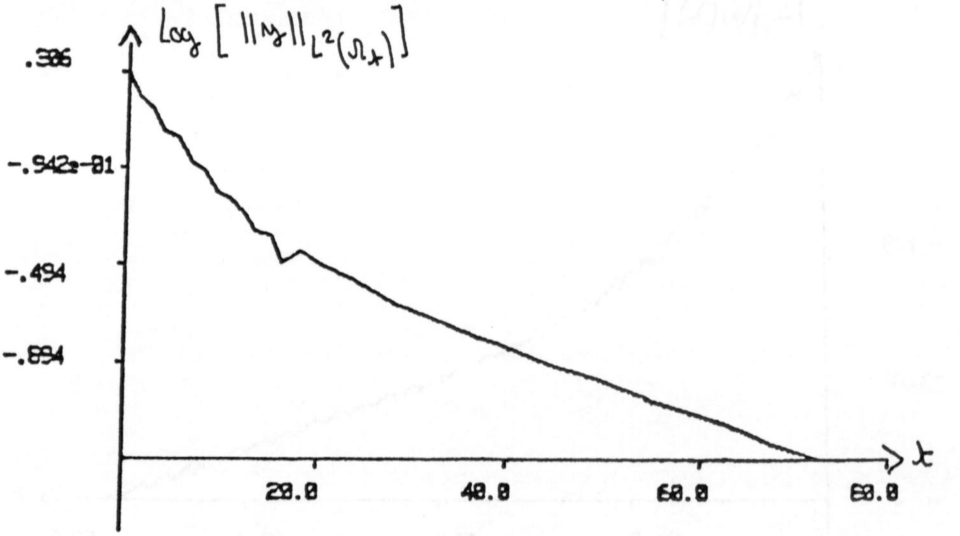

$\lambda = 0.5 \qquad \mu = 0.2 \qquad dx_1 = dx_2 = 0.5 \qquad \Delta t = 0.05$

Initial uniform force $f(x_1, x_2) = \cos(2\, x_2)$

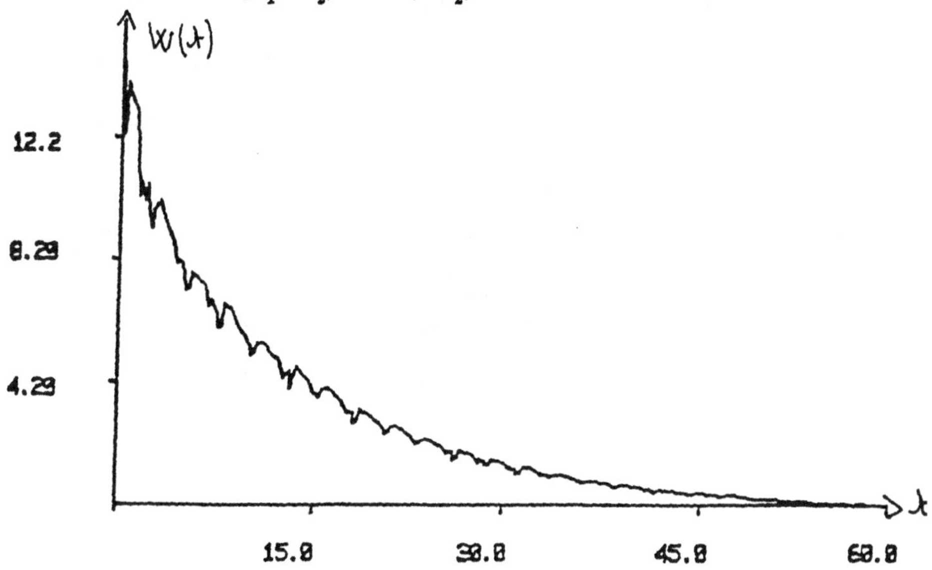

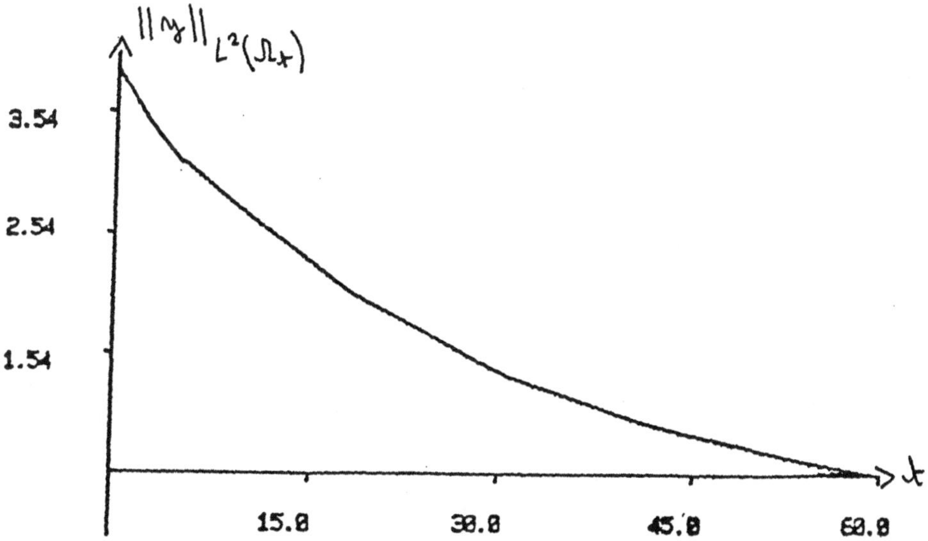

So, in this last section, we have proved, numerically, the exponential decay of the elastic energy $W(t)$ and of the norm of the deflection $\|y\|_{L^2(\Omega_t)}$.

References

[1] John P. Quinn and David L. Russel : "*Asymptotic stability and energy decay rates for solutions of hyperbolic equations with boundary damping* " , Proceedings of the Royal Society of Edinburgh , 77A , pp 97 - 127 , 1977 .

[2] J.P. Zolesio : "*Identification de domaine par déformation* " , Thèse de Doc. d'état , Nice (1979).

[3] J.P. Zolesio : "*The speed method in shape optimization, in Optimization of Distributed Parameter structures , Haug, E., Céa, J. (eds), Sijthoff and Noordhoff, 1981.*

[4] Ch. Truchi : "*Stabilisation par variation du domaine*" , Thèse de Doctorat , Nice (April 1987).

[5] J.P. Zolesio : "*Shape Stabilization of Flexible Structures*" , Lecture Notes in Control and Information Sciences , Edited by M. Thoma . Distributed Parameter Systems . Proceedings of the 2nd International Conference Vorau , Austria (1984) . Edited by F. Kappel , K. Kunish , W. Schapacher , Springer-Verlag . Berlin - Heidelberg - New York - Tokyo .

Lecture Notes in Control and Information Sciences

Edited by M. Thoma and A. Wyner

Vol. 62: Analysis and Optimization
of Systems
Proceedings of the Sixth International
Conference on Analysis and Optimization
of Systems
Nice, June 19–22, 1984
Edited by A. Bensoussan, J. L. Lions
XIX, 591 pages. 1984.

Vol. 63: Analysis and Optimization
of Systems
Proceedings of the Sixth International
Conference on Analysis and Optimization
of Systems
Nice, June 19–22, 1984
Edited by A. Bensoussan, J. L. Lions
XIX, 700 pages. 1984.

Vol. 64: Arunabha Bagchi
Stackelberg Differential Games
in Economic Models
VIII, 203 pages, 1984

Vol. 65: Yaakov Yavin
Numerical Studies
in Nonlinear Filtering
VIII, 273 pages, 1985.

Vol. 66: Systems and Optimization
Proceedings of the Twente Workshop
Enschede, The Netherlands, April 16–18, 1984
Edited by A. Bagchi, H. Th. Jongen
X, 206 pages, 1985.

Vol. 67: Real Time Control of Large Scale Systems
Proceedings of the First European Workshop
University of Patras, Greece, Juli 9–12, 1984
Edited by G. Schmidt, M. Singh, A. Titli,
S. Tzafestas
XI, 650 pages, 1985.

Vol. 68: T. Kaczorek
Two-Dimensional Linear Systems
IX, 397 pages, 1985.

Vol. 69: Stochastic Differential Systems –
Filtering and Control
Proceedings of the IFIP-WG 7/1 Working Conference
Marseille-Luminy, France, March 12-17, 1984
Edited by M. Metivier, E. Pardoux
X, 310 pages, 1985.

Vol. 70: Uncertainty and Control
Proceedings of a DFVLR International Colloquium
Bonn, Germany, March, 1985
Edited by J. Ackermann
IV, 236 pages, 1985.

Vol. 71: N. Baba
New Topics in Learning Automata
Theory and Applications
VII, 231 pages, 1985.

Vol. 72: A. Isidori
Nonlinear Control Systems:
An Introduction
VI, 297 pages, 1985.

Vol. 73: J. Zarzycki
Nonlinear Prediction
Ladder-Filters for Higher-Order
Stochastic Sequences
V, 132 pages, 1985.

Vol. 74: K. Ichikawa
Control System Design based on
Exact Model Matching Techniques
VII, 129 pages, 1985.

Vol. 75: Distributed Parameter
Systems
Proceedings of the 2nd International
Conference, Vorau, Austria 1984
Edited by F. Kappel, K. Kunisch,
W. Schappacher
VIII, 460 pages, 1985.

Vol. 76: Stochastic Programming
Edited by F. Archetti, G. Di Pillo,
M. Lucertini
V, 285 pages, 1986.

Vol. 77: Detection of
Abrupt Changes in Signals
and Dynamical Systems
Edited by M. Basseville,
A. Benveniste
X, 373 pages, 1986.

Vol. 78: Stochastic
Differential Systems
Proceedings of the 3rd Bad Honnef
Conference, June 3–7, 1985
Edited by N. Christopeit, K. Helmes,
M. Kohlmann
V, 372 pages, 1986.

Vol. 79: Signal
Processing for Control
Edited by K. Godfrey, P. Jones
XVIII, 413 pages, 1986.

Vol. 80: Artificial Intelligence
and Man-Machine Systems
Edited by H. Winter
IV, 211 pages, 1986.

Lecture Notes in Control and Information Sciences

Edited by M. Thoma and A. Wyner

Vol. 81: Stochastic Optimization
Proceedings of the International Conference, Kiew, 1984
Edited by I. Arkin, A. Shiraev, R. Wets
X, 754 pages, 1986.

Vol. 82: Analysis and Algorithms of Optimization Problems
Edited by K. Malanowski, K. Mizukami
VIII, 240 pages, 1986.

Vol. 83: Analysis and Optimization of Systems
Proceedings of the Seventh International Conference of Analysis and Optimization of Systems
Antiba, June 26-27, 1986
Edited by A. Bensoussan, J. L. Lions
XVI, 901 pages, 1986.

Vol. 84: System Modelling and Optimization
Proceedings of the 12th IFIP Conference
Budapest, Hungary, September 2–6, 1985
Edited by A. Prékopa, J. Szelezsán, B. Strazicky
XII, 1046 pages, 1986.

Vol. 85: Stochastic Processes in Underwater Acoustics
Edited by Charles R. Baker
V, 205 pages, 1986.

Vol. 86: Time Series and Linear Systems
Edited by Sergio Bittanti
XVII, 243 pages, 1986.

Vol. 87: Recent Advances in System Modelling and Optimization
Proceedings of the IFIP-WG 7/1 Working Conference
Santiago, Chile, August 27-31, 1984
Edited by L. Contesse, R. Correa, A. Weintraub
IV, 199 pages, 1987.

Vol. 88: Bruce A. Francis
A Course in H_∞ Control Theory
XI, 156 pages, 1987.

Vol. 88: Bruce A. Francis
A Course in H_∞ Control Theory
X, 150 pages, 1987.
Corrected - 1st printing 1987

Vol. 89: G. K. H. Pang/A. G. J. McFarlane
An Expert System Approach to Computer-Aided Design of Multivariable Systems
XII, 223 pages, 1987.

Vol. 90: Singular Perturbations and Asymptotic Analysis in Control Systems
Edited by P. Kokotovic, A. Bensoussan, G. Blankenship
VI, 419 pages, 1987.

Vol. 91 Stochastic Modelling and Filtering
Proceedings of the IFIP-WG 7/1 Working Conference
Rome, Italy, Decembre 10-14, 1984
Edited by A. Germani
IV, 209 pages, 1987.

Vol. 92: L. T. Grujić, A. A. Martynyuk, M. Ribbens-Pavella
Large-Scale Systems Stability Under Structural and Singular Perturbations
XV, 366 pages, 1987.

Vol. 93: K. Malanowski
Stability of Solutions to Convex Problems of Optimization
IX, 137 pages, 1987.

Vol. 94: H. Krishna
Computational Complexity of Bilinear Forms
Algebraic Coding Theory and Applications to Digital Communication Systems
XVIII, 166 pages, 1987.

Vol. 95: Optimal Control
Proceedings of the Conference on Optimal Control and Variational Calculus
Oberwolfach, West-Germany, June 15-21, 1986
Edited by R. Bulirsch, A. Miele, J. Stoer and K. H. Well
XII, 321 pages, 1987.

Vol. 96: H. J. Engelbert/W. Schmidt
Stochastic Differential Systems
Proceedings of the IFIP-WG 7/1 Working Conference
Eisenach, GDR, April 6-13, 1986
XII, 381 pages, 1987.

Lecture Notes in Control and Information Sciences

Edited by M. Thoma and A. Wyner

Vol. 97: I. Lasiecka/R. Triggiani (Eds.)
Control Problems for Systems
Described by Partial Differential Equations
and Applications
Proceedings of the IFIP-WG 7.2
Working Conference
Gainesville, Florida, February 3-6, 1986
VIII, 400 pages, 1987.

Vol. 98: A. Aloneftis
Stochastic Adaptive Control
Results and Simulation
XII, 120 pages, 1987.

Vol. 99: S. P. Bhattacharyya
Robust Stabilization Against
Structured Perturbations
IX, 172 pages, 1987.

Vol. 100: J. P. Zolésio (Editor)
Boundary Control and Boundary Variations
Proceedings of the IFIP WG 7.2 Conference
Nice, France, June 10-13, 1987
IV, 398 pages, 1988.

Vol. 101: P. E. Crouch, A. J. van der Schaft
Variational and Hamiltonian
Control Systems
IV, 121 pages, 1987.